METHODS IN MOLECULAR BIOLOGY

Series Editor
John M. Walker
School of Life and Medical Sciences
University of Hertfordshire
Hatfield, Hertfordshire, UK

For further volumes:
http://www.springer.com/series/7651

For over 35 years, biological scientists have come to rely on the research protocols and methodologies in the critically acclaimed Methods in Molecular Biology series. The series was the first to introduce the step-by-step protocols approach that has become the standard in all biomedical protocol publishing. Each protocol is provided in readily-reproducible step-by step fashion, opening with an introductory overview, a list of the materials and reagents needed to complete the experiment, and followed by a detailed procedure that is supported with a helpful notes section offering tips and tricks of the trade as well as troubleshooting advice. These hallmark features were introduced by series editor Dr. John Walker and constitute the key ingredient in each and every volume of the Methods in Molecular Biology series. Tested and trusted, comprehensive and reliable, all protocols from the series are indexed in PubMed.

Manuscript Instructions https://drive.google.com/file/d/1K4Az5vpAFVzYadJ3jKx2-QSo7Dqa_SNA/view?usp=sharing

Biomedicine and Biotechnology in Space

Molecular Methods

Edited by

Mariano Bizzarri

Department BeSSA - Wellbeing, Health and Environmental Sustainability, Sapienza University of Rome, Rome, Italy

Humana Press

Editor
Mariano Bizzarri
Department BeSSA - Wellbeing
Health and Environmental Sustainability
Sapienza University of Rome
Rome, Italy

ISSN 1064-3745 ISSN 1940-6029 (electronic)
Methods in Molecular Biology
ISBN 978-1-0716-5173-5 ISBN 978-1-0716-5174-2 (eBook)
https://doi.org/10.1007/978-1-0716-5174-2

This Humana imprint is published by the registered company Springer Science+Business Media, LLC, part of Springer Nature.
The registered company address is: 1 New York Plaza, New York, NY 10004, U.S.A.

Preface

As humanity continues to push toward interplanetary travel and beyond, short- and long-term adaptation of the human body to extreme environmental conditions in space remains a fundamental concern. Astronauts experience physically and mentally enduring situations like microgravity, extreme heat or cold, radiation, isolation, and other unforeseen technical challenges. Indeed, a variety of medical conditions can affect astronauts during prolonged stays in space, including blood clots, kidney stones, radiation-related effects, and a range of other illnesses. To date, a list of 100 spaceflight medical conditions that have occurred in spaceflight are of significant concern for affecting human well-being and/or can significantly affect the probability of mission's success. Therefore, as NASA and European agencies aim to send astronauts farther into space for longer durations, managing and maintaining their health and performance has become a top priority.

In the context of human space exploration, the assessment of clinical conditions, i.e., the specific task of clinical pathology, and consequently the development of countermeasures must consider an individual-focused precision medicine, as well as specific mission architecture and space conditions in which each individual astronaut is operating, including duration, radiation dose, and operational requirements.

This integrated approach requires performing deep quantitative molecular analysis of an individual's genotype, coupled with assessment of molecular, physiological, morphological, environmental, and behavioral phenotypes in order to develop individualized, precision countermeasures.

Defining a strategy to establish a sound analytical approach is among the topics of the present volume. Specifically, we aim to assess those parameters useful for identifying the "fingerprint" of specific biological (complex) processes, including metabolic, immunologic, protein/genetic, and signaling pathways, in order to depict a comprehensive and systemic model able to provide clear explanations and preventative health scenarios. Such studies are challenged by extraordinary complexities related to the acquisition of biological samples, lack of conventional analytical devices, and restrictions due to size-limits onboard of either the International Space Station or the space vehicle. However, the uncertainties cannot be restricted only to technological challenges, given that we are confronting the boundaries of medicine in addressing very unusual (and unexpected) issues. Indeed, biological and medical investigations carried out in a space environment have forced us to cope with a true conceptual revolution in which the relationships between physics and biology should be reframed in full. This approach requires a relevant effort to translate findings obtained at the molecular level (usually recorded in vitro) to higher levels (tissues and organism) through integration with physiological (systemic) markers. Therefore, we need to develop and implement new technologies for widening data recording at these different tiers of observation. Precisely, this attempt focuses on five key targets: (a) genetic variants that may drive unique risk profiles of astronauts in the space environment; (b) genetic pharmagenomic profiles; (c) adaptive changes in metabolism (including the microbiota); (d) post-translational molecular changes (that can significantly affect biochemical reactions and enzyme kinetics); and (e) synergy between the above-mentioned factors and the medical history of the astronaut.

To address such a challenge, an impressive technological endeavor has been carried out to develop new diagnostic tools, medical devices, and infrastructures for developing remote sensing and medical support, given that medical capabilities are inherently limited during spaceflights. Improvements in telemedicine, realization of multi-sensor applications, innovation in -omic-based approaches for analytical discoveries, and advances in image analysis would never have been discovered if it had not been necessary for space exploration. Briefly, space biomedicine has represented—and still represents—a major stimulus for the development of advanced biomedical technologies.

Studies included in this volume describe several advanced techniques specifically designed to address difficulties and complexities associated with molecular biological inquiry in microgravity conditions. Specifically, the chapters deal with the specific space environment (J.J.W.A. van Loon, *Hypergravity, Microgravity, and Animal Research* and O. Naimark, *Assessing the Effects of Microgravity on Eukaryotic Cells Using a Statistical Thermodynamic Approach*) and unique systems biology constraints in space (G. Hardiman, *Multiscale Modeling and Systems Biology in Microgravity Investigations*). Cell and pathways studies are extensively investigated in chapters from C.R.G. Willis (*Transcriptomics in Space: A Basic R Pipeline for Analyzing Bulk RNA-Sequencing Data*), A. Beheshti (*A Systems Biology Approach to Discovery and Validation of MicroRNA Signature Health Risks Associated with the Space Environment*), A. Catizone (*The Cytoskeleton in Space*), and A. Fuso (*Epigenetics in Space: The Dynamic Cellular Response*).

Tissues and physiological apparatus are analyzed in several contributions, including R. Negri (*Modulation of Cell-to-Cell Adhesion and Interaction with the Extracellular Matrix in Microgravity Conditions*), D. Atiakshin (*Histochemistry and Immunohistochemistry in Space Biology: Analysis of the Functional Morphology of the Tissue Microenvironment*), N. Monti (*Ovarian Reproductive and Endocrine Function Investigated in Microgravity Environment*), and B.S. Samuel (*Defining Microbiome Impact on Host Physiology During Spaceflight Using Caenorhabditis elegans*).

Specific technological approaches are addressed by L. De Sio (*3D Bioprinted Tumor Constructs for Photothermal Therapy: Emerging Opportunities in Microgravity Research*) and C. Marchese (*Regenerative Medicine in Space: Advancing Bone Repair with Adipose Stem Cells and Bone-on-Chip Technology*), while personalized-based approaches are discussed in the chapters from S.V. Costes (*Integrated Platforms to Further Advance Space Biology Research*) and M.A. Schmidt and C.E. Mason (*Precision Medicine in Astronauts: Principles and Targets in Molecular Assessment*).

Hopefully, these contributions can provide useful insights into this very complex field of scientific research.

Rome, Italy Mariano Bizzarri

Contents

Contributors

ANTONIO ANGELONI • *Department BeSSA - Wellbeing, Health and Environmental Sustainability - National Space Biomedicine Laboratory, Systems Biology Group, Sapienza University of Rome, Rome, Italy*

DMITRII ATIAKSHIN • *Research and Educational Resource Center for Immunophenotyping, Digital Spatial Profiling and Ultrastructural Analysis Innovative Technologies, Peoples' Friendship University of Russia (RUDN University), Moscow, Russia; Research Institute of Experimental Biology and Medicine, Burdenko Voronezh State Medical University, Voronezh, Russia*

YURI V. BAYANDIN • *Institute of Continuous Mechanics UrB RAS, Perm, Russia*

CLAUDIA BEARZI • *Institute for Biomedical Technologies, National Research Council (ITB-CNR), Segrate, Milan, Italy; Fondazione Istituto Nazionale di Genetica Molecolare, Milan, Italy*

AFSHIN BEHESHTI • *Center for Space Biomedicine, Trivedi Institute for Space and Global Biomedicine, University of Pittsburgh School of Medicine, Pittsburgh, PA, USA; Stanley Center for Psychiatric Research, Broad Institute of MIT and Harvard, Cambridge, MA, USA; McGowan Institute for Regenerative Medicine, Department of Surgery, University of Pittsburgh School of Medicine, Pittsburgh, PA, USA; Computational and Systems Biology, University of Pittsburgh School of Medicine, Pittsburgh, PA, USA*

MARIKA BERARDINI • *Department of Anatomy, Histology, Forensic-Medicine and Orthopedics, "Sapienza" University of Rome, Rome, Italy*

MARIANO BIZZARRI • *Department BeSSA - Wellbeing, Health and Environmental Sustainability, Head of the National Space Biomedicine Laboratory, Systems Biology Group, Sapienza University of Rome, Rome, Italy*

DANA BLACKBURN • *Alkek Center for Metagenomics and Microbiome Research and Department of Molecular Virology and Microbiology, Baylor College of Medicine, Houston, TX, USA*

IGOR B. BUCHWALOW • *Research and Educational Resource Center for Immunophenotyping, Digital Spatial Profiling and Ultrastructural Analysis Innovative Technologies, Peoples' Friendship University of Russia (RUDN University), Moscow, Russia; Institute for Hematopathology, Hamburg, Germany*

EMANUELE CACCI • *Department of Biology and Biotechnologies "Charles Darwin", Sapienza University of Rome, Rome, Italy*

THOMAS CAHILL • *School of Biological Sciences and Institute for Global Food Security, Queen's University Belfast, Belfast, UK*

ANGELA CATIZONE • *Department of Anatomy, Histology, Forensic-Medicine and Orthopedics, "Sapienza" University of Rome, Rome, Italy*

MAILA CHIRIVÌ • *Department of Molecular Medicine, Sapienza University, Rome, Italy; UOC Neurology, Fondazione Ca'Granda, Ospedale Maggiore Policlinico, Milan, Italy*

SYLVAIN V. COSTES • *Blue Marble Space Institute of Science, Seattle, WA, USA*

GIORGIA CUCINA • *Department BeSSA - Wellbeing, Health and Environmental Sustainability - National Space Biomedicine Laboratory, Systems Biology Group, Sapienza University of Rome, Rome, Italy*

Luciano De Sio • *Department of Medical-Surgical Sciences and Biotechnologies, Sapienza University of Rome, Latina, Italy*
Monica Driscoll • *Department of Molecular Biology and Biochemistry, Rutgers University, New Brunswick, NJ, USA*
Francisco J. Enguita • *Faculdade de Medicina, Universidade de Lisboa, Lisboa, Portugal*
Andrea Fuso • *Deptartment of Experimental Medicine, Sapienza University of Rome, Rome, Italy; CRiN, Center for Research in Neurobiology "D. Bovet", Sapienza University of Rome, Rome, Italy*
Emanuele Galante • *Department BeSSA - Wellbeing, Health and Environmental Sustainability - National Space Biomedicine Laboratory, Systems Biology Group, Sapienza University of Rome, Rome, Italy*
Gary Hardiman • *School of Biological Sciences and Institute for Global Food Security, Queen's University Belfast, Belfast, UK*
Jeffrey A. Jones • *Center for Space Medicine, Baylor College of Medicine, Houston, TX, USA*
Ana Lúcia Leitão • *Departamento de Química, Faculdade de Ciências e Tecnologia, NOVA FCT, Universidade NOVA de Lisboa, Caparica, Portugal*
Guillermo H. López-Campos • *Wellcome-Wolfson Institute for Experimental Medicine at Queen's University Belfast, Belfast, UK*
Giuseppe Lupo • *Department of Biology and Biotechnologies "Charles Darwin", Sapienza University of Rome, Rome, Italy*
Cinzia Marchese • *Department of Experimental Medicine, Sapienza University of Rome, Rome, Italy*
Christopher E. Mason • *Department of Physiology and Biophysics, Weill Cornell Medicine, New York, NY, USA*
Noemi Monti • *Department BeSSA - Wellbeing, Health and Environmental Sustainability - National Space Biomedicine Laboratory, Systems Biology Group, Sapienza University of Rome, Rome, Italy*
Oleg B. Naimark • *Institute of Continuous Mechanics UrB RAS, Perm, Russia*
Rodolfo Negri • *Department of Biology and Biotechnologies "Charles Darwin", Sapienza University of Rome, Rome, Italy*
Aleksandr S. Nikitiuk • *Institute of Continuous Mechanics UrB RAS, Perm, Russia*
Elena Di Nisio • *Department of Biology and Biotechnologies "Charles Darwin", Sapienza University of Rome, Rome, Italy*
Gbolaga O. Olanrewaju • *Department of Biomedical Sciences, Heritage College of Osteopathic Medicine, Ohio University, Athens, OH, USA; Ohio Musculoskeletal and Neurological Institute, Ohio University, Athens, OH, USA*
Mariagiovanna Pais • *School of Biological Sciences and Institute for Global Food Security, Queen's University Belfast, Belfast, UK*
Audrey J. Parish • *Alkek Center for Metagenomics and Microbiome Research and Department of Molecular Virology and Microbiology, Baylor College of Medicine, Houston, TX, USA*
Giovanni Perotto • *Smart Materials, Istituto Italiano di Tecnologia, Genoa, Italy*
Francesca Petronella • *Institute of Crystallography, National Research Council, (CNR-IC), Monterotondo, Rome, Italy*
Paola Pontecorvi • *Department of Experimental Medicine, Sapienza University of Rome, Rome, Italy*

ALESSANDRO QUERQUI • *Department BeSSA - Wellbeing, Health and Environmental Sustainability - National Space Biomedicine Laboratory, Systems Biology Group, Sapienza University of Rome, Rome, Italy*
BUSHRA RAHMAN • *Department of Chemical Engineering, Texas Tech University, Lubbock, TX, USA*
GIULIA RICCI • *Department of Experimental Medicine, Università degli Studi della Campania "Luigi Vanvitelli", Naples, Italy*
ROBERTO RIZZI • *Department of Medical-Surgical Sciences and Biotechnologies, Sapienza University of Rome, Latina, Italy*
PAOLO ROSA • *Department of Medical-Surgical Sciences and Biotechnologies, Sapienza University of Rome, Latina, Italy*
BUCK S. SAMUEL • *Alkek Center for Metagenomics and Microbiome Research and Department of Molecular Virology and Microbiology, Baylor College of Medicine, Houston, TX, USA*
LAUREN M. SANDERS • *Blue Marble Space Institute of Science, Seattle, WA, USA*
ATIYYA P. SAROYIA • *Department of Chemical Engineering, Texas Tech University, Lubbock, TX, USA*
CALEB M. SCHMIDT • *Sovaris Aerospace, Boulder, CO, USA; Department of Systems Engineering, Colorado State University Fort Collins, Fort Collins, CO, USA*
JULIAN C. SCHMIDT • *Sovaris Aerospace, Boulder, CO, USA*
MICHAEL A. SCHMIDT • *Sovaris Aerospace, Boulder, CO, USA; Aerospace Medicine Program, Department of Medicine, University of Central Florida College of Medicine, Orlando, FL, USA; Center for Aerospace and Extreme Environments Medicine, University of Central Florida College of Medicine, Orlando, FL, USA*
MARIA LAURA SFORZA • *Department of Medical-Surgical Sciences and Biotechnologies, Sapienza University of Rome, Latina, Italy*
VIKTORIYA SHISHKINA • *Research Institute of Experimental Biology and Medicine, Burdenko Voronezh State Medical University, Voronezh, Russia*
NATHANIEL J. SZEWCZYK • *Ohio Musculoskeletal and Neurologic Institute and Department of Biomedical Sciences, Ohio University, Athens, OH, USA; Department of Biomedical Sciences, Heritage College of Osteopathic Medicine, Ohio University, Athens, OH, USA; Ohio Musculoskeletal and Neurological Institute, Ohio University, Athens, OH, USA*
MARCO TAFANI • *Department of Experimental Medicine, "Sapienza" University of Rome, Rome, Italy*
MARKUS TIEMANN • *Institute for Hematopathology, Hamburg, Germany*
JACK J. W. A. VAN LOON • *Department Oral & Maxillofacial Surgery/Pathology, Amsterdam Movement Sciences & Amsterdam Bone Center (ABC), Amsterdam University Medical Center Location, Vrije Universiteit Amsterdam & Academic Center for Dentistry Amsterdam (ACTA), Amsterdam, The Netherlands; TEC-MMG-LIS Lab, European Space Agency (ESA), European Space Research and Technology Center (ESTEC), Noordwijk, The Netherlands*
SIVA A. VANAPALLI • *Department of Chemical Engineering, Texas Tech University, Lubbock, TX, USA*
CRAIG R. G. WILLIS • *Research Software and Analytics, University of Exeter, Exeter, UK*

Part I

Introduction

Check for updates

Chapter 1

Biomedicine and Biotechnology in Space: An Introduction

Mariano Bizzarri

Abstract

As humanity continues to push toward interplanetary travel and beyond, short- and long-term adaptation of the human body to extreme environmental conditions in space remains a fundamental concern. An astronaut experiences physically and mentally enduring situations like microgravity, extreme heat or cold, radiation, isolation, and other unforeseen technical challenges. The assessment of clinical conditions—the specific task of Clinical Pathology—during spaceflights is becoming a relevant priority. Defining a strategy to establish a sound analytical approach is among the topics of the present volume. Namely, we are trying to assess those parameters to identify the "fingerprint" of specific biological (complex) processes, including metabolic, immunologic, protein/genetic and signaling pathways, in order to depict a comprehensive, systemic model able to provide clear explanations and preventative health scenarios. To address such a challenge, an impressive technological endeavor is going on to develop innovative diagnostic tools, medical devices and infrastructures for developing remote sensing and medical support, given that medical capabilities are inherently limited during spaceflights.

Keywords Spaceflight, Biotechnology, Microgravity, Omics technologies, Laboratory medicine in space

Space Biomedicine is gaining momentum, and it promises to provide an invaluable body of knowledge and technological advancements. Innovation in space medicine is mandatory for "preparing astronauts for long-duration missions into the solar system and provide lasting benefits to life on Earth." Undoubtedly, human health is severely threatened by the space environment, and the development of specific medical tools and pharmacological countermeasures is both obligatory to ensure a minimum of safety conditions. Undoubtedly, the space environment is an "extreme" environment, in which humans are facing three main deadly challenges: (1) impressive changes in fundamental physical forces (modified gravity and electromagnetic fields), affecting every level of the organism (from molecules to the entire human body); (2) Exposure to cosmic rays and radiation hazard, with consequent increased risk of carcinogenesis and mutagenesis; (3) Endocrine and psychosocial disturbances, following the physical isolation and the disruption of basic chronobiological rhythms [1] (Fig. 1). The consequent health risk, depending on the

Mariano Bizzarri (ed.), *Biomedicine and Biotechnology in Space*, Methods in Molecular Biology, vol. 3000,
https://doi.org/10.1007/978-1-0716-5174-2_1,

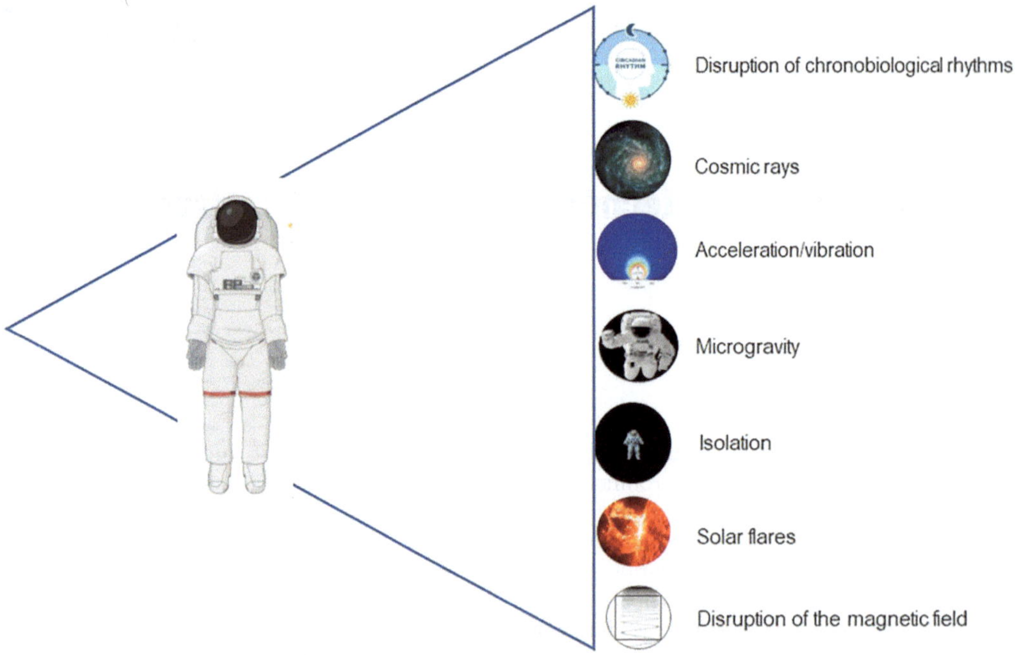

Fig. 1 Main environmental stress factors to which human beings are exposed during spaceflight

duration of the spaceflight and on the distance reached from the Earth, is significant and cannot be underestimated [2].

As humanity continues to push toward interplanetary travel and beyond, short- and long-term adaptation of the human body to extreme environmental conditions in space remains a fundamental concern. An astronaut is subjected to physically and mentally enduring situations like microgravity, extreme heat or cold, radiation, isolation, and other unforeseen technical challenges. Indeed, a variety of medical conditions can affect astronauts who live for prolonged periods in space, including blood clots, kidney stones, radiation exposure, and a range of other illnesses and injuries [3]. NASA's Human Research Program currently recognizes five principal hazards of human spaceflight (radiation, psychosocial isolation, distance from Earth, microgravity fields, and closed environment), although the cumulative and interactive challenges posed by these conditions, as well as their potential impact upon human health, are only beginning to be understood by the scientific community. To date, a list of 100 spaceflight medical conditions that have occurred in spaceflight are of significant concern for affecting human well-being and/or can significantly affect the probability of mission's success [4].

Medical risk is roughly dependent on the duration of space flight missions, although critical issues can be observed even after few days in orbit [5]. However, on longer missions—such as a mission to Mars extended for 3 years—health challenges and risks increase exponentially. No human has yet spent such a consecutive length of time in space. This extended stay in a hostile environment increases the likelihood of the occurrence of medical events, both minor and severe [6]. These events include, but are not limited to, disease, trauma, decompression sickness, burns, toxic exposures, overheating or overcooling, life support failure, depressurization, and meteorite impacts. It has been calculated that for a 1195-day mission, the crew health index (the percentage of quality-adjusted time lost due to in-flight medical events) would range from 8% to 15% depending on the medical capability provided. To manage the physiological impact of these situations, the use of real-time diagnostic laboratory medicine will be necessary for the health of the astronaut. Therefore, as NASA and European agencies aim to send astronauts farther into space for longer durations, managing and maintaining their health and performance is becoming a top priority.

From a theoretical standpoint, an astronaut's genotype and molecular phenotype bear heavily on how each individual will uniquely interact with the spaceflight environment. This evidence requires an individual-focused precision medicine. Precision medicine (PM) can be defined as "the science and practice of providing methods of treatment and prevention tailored to an individual's molecular, physiological, morphological, and behavioral characteristics" [7]. In the context of human space exploration (HSE), the assessment of clinical conditions—the specific task of Clinical Pathology—and consequently the countermeasures development must also consider the specific mission architecture and space conditions, in which each individual astronaut is operating, including duration, radiation dose, and operational requirements. This integrated approach requires performing deep quantitative molecular analysis of an individual's genotype, coupled with assessment of the molecular, physiological, morphological, environmental, and behavioral phenotypes, in order to develop individualized, precision countermeasures. Although a "personalized" approach in assessing astronaut health has been the rule for clinicians since the beginning of space travel, in the past such a task has been fulfilled mostly through "empirical" and semiquantitative approaches. Currently, the time has come for PM to rely on an extended analytical array of molecular and genetic parameters. Yet, those data still need to be integrated into a robust systems biology model in order to extract meaningful medical records. A systems biology approach implies that we can manage and integrate concepts like personalization and stratification. Personalization deals with the assessment of *pattern of variance* (related to significant molecular/physiological parameters) of each single individual and would provide (at least in principle) tailored countermeasures. Stratification means that individuals sharing common biochemical, physiological, and genetic features, including the capability to respond to stressors and drugs administration, should be grouped all together. Furthermore, systems biology should aim at unveiling unsupervised targets in order to detect off-target or unanticipated effects that would enable the discovery of novel pathways and (unsuspected) mechanisms.

This approach requires a relevant effort in translating findings obtained at the molecular level (usually recorded in vitro studies) to higher levels (tissues, organism) through an integration with physiological (systemic) markers. Therefore, we need to develop and implement new technologies for widening data recording at these different tiers of observation. Precisely, this attempt focuses on five key targets: (a) genetic variants that may drive unique risk profiles of astronauts in the space environment; (b) genetic pharmagenomic profiles that may modulate drug efficacy/metabolism; (c) adaptive changes in metabolism (including the microbiota) that can impinge on astronaut's risk; (d) posttranslational molecular changes that can significantly affect biochemical reactions and enzyme kinetics; and (e) synergy between the above mentioned factors and the medical history of astronaut. A clear-cut example of this approach has been provided in oncology by the use of integrated programs—like *Oncotype Dx®* [8]—in order to draw a therapeutic program for breast cancer patients. Current development in space biomedicine lets us foresee similar tools to identify medical risks for astronauts.

This is a critical step, which deserves special attention, given that "variation" does not necessarily imply "disease," and medical recordings are not settled to perform diagnoses. Variance observed upon diverse constraints, under the activity of different "fields" (gravitational, magnetic, and radiation-related), does not imply "illness," and can merely suggest adaptability. Variance does (can)not mean disease. The question is, therefore, when and where variability end up becoming disease. There is a great debate about the relative meaning of "variance" and "disease," and it is still difficult to resolve such issue [9]. However, a dynamic parameter, usually neglected/under-rated on terrestrial conditions, can help in assigning a physiological value to the changes observed in spaceflight conditions: the optimal performance. Optimal performance should be adequately quantitated and identified for spaceflight missions (for astronauts as well as for civilian travelers), as it has a relevant impact on both success and safety of a mission. Physiological recording would, therefore, be coupled with extensive pattern analysis, aimed at characterizing biochemical networks, in order to identify the respective borders of optimal and poor performance, the latter non-necessarily coincident with an illness state. In general, we should seek to align molecular patterns to different phenotypes, including the performance, clinical, and behavioral ones. Noticeably, this is the preliminary step to identify tailored, precision countermeasures to the individual astronaut or stratified group within a crew [10].

Defining a strategy to establish a sound analytical approach is among the topics of the present volume. Namely, we are trying to assess those parameters able to identify the "fingerprint" of specific biological (complex) processes, including metabolic, immunologic, protein/genetic, and signaling pathways, in order to depict a comprehensive, systemic model able to provide clear explanations and preventative health scenarios (Fig. 2). Changes in these molecular patterns should be compared also with the intended outputs. In addition, genetic studies are expected to provide meaningful insights into single nucleotide polymorphisms (SNPs) and variants that are of potential relevance for establishing proper pharmacogenomics, tailored drug prescription and personalized health/performance assessment. For instance, it has been observed that a high penetrance hemochromatosis genotype could improve athletes' performance on Earth [11]. However, the same genotype (characterized by increased iron storage) may increase the oxidative iron catalyst under the activity of radiation in space, therefore, raising some nontrivial concerns. The consequent countermeasure should limit iron assumption with the diet in astronauts carrying that genotype. Noteworthy, it should be kept on mind that the space environment exerts several effects not limited to a small number of genes or a single subcellular component, but instead involving many living compartments and functions. The "space field" induces indeed field-dependent consequences that do not recognize single, specific targets but instead involve the organism as a whole [12]. Consequently, these effects must be investigated and can be understood only by taking a global, integrative approach that parallels the approach used to deal with consequences of stresses

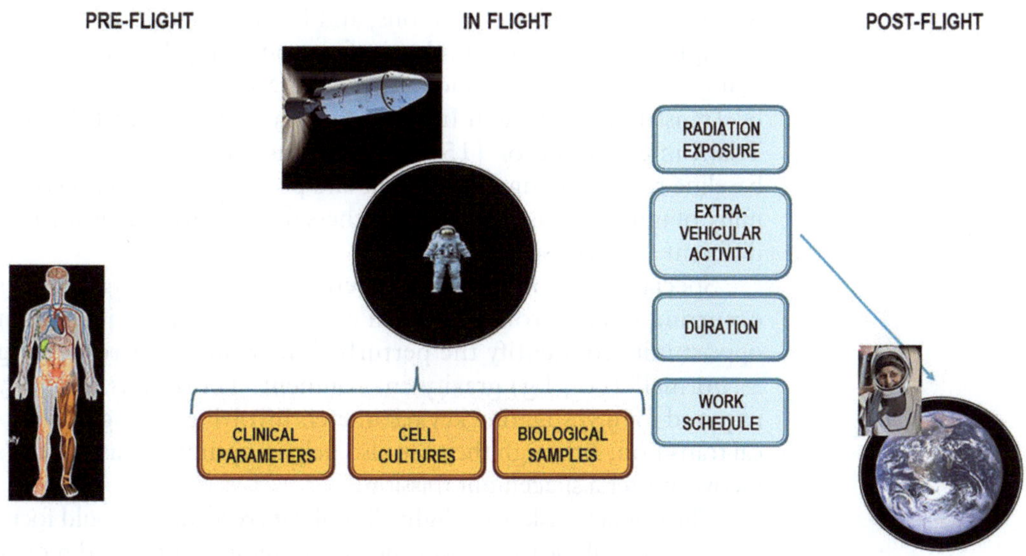

Fig. 2 Flight, environment, and duration of exposure dependence of human health risk during spaceflight

experienced on the Earth, where the antiphysiological menace is multifactorial and systemic [13].

Currently, the hostile space environment severely narrows our ability in planning future human exploration beyond the low Earth orbit (LEO), in which the majority of studies have been hitherto performed. Such studies are challenged by extraordinary complexities related to the acquisition of biological samples, lack of conventional analytical devices, restrictions due to size-limits on board of either the International Space Station (ISS) or the space vehicle. However, the uncertainties cannot be restricted only to technological challenges, given that we are confronting with the boundaries of medicine in addressing very unusual (and unexpected) issues. Indeed, biological and medical investigations carried out in a space environment have forced us to cope with a true conceptual revolution in which the relationships in between physics and biology should be reframed in full. This is probably the most exciting and promising field of enquiry, as these relationships entail a different theoretical viewpoint, and can no longer be restricted to technical/technological approaches aiming at studying biological phenomena with increased precision [14].

Careful consideration of specific environmental cues (radiation, solar flares, and galactic ions) as well as proper evaluation of time exposure is required in planning a versatile PM. Indeed, almost three principal conditions deserve to be investigated over time: preflight, in-flight, and postflight (Fig. 2). To interpret changes during those steps, multiple baseline levels early in the process of astronaut training are warranted. Those findings would be critical for properly identifying the pattern of variance, before and during spaceflight, in order to optimize the astronaut performance. Currently, astronauts undergo extensive medical assessments before launch to establish baseline health metrics. These include cardiovascular fitness, cognitive testing, psychological evaluations, and body composition analysis. Preflight data sets the foundation for monitoring health changes that occur during the mission. The EXPAND database, developed by Translational Research Institute for Space Health (TRISH) and TrialX in collaboration [15], integrates and securely stores these baseline records alongside in-flight and postflight data. This continuity of information enables researchers to compare changes across different stages of spaceflight.

Special attention must be directed toward assessing changes occurring once astronauts return to Earth. This is an important opportunity to identify the perturbation arising after recovering a "normal" (i.e., 1 g) gravity environment. This task is primarily required for investigating key parameters involving during the critical transition, although they can also help in optimizing astronaut's recovery after a spaceflight mission.

The primary task of in-flight clinical data recording should focus on a few critical measures. Key molecular findings, obtained mostly through salivary and urinary analyses, could be as informative as possible in order to reconstruct the physiological status. In addition, those markers should indicate a rationale for the adoption of mandatory (and appropriate) countermeasures. Today and in the future, personalized medicine approaches will afford us the capability to individualize countermeasures in an effort to improve astronaut performance and safety [16].

To fulfil such task, we should rely on developing high-throughput techniques for monitoring the identity and activity of genes, proteins, and metabolites in organisms and their cellular and subcellular components, and interpreting their behavior according to a system biology view in order to grasp the complex interactions between living systems and environmental stressors [17]. Accordingly, systems biology can be defined as the discipline investigating how functions arise from the interactions among the system's components. These techniques belong to different research areas (including genomics, transcriptomics, proteomics, and metabolomics) known as "omics" [18], and require a profound shift in the conceptual paradigm that must be supported by a robust technological advancement. Multivariate analysis and modeling methods are applied to the high-dimensional data sets derived from these studies, in order to describe systems, derive meaning, and generate new working hypotheses.

This objective requires improved analytical devices and innovative technologies in order to acquire and manage a huge amount of clinical data. In addition, deploying "omics" tools onboard spacecraft, even those that are based on mature technologies, poses significant technical difficulties and requires substantial reengineering of their ground-based counterparts. Despite the impressive progress made in the last 5 years, miniaturization, automation, compatibility of protocols, and materials with spacecraft or microgravity need to be further improved.

Acquiring biological data—from both in vitro experiments and astronauts—is, however, complicated due to the environmental constraints, which require the astronaut to operate mostly in autonomy. This is because both medical equipment, reliable technologies, available (operating) space, and specialized personnel are limited. Furthermore, chemical apparatus and diagnostic devices for in-flight application should be carefully designed to deal with various physical limitations (reduced volume and power consumption) and to accommodate with the specific biophysical limits imposed by microgravity, which modulates in a significant manner fluid dynamics (sedimentation, shear stress, convection, buoyancy). In addition, acquisition and management of biological samples (blood, urine, and saliva) is critical, particularly for ensuring safe withdrawal and proper preanalytical/analytical processing. It should be outlined how invaluable the opportunity offered by a space station (i.e., the International Space Station, ISS) is in testing new technologies and devices (as biochemical sensors) for investigating biological processes and monitoring astronauts' health. For several key parameters—especially those belonging to the area of interest of Clinical Pathology—biosensor-based devices provide a useful technological answer in ensuring a reliable analytical performance.

Diagnostic biosensors have already found application in areas such as emergency medicine and resource-limited settings [19], given that they do not require complicated sample pretreatment and can provide semiquantitative or quantitative results in a few minutes. However, sensor devices for in-flight application should satisfy specific criteria, usually more stringent than those that are needed for terrestrial applications. Noticeably, some sensors are already in use on board of the ISS [20]. For space vehicles, "operational constraints" include volume, weight, power, reagent stability, waste generation/disposal, and microfluidics in microgravity. There are multiple international efforts underway to develop miniaturized, low power, microgravity-compatible laboratory instruments. Any piece of hardware going to space must undergo testing for ruggedness and be able to withstand the vibrations of the launch. All hardware must be tested for chemical off-gassing and trace contaminants, to minimize the strain on the ISS air handlers and trace contaminant control systems. Many laboratory analyzers cannot sustain the rigors of this ruggedness testing, and therefore, are not suitable for spaceflight. In addition, the behavior of fluids in microgravity complicates laboratory testing. Liquid reagent vials or syringes will show even-dispersal of air and liquid, with disappearance of the air–fluid interface and resultant formation of bubbles. This challenges the functionality of the instrument and may result in damage to sensitive equipment. Liquid droplets can float freely in the air without a container, which can potentially create a hazard for crew and equipment.

Therefore, NASA has made a concerted effort to test medical technologies over the past decades to advance human health and diagnostic performance in space. As an example, the Exploration Medical Capability program (ExMC) recently adapted a Reusable Handheld Electrolyte and Laboratory Technology for Humans (rHEALTH One) flow cytometer for flight on the space station. Before launch to the orbiting laboratory, the analyzer was modified to function properly in microgravity. This included designing a water bag assembly to prevent air bubbles from entering the lines and to prevent any leakage, ensuring the samples would flow correctly through the device in the absence of gravity. After modification, the rHEALTH One analyzer was launched in February 2022. ESA astronaut Samantha Cristoforetti put it through a series of tests on the space station over 2 days. The results of the microgravity tests will be compared to postflight biological samples to confirm rHEALTH One correctly analyzed and detected any changes. Currently, space diagnostic technology for blood analysis has been developed (as the HemoCue) [21] that requires small sample volume, performs multiple clinical panels, and fits in your hand. However, challenges including limited shelf life of 6 months, refrigeration needs and data reliability issues still exist.

To address such a challenge, an impressive technological endeavor has been carried out to develop new diagnostic tools, medical devices and infrastructures for developing remote sensing and medical support, given that medical capabilities are inherently limited during spaceflights [22]. Improvement in telemedicine, realization of multisensor applications, and innovation in–omic-based approach for analytical discoveries, as well as advances in image analysis would never have been discovered, if it had not been necessary for space exploration. In a nutshell, space biomedicine has represented—and still represent—a major stimulus for the development of advanced biomedical technologies [23, 24].

Studies included in this volume describe several advanced techniques specifically designed to address difficulties and complexities associated with molecular biological inquiry in microgravity conditions.

The chapter authored by Oleg Naimark—*Cell Mechanobiology in Microgravity*—describes a very unusual behavior of living cells within a microgravity field that must mandatory be kept in mind when we perform biochemical studies in space. As previously reported by other reports [25], exposure to microgravity results in the disruption of cell shape uniformity, leading to the emergence of diverse morphological phenotypes downstream of mitosis. Cells are partitioned into two distinct populations, with different morphological and functional characteristics. Intriguingly, the two emergent phenotypes do not involve appreciable change in their respective genomic expression pattern, validating the concept that the same genotype can sustain different phenotypes according to the environmental milieu. The first population consists of flattened spindle-shaped adherent cells, while the second population comprises floating-clump cells, constituted by rounded cells growing as 3D organoid-like structure. These populations grow as stable clusters in microgravity, although they regain their native phenotype when replaced in normal (1 g) gravity. These findings highlight how a change in the field of physical forces can affect significantly the structure and properties of cells and tissues on relatively large scales, through the integration of large-scale correlations between chemical and physical factors. Undoubtedly, these premises suggest that physical factors should be considered true "principles of organization" in biology. In other words, the lack/reduction in gravity acts as an organizing factor and consequently, this implies that living organisms in weightlessness undergo a profound reprogramming of their structures and biological functions. Indeed, as forecasted by the nonequilibrium thermodynamics, dissipative systems experience new self-organizing states from a homogeneous conformational state, downstream of a sharp symmetry breaking [26]. Briefly, cells are usually insensitive to weak forces, when they are in a condition of quasiequilibrium thermodynamic, in which the energy of these influences is less than the energy of thermal motion. However, for out-of-equilibrium dissipative systems, the situation changes radically when the system encounters a critical transition. In such cases, the interaction between structural elements, besides external fields, provides a means of overcoming thermal fluctuations. This renders intracellular structures—specifically the cellular cytoskeleton—susceptible to changes in gravity.

The aforementioned considerations are critical in modeling biological investigations, as discussed by Gary Hardiman et al., in the chapter *Multiscale Modeling and Systems Biology in Microgravity Investigations.* Studies performed in vitro (2D cell monolayers, 3D spheroids, or organ-on-a-chip platforms) offer precise control over culture conditions, high throughput, and detailed molecular readouts, whereas in vivo models (e.g., rodents flown to the ISS or housed in specialized ground-based clinostat cages) capture systemic, multiorgan responses to microgravity, but involve greater logistical complexity and ethical considerations. Together, these platforms and techniques form an integrated toolkit for unraveling the cellular and organismal adaptations to reduced-gravity environments. Similarly, metabolomics workflows require careful metabolite quenching (e.g., cold methanol) and extraction from cell lysates, with both nuclear magnetic resonance (NMR) spectroscopy and high-resolution LC–MS methods used to profile small-molecule alterations in central carbon, lipid, and amino acid pathways. Complementing these empirical approaches, molecular dynamics simulations recreate the microgravity environment in silico by modifying force-field parameters, employing explicit solvent models, and introducing pseudogravity constraints (or gravity vector averaging algorithms) to probe how proteins and membranes behave under reduced gravitational forces. Trajectory analyses—measuring root-mean-square deviation (RMSD), hydrogen-bond lifetimes, and solvent density—reveal microgravity-induced conformational shifts that correlate with experimental omics findings. Together, these methodologies provide a comprehensive, multiscale view of molecular adaptations to spaceflight. Pathway analysis complements network approaches by functionally categorizing differentially expressed genes into established biological processes, which can be implemented using customized data integration workflows in Top-Down, Bottom-Up, and Middle-Out Approaches.

Special attention should be paid to techniques aimed at investigating transcriptomic processes. In the chapter *Transcriptomics in Space: a Basic R Pipeline for Analyzing Bulk RNA-Sequencing Data*, authored by Craig R. G. Willis et al., this issue is largely debated. Spaceflight experiments are currently generating large-scale molecular datasets across different tissues and organisms, demonstrating how large-scale study of gene expression ("transcriptomics") can be useful for producing new insights into the signalling pathways in spaceflight conditions. Technologically, bulk RNA-sequencing provides a simple yet powerful platform for studying transcriptome-wide dysregulation in space, as described by the proposed basic pipeline, based on bulk RNA-sequencing, owing to the wide range of transcriptomics-centric R packages that exist, and the ability to download and execute external transcriptomics tools within R itself.

This issue is further investigated in the chapter, authored by A. Beheshti et al., *From Transcriptomic Data to miRNAs.* Noticeably, transcriptomic data and pathway analyses can provide relevant information about key genes driving specific physiological states, which can in turn be used to predict the miRNAs regulating those states. miRNAs that act as upstream regulators of these driver genes are selected as candidates for a spaceflight-associated miRNA signature. The miRNA regulatory action exerts a "buffer" effect that maintains gene expression levels between homeostatic intervals, and any imbalance in miRNA levels will modify the physiological patterns of the genomic output. Cells can actively secrete miRNAs to the extracellular environment, and consequently, miRNAs can be detected in all biofluids. Circulating miRNAs have emerged as minimally invasive biomarkers capable of capturing complex physiological changes, including those driven by spaceflight. Additionally, beyond their role as biomarkers, miRNAs are increasingly recognized as therapeutic targets. Analytical techniques have been recently implemented with droplet digital polymerase chain reaction (ddPCR), which, compared to conventional PCR methods, enables absolute quantification of miRNA copy numbers and overcomes several limitations associated with traditional miRNA expression assays. Harnessing the power of miRNAs not only deepens our understanding of space biology, but also likely facilitates the development of targeted therapeutic interventions.

The cytoskeleton (CSK) is a prominent example of gravity-sensitive structure. In the chapter, authored by A. Catizone et al.—*The Cytoskeleton in Space*—it is explained how human cells can perceive mechanical forces through field-dependent effects exerted upon CSK. Findings gathered in the last 20 years have given rise to the branch of cell biology called "mechanobiology." The mechanosensitive cytoskeletal structures play a key role in cell gravity sensing, allowing them to adapt to changes in gravitational environments and, subsequently, to trigger adaptive modifications on a number of enzymatic and genomic pathways. Investigating CSK in space requires a concerted effort in conjugating molecular and imaging (confocal microscopy) techniques, thoroughly described herein.

It is worth noting that several of these adaptive mechanisms involve epigenetic modifications. In the chapter, authored by A. Fuso et al.—*Epigenetics in Space: the Dynamic Cellular Response*—it is well described how, unexpectedly, despite their fundamental involvement in the regulation of gene expression, epigenetic modifications have been rarely investigated in space environment. Epigenetic modifications—including DNA methylation, histones modifications, chromatin remodeling, and noncoding RNAs—participate in modulating cell response to space environment (specifically downstream microgravity exposure), leading to several adaptive changes of the epigenome, and consequently to modifications in cell phenotype. Findings gathered up to now suggest that this three-way interaction—environment-epigenome-cell—is associated with cell and tissue altered homeostasis.

In the chapter *Modulation of Cell-to-Cell Adhesion and Interaction with the Extracellular Matrix in Microgravity Conditions*, authored by R. Negri et al., the authors investigated how cells and their microenvironment cooperate each other through complex interactions that, ultimately, shape the phenotype and functional properties of cells/tissues in microgravity. The suggested protocol encompasses morphological and transcriptomic analyses on human neural progenitor cells (hNSPCs) cultured in flasks coated with poly-L-ornithine and laminin matrix at low density and subjected to microgravity simulation. The control group is represented by cultures growing in normal gravity or at higher-density but lacking a matrix.

Dmitrii Atiakshin et al. further address this issue—i.e., the relationship between cells and the extracellular matrix in space conditions—in the chapter *Histochemistry and Immunohistochemistry in Space Biology: Analysis of the Functional Morphology of the Tissue Microenvironment.* Namely, the chapter highlights how the use of multiplex histochemical and immunohistochemical technologies for detecting tissue structures on a histological sample, and artificial intelligence technologies in bioinformatics analysis of big data can provide significant advantages in the multidimensional assessment of tissue microenvironment. Technologies for multiplex staining of histological sections, spatial phenotyping, and digital profiling provide unique information on the state of the immune and stromal landscapes in situ, the extracellular matrix remodeling, as well as the role of intercellular interactions in the implementation of key mechanisms of morphogenesis in space flight conditions.

Integrated, complex functions—like that supported by the reproductive system—may pose a technical and conceptual challenge in space conditions. This issue is addressed in the chapter *Ovarian Reproductive and Endocrine Function Investigated in Microgravity Environment*, authored by N. Monti et al. The paper describes how folliculogenesis and estrogen production are severely altered in space environment. Any disruption in the delicate endocrine cross talk occurring involving gonadotropins, ovarian control factors and steroid hormones may impair oocyte quality, ovulatory function, and overall reproductive health. Microgravity interferes with this delicate hormonal balance, potentially reducing estrogen synthesis and impairing follicular development mostly by downregulating aromatase activity. As the regulation of ovarian function is distributed across different tiers, investigation is particularly delicate and requires studies spanning through different levels and should be compared with data obtained from in vitro cultures.

Regulation of cells/tissues functions is dependent on complicated interactions with the microenvironment, as well as with the microbiome. Recently, this latter question is attracting a lot of attention, as microbiome impact on human health in space conditions cannot be underestimated. The chapter *Defining Microbiome Impact on Host Physiology During Spaceflight Using Caenorhabditis elegans*, authored by S. Buck et al., provides an elegant investigation protocol for addressing such issue, based on the development of a biocompatible platform, which enables the cultivation of a nematode (*C. elegans*) with live microbiome over extended periods. The proposed system (based on polyethylene bags associated with a microfluidic platform) enables the growing of a large nematode population and subsequent extended molecular inquiries, thereby facilitating systems-level investigations of host–microbiome interactions in microgravity.

Complexities embedded in cell interactions and involving the microenvironment require new, advanced experimental tools for appropriate studies. The development of 3D-printed construct allow now the culture of complex three-dimensional organoids, which can proficiently be used for space research. A comprehensive review—*3D Bioprinted Tumor Constructs for Photothermal Therapy: Emerging Opportunities in Microgravity Research*—authored by L. De Sio et al., addresses this question in depth.

Molecular investigations are critical for planning effective regenerative strategies in space, as discussed by C. Marchese et al., in the chapter *Regenerative Medicine in Space: Advancing Bone Repair with Adipose Stem Cells and Bone-On-Chip Technology*. Regenerative medicine is specifically critical for counteracting astronaut bone loss in microgravity, where reduced mechanical loading accelerates osteopenia and impairs fracture healing. Bone-on-chip (BOC) technology, a microfluidic system mimicking bone physiology, enables precise studies of bone formation and degeneration in space. Adipose-derived stem cells (ASCs) committed to osteogenic differentiation (osASCs), offer a promising alternative to primary osteoblasts due to their ease of isolation, expansion potential, and strong regenerative properties. Integrating osASCs into BOC models facilitates research on bone loss mechanisms and novel osteoporosis treatments.

Overall, these contributions outline the need for integrated platforms capable of managing complex datasets. The chapter *Biomedicine and Biotechnology in Space—Methods and Protocols in Systems Biology*, authored by S. Costes et al., introduces the NASA Open Science Data Repository (OSDR), designed to unify multiomics, physiological, environmental telemetry, and radiation exposure data from spaceflight and analog experiments. By strictly adhering to FAIR (Findable, Accessible, Interoperable, Reusable) principles, OSDR ensures standardized data curation and facilitates integrated analyses critical for understanding space-induced biological responses. The integration of AI technologies and federated learning frameworks will transform OSDR into a dynamic predictive modeling platform, enhancing discoveries and astronaut health research.

Finally, the chapter *Precision Medicine in Astronauts: Principles and Targets in Molecular Assessment*, authored by M. A. Schmidt and C. E. Mason, summarizes many relevant technical topics by specifically focusing on Precision Medicine for astronauts. That objective can be attained only by expanding the characterization of individual genotype/phenotype in space conditions, by identifying a network of molecular targets. This approach is required for planning specific countermeasures tailored to individuals and crews.

References

1. Afshinnekoo E, Scott RT, MacKay JM, Pariset E, Cekanaviciute E, Barker R, Gilroy S, Hassane D, Smith SM, Zwart SR, Nelman-Gonzalez M, Crucian BE, Ponomarev SA, Orlov OI, Shiba D, Muratani M, Yamamoto M, Richards SE, Vaishampayan PA, Meydan C, Foox J, Myrrhe J, Istasse E, Singh N, Venkateswaran K, Keune JA, Ray HE, Basner M, Miller J, Vitaterna MH, Taylor DM, Wallace D, Rubins K, Bailey SM, Grabham P, Costes SV, Mason CE, Beheshti A (2020) Fundamental biological features of spaceflight: advancing the field to enable deep-space exploration. Cell 183:1162–1184. https://doi.org/10.1016/j.cell.2020.10.050
2. Antonsen EL, Myers JG, Boley L, Arellano J, Kerstman E, Kadwa B, Buckland DM, Van Baalen M (2022) Estimating medical risk in human spaceflight. NPJ Microgravity 8(1):8. https://doi.org/10.1038/s41526-022-00193-9
3. Tomsia M, Cieśla J, Śmieszek J, Florek S, Macionga A, Michalczyk K, Stygar D (2024) Long-term space missions' effects on the human organism: what we do know and what requires further research. Front Physiol 15:1284644. https://doi.org/10.3389/fphys.2024.1284644
4. Antonsen E et al (2017) Risk of adverse health outcomes and decrements in performance due to in-flight medical conditions. https://ntrs.nasa.gov/archive/nasa/casi.ntrs.nasa.gov/20170004604.pdf
5. Skibba R (2024) Astronauts face health risks-even on short trips in space. Science 384(6701):1159–1160. https://doi.org/10.1126/science.adr0287
6. Bizzarri M, Gaudenzi P, Angeloni A (2023) The biomedical challenge associated with the Artemis space program. Acta Astronaut 212:14–28. https://doi.org/10.1016/j.actaastro.2023.07.021
7. Marques L, Costa B, Pereira M, Silva A, Santos J, Saldanha L, Silva I, Magalhães P, Schmidt S, Vale N (2024) Advancing precision medicine: a review of innovative in silico approaches for drug development, clinical pharmacology and personalized healthcare. Pharmaceutics 16(3):332. https://doi.org/10.3390/pharmaceutics16030332
8. Tanaka H (2010) Omics-based medicine and systems pathology. A new perspective for personalized and predictive medicine. Methods Inf Med 49(2):173–185. https://doi.org/10.3414/ME9307
9. Bizzarri M (2020) Approaching complex diseases network-based pharmacology and systems approach in bio-medicine: network-based pharmacology and systems approach in bio-medicine. In: Bizzarri M (eds) Springer-Nature. https://doi.org/10.1007/978-3-030-32857-3
10. Schmidt MA, Jones JA, Mason CE (2023) Optimizing human performance in extreme environments through precision medicine: from spaceflight to high-performance operations on Earth. Cambridge Prisms: Precis Med 1:e27. https://doi.org/10.1017/pcm.2023.16
11. Schmidt MA, Schmidt JS (2025) Using functionally characterized networks in the application of precision medicine in space. In: Legato MJ, Schmidt MA (eds) Building a space faring civilization: advancing the renaissance of science, medicine, and human performance in civilian spaceflight. Elsevier, New York

12. Masiello MG, Verna R, Cucina A, Bizzarri M (2018) Physical constraints in cell fate specification. A case in point: microgravity and phenotypes differentiation. Prog Biophys Mol Biol 134:55–67. https://doi.org/10.1016/j.pbiomolbio.2018.01.001
13. Buescher JM, Driggers EM (2016) Integration of omics: more than the sum of its parts. Cancer Metab 4:4. https://doi.org/10.1186/s40170-016-0143-y
14. Bizzarri M, Fedeli V, Piombarolo A, Angeloni A (2022) Space biomedicine: a unique opportunity to rethink the relationships between physics and biology. Biomedicines 10(10):2633. https://doi.org/10.3390/biomedicines10102633
15. https://trialx.com/how-health-data-is-collected-in-space-missions/
16. Schmidt MA, Goodwin TJ (2013) Personalized medicine in human space flight: using omics based analyses to develop individualized countermeasures that enhance astronaut safety and performance. Metabolomics 9(6):1134–1156. https://doi.org/10.1007/s11306-013-0556-3
17. Bizzarri M, Palombo A, Cucina A (2013) Theoretical aspects of systems biology. Prog Biophys Mol Biol 112(1–2):33–43. https://doi.org/10.1016/j.pbiomolbio.2013.03.019
18. Karouia F, Peyvan K, Pohorille A (2017) Toward biotechnology in space: high-throughput instruments for in situ biological research beyond Earth. Biotechnol Adv 35(7):905–932. https://doi.org/10.1016/j.biotechadv.2017.04.003
19. Cummins BM, Ligler FS, Walker GM (2016) Point-of-care diagnostics for niche applications. Biotechnol Adv 34(3):161–176. https://doi.org/10.1016/j.biotechadv.2016.01.005
20. Roda A, Mirasoli M, Guardigli M, Zangheri M, Caliceti C, Calabria D, Simoni P (2018) Advanced biosensors for monitoring astronauts' health during long-duration space missions. Biosens Bioelectron 111:18–26. https://doi.org/10.1016/j.bios.2018.03.062
21. Crucian B, Valentine R, Calaway K, Miller R, Rubins K, Hopkins M, Salas Z, Krieger S, Makedonas G, Nelman-Gonzalez M, McMonigal K, Perusek G, Lehnhardt K, Easter B (2021) Spaceflight validation of technology for point-of-care monitoring of peripheral blood WBC and differential in astronauts during space missions. Life Sci Space Res (Amst) 31:29–33. https://doi.org/10.1016/j.lssr.2021.07.003
22. Antonsen EL, Mulcahy RA, Rubin D, Blue RS, Canga MA, Shah R (2018) Prototype development of a tradespace analysis tool for spaceflight medical resources. Aerosp Med Hum Perform 89(2):108–114. https://doi.org/10.3357/AMHP.4959.2018
23. Miklós V Everyday things you didn't know were invented by NASA. https://gizmodo.com/everyday-things-you-didnt-know-were-invented-by-nasa-1663622199
24. Guarnieri JW, Kim J, Wallace DC, Mason CE, Muratani M, Beheshti A (2025) Methodologies for mitochondrial omic profiling during spaceflight. Methods Mol Biol 2878:273–291. https://doi.org/10.1007/978-1-0716-4264-1_15
25. Po A, Giuliani A, Masiello MG, Cucina A, Catizone A, Ricci G, Chiacchiarini M, Tafani M, Ferretti E, Bizzarri M (2019) Phenotypic transitions enacted by simulated microgravity do not alter coherence in gene transcription profile. NPJ Microgravity 5:27. https://doi.org/10.1038/s41526-019-0088-x.
26. Kondepudi DK, Prigogine I (1983) Sensitivity of non-equilibrium chemical systems to gravitational field. Adv Space Res 3:171–176

Part II

Conceptual Breakthroughs and Research Strategies

Check for updates

Chapter 2

Hypergravity, Microgravity, and Animal Research

Jack J. W. A. van Loon

Abstract

Gravity plays a significant role in both life and physical science processes. In life sciences, at the level of a whole organism, gravity as unit gravity on Earth ($9.81\,ms^{-2}$), the lack of gravity during free fall, or the increase of gravity while, e.g., during centrifugation is reflected in an organism's metabolism, anatomy, and behavior. In this chapter, the impact of near weightlessness and hypo- and hypergravity on animals related to buoyancy, sedimentation, and hydrostatic pressure will be addressed. It will provide a concise historical overview of these processes for in-flight and ground-based studies on animals and the experimental modalities that have been or could be used in future investigations.

Keywords Animal, Spaceflight, Centrifuges, Hypergravity, Microgravity simulation

1 Introduction

There are four basic forces in nature: the weak nuclear force, the strong nuclear force, the electromagnetic force, and gravity where gravity, g (with lower case "g" not to be mistaken with the capital "G" for the gravitational constant!), is the one that has a very extended range of impact going from the level of molecules (size order $7.4 \times 10^{-11}\,m$) to larger than the Milky Way (size order ~$2.4 \times 10^{20}\,m$) that spans more than 30 orders of magnitude. Within that range, life on Earth developed at a relatively stable gravity level of one "g" of around $9.8\,m/s^2$. This chapter will not address the origin of life on our planet (for that, see a comprehensive publication edited by Seckback and Walsh [1] and also many other works looking into this subject). In the very early days of life development, a direct role of gravity at the level of life-related molecules or very small single-cellular structures is most likely negligible (see e.g., [2–4]). Indirect effects of the force of gravity can be found in remnants of very early life dating from the Precambrian times found as preserved in fossilized stromatolites (see e.g., [5]). In this case, gravity had an extracellular effect of fluid convection. However, there surely are single-cell organisms that can orient themselves related to the gravity

Mariano Bizzarri (ed.), *Biomedicine and Biotechnology in Space*, Methods in Molecular Biology, vol. 3000,
https://doi.org/10.1007/978-1-0716-5174-2_2,

vector. Organisms like the single-cell flagellate eukaryote *Euglena* [6], the ciliated protozoon *Loxodes* [7], or the magnetotactic bacterium *Magnetospirillum magnetotacticum* [8] can directly sense and make use of the direction of the gravity vector.

We will focus on the type of studies that have been performed from the very early days of spaceflight and altered gravity related research, limited to the animal kingdom and discussing the current approach in order to reveal and understand the role of weight in living systems.

Probably one of the very first documented publications mentioning a role of gravity in the anatomy of living systems was by the Italian homo universalis Galileo Galilei who stated in his publication "Two New Sciences" published in 1638 that *Man cannot build a house, nor nature can construct an animal beyond a certain size without altering the design or material* (see Fig. 1A). And although this effect is clearly visible at a systemic level of the skeleton, it has also been shown that during bone development the mineralization of the cartilaginous matrix is directly impacted by gravity at a local tissue level. In vitro studies using embryonic mouse metatarsal long bones showed decreased mineralization, and increased mineral resorption, when developed under near weightlessness conditions [9], and this reduced mineralization can be reversed by exposing these explants in-flight to at least 6 h of 1 g in an on-board centrifuge [10].

So there is a direct impact of gravity on living systems, especially when they are larger in size (i.e., mass). There are some general phenomena that depend on or are driven by gravity, which are relevant to animals: buoyancy, sedimentation, hydrostatic pressure, and convection. We will address the first three in more detail. For other

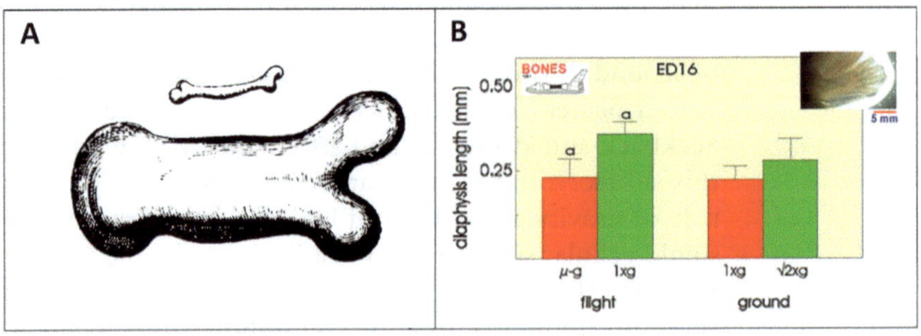

Fig. 1 **(A)** Image from Galileo Galilei's work "Two New Sciences" stated that "…bones of a large animal should be proportionally much larger than the similar bones of a small animal. And I discovered that this happens because the weight grows faster than the resistance of the bones: one animal twice as much as another has each bone twice as much more resistant, but the animal will be eight times heavier…." **(B)** Mineralization of mouse metatarsal long bones under in-flight μg or in-flight 1 g control conditions. Decreased mineralization under microgravity conditions can be clearly seen. This in vitro study was performed in ESA Type-I experiment containers incubated in the Biorack facility [9]

forces related to life, I would recommend the reader to explore, e.g., the works by Steven Vogel [11].

2 Buoyancy

The musculoskeletal system plays a crucial role in the development of animals with respect to gravity. As life has most likely begun in water, its development was aided by the gravity-dependent buoyant environment. The support of this mechanically supportive environment is reflected in more white muscles with fast-twitch fibers, lower myoglobin content, and fewer mitochondria suited for intense bursts of activity found in fish compared to the slow-twitch fibers, higher myoglobin content, and abundant mitochondria seen in red muscles of land living animals, which are involved in more endurance activities coping with gravity in a non-buoyant environment [12].

So when around 375 million years ago the *Tiktaalik* (Fig. 2) went from a buoyant to a nonbuoyant habitat, it meant that its musculoskeletal system, its joints, its bone support capacity, and its muscle type were then capable of supporting its own weight on land [13]. The whole locomotory and skeletal system comes under a completely new set of forces when the support of the water is withdrawn, and the effects of gravity become insistent. In addition, the skin must be

Fig. 2 Artist impression of the Tiktaalik, the animal that was able to go from a buoyant environment onto land some 375 million years ago(See Shubin et al. [152], Credit: Zina Deretsky, National Science Foundation)

changed to resist desiccation, the respiratory system must be adapted to use gaseous oxygen, and receptors must be adapted to signal the strange new configurations of stimuli [14].

In nearly all animals, the initial embryonic development starts from within a buoyant environment of the amniotic fluid in, e.g., amphibian and bird eggs or in the uterus of mammals [15, 16]. Although buoyancy scales linearly with the specific density ergo weight of a structure, an increase in gravity might have an effect on animal development during embryogenesis.

It was for a long time that one presumed that gravity was involved in the establishment of the bilateral axis in embryo development, especially in yolk-rich eggs. However, from the first in-flight fertilization of amphibian eggs of *Xenopus laevis,* as was performed in a sounding rocket in 1988, it was shown that fertilization is possible in a near weightlessness environment [17], but also later stages of development, up to gastrulae, showed that gravity is not required to establish the dorso-ventral axis in these species [18, 19]. See also the interesting work from Houston on this subject [20].

The significance of living in a buoyant environment is also nicely demonstrated in the amphibious fish *Kryptolebias marmoratus,* which lives parts of the year on land. Turko and his co-workers found that during their period on land, these fish increased their gill arch stiffness for some 60% compared to their buoyant living period. The increased mechanical properties are related to an increase in bone-related proteins, the same proteins as seen in bone formation in tetrapods [21].

3 Sedimentation

Sedimentation is another phenomenon that is gravity-dependent. We are all familiar with the concept that heavy/larger particles settle faster in a liquid than lighter/smaller particles (of the same density). However, this is only true when this observation is made in a gravitational field. Under weightlessness, these particles will not settle but float in the liquid driven only by Brownian motion.

Within the living nature, we see numerous multicellular organisms that make use of this settling phenomenon. One of the best-known examples is the case of starch grains in, especially, the columella cells of a plant. Settling of these heavy particles in the plant will direct its growth (see other chapters of this publication with an extensive explanation of this plant gravitropic process). However, this sensory mechanism is not much different from the heavy particles in our inner ear vestibular system. Also here the heavy particles, otoconia, in the saccule provide, via the stereocilia on the vestibular hair cells, our brain information regarding the direction and magnitude of the gravity vector in order to facilitate orientation and postural control.

During evolution, the first signs of an acceleration sensory system were reported from jellyfish, a part of the *Cnidaria* phylum whose fossils date back to the Middle Cambrian, some 500 million years ago [22]. Although the composition of the minerals in the various vestibular systems is nearly the same, all calcium carbonate ($CaCO_3$), the type of crystals formed depends on the evolutionary position ranging from primitive to advanced fish and higher vertebrates, evolving from apatite to aragonite and eventually calcite, respectively [23]. And while mammals have two compartments for sensing accelerations, the utricle and the saccule, for horizontal and vertical accelerations, respectively, non-mammalian vertebrates, like fish, have a third compartment to sense accelerations, often combined with a hearing function, the lagena [24]. On the other hand, other species like the unsegmented pseudocoelomate *C. elegans* do not have a vestibular system, as mentioned for fish or mammals, but are capable of sensing gravity mediated by active neural processes that involve sensory cilia and dopamine [25].

Obviously, there have been numerous studies performed to explore the effects of gravity on the vestibular system as well as on the possible role of the vestibular system on other parts of an organism.

In general, we can say that the development and maintenance of the vestibular organs are gravity-dependent. The exposure of zebrafish (*Danio rerio*) to moderate hypergravity (HG) of 3 g for 5 days in a large centrifuge [26] resulted in a significantly larger head and a significant increase in bone formation for a subset of the cranial bones (see Fig. 3A–C), while mineralization of the otoconia is significantly diminished [27]. So we see in the same animal, both an increase and in other locations a decrease in mineralization most likely related to its function. The mandibula in a higher mechanical load environment of 3 g needs to be better capable of withstanding such loads, while for otoconia, there might have been some developmental feedback mechanism regarding the level of mineralization. It might be that the level of the efferent signals from the sensing organ toward the brain is sufficient with a lower mass of the otolith at 3 g. A similar result has been reported for mice that have been exposed to microgravity on board the ISS for 90 days. They seem to have an increase in otoconia size compared to controls, while mice exposed to 2.38 g for 3 months showed an ablation or thinning of the outer shell and cavitation of the inner core of the otoconia [28]. This phenomenon is not only seen in vertebrates but also in *Aplysia* (sea hares), which do not have a complex developed vestibular system as in vertebrates, but a more rudimentary statocyst. However, the principle of a heavy structure (statolith) resting on the sensory epithelium (setae) is very much similar in this balance sensory receptor. They are reported to have a decrease in statolith diameter when exposed to 2 g [29], but also, when compensated for body diameter, the effect was larger at $5 \times g$

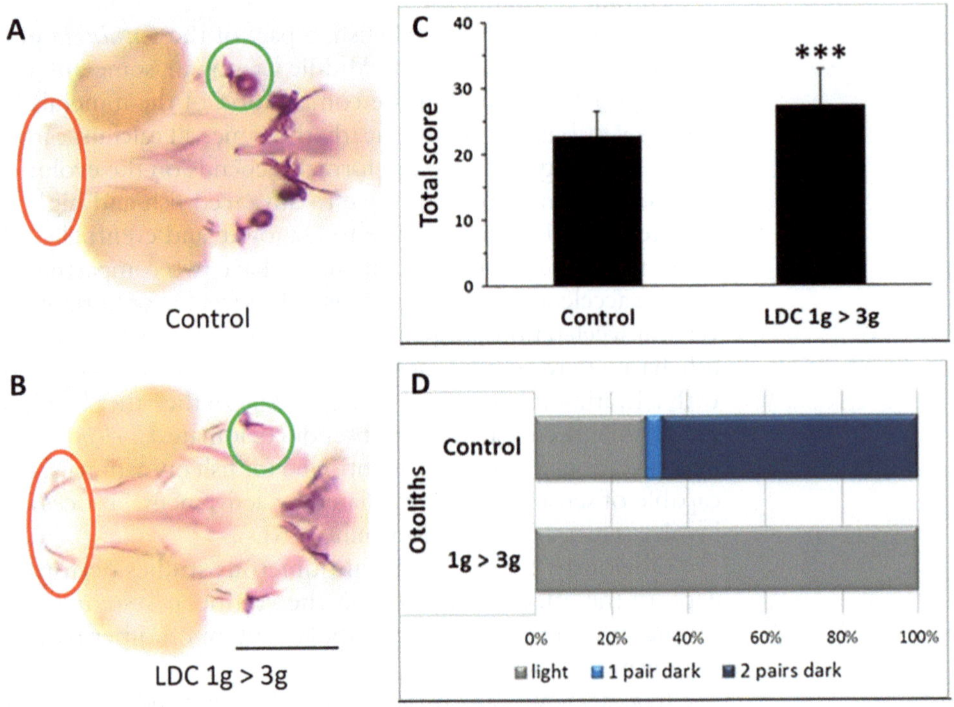

Fig. 3 Zebrafish after exposure to 3 × g at the LDC centrifuge facility at ESA-ESTEC in the Netherlands. Alizarin red staining of 9 days post fertilization control larvae (**A**) and larvae treated for 4 days in 3 g hypergravity (**B**). (**C**) Global score for bone formation in control and 3 g-treated larvae. (**D**) Comparison of cumulated frequencies of, respectively, light, one pair dark, or two pairs dark otoliths in control and 3 g-treated larvae. The red ellipse in A and B indicates the increased mineralization of the mandibulae at 3 g, while the green circle indicates the reduced mineralization in the otolith at 3 g(From Aceto et al. [27] with permission)

than at 3 × g compared to 1 × g controls [30]. However, there seems to be no direct effects on statolith diameter in the latter study.

Other extensive research regarding the vestibular system is performed on fish exposed to near weightlessness in parabolic flights with guppies (*Poecilia reticulata*), blind cavefish (*Astyanax mexicanus*), and angelfish (*Pterophyllum*) [31] or even in drop towers using cichlid fish (*Oreochromis mossambicus*) [32]. At one of the Skylab missions in the 1970s, two mummichog fish (*Fundulus heteroclitus*) were exposed to chronic near-weightlessness. Upon video observation at day 3 in flight, it was found that both fish swam in tight circles for a considerable fraction of the observed time, resembling the "looping behavior" as observed in previous parabolic aircraft flights. However, the frequency of looping diminished slowly after the third day until normal swimming was prevalent. On flight day 22, both fish swam normally with their backs turned toward the light source, indicating that the animals had adapted to the weightless conditions after some time [33]. An interesting observation was made by Hilbig and colleagues, who noted some typical looping and spinning movements [34]. It appeared that animals showing this response had otolith asymmetry (i.e., differently weighed statoliths on the right and the left side of the head), which was displayed during the microgravity phase of a parabolic flight.

Although the vestibular system is still regarded as only involved in balance and orientation, there have been many animal studies that show that the vestibular system might have an impact on many other physiological systems. One useful model in that respect is the *het*-mouse, a mouse strain that lacks macular otoconia. When wild-type animals were exposed to 2 g for 8 weeks, they showed a decrease in total body temperature and an attenuated total body circadian rhythm temperature, while there was hardly a response in the *het*-mice. Changes seen in WT at hyper-g mice like ambulatory activity, heart rate, feeding, and adiposity were diminished or lacking in *het*-mice [35]. Also the baroreflex regulation, as measured by heart rate (HR) and mean arterial pressure (MAP), in the otoconia-deficient C57BL/6JEi head tilt (*het*) mice exposed to phenylephrine (PE) or sodium nitroprusside (SNP) was respectively increased and decreased in *het*-mice compared to wild-type animals [36]. The vestibular system even seems to have a direct link with the musculoskeletal system. When mice are exposed to 3 g for 4 weeks, they show an increase in bone parameters, such as a decrease in RANKL and RANKL/OPG, as well as increased soleus muscle tissue weight. However, these phenotypic adaptations to this increased mechanical environment are not expressed when the mice have their vestibular system mechanically removed [37].

Besides the more functional or phenotypical approaches, recent studies have also focused on the mechanisms, proteins, and genes involved in the development of otoconia to better understand and treat pathologies such as imbalance and vertigo [38].

4 Hydrostatic Pressure

The impact of the gravity-driven hydrostatic pressure is mainly relevant for the cardiovascular system especially in bipedal animals, so the impact of gravity for this system is less of a concern for rats and mice as it is for humans. Therefore, rodents might not always be the best translational model to explore possible effects of chronic near weightlessness hematopoietic responses [39, 40]. Circulatory-hematopoietic responses related to chronic hypergravity can be different in rats compared to chickens. Extrapolated plasma volume data from 3 g chronically accelerated chicken indicated that the chick, with its upright posture in the gravitational field, would be a very comparable model for human regarding this parameter [41, 42]. However, over the years and due to the possibility for genetic manipulation, the mouse gained more and more terrain in the field of translational studies.

The impact of hydrostatic pressure is clearly reflected in some anatomical adaptation in various animals. One of the tallest animals, the giraffe, has to cope with a large hydrostatic pressure difference in its cardiovascular system. Its arterial pressure (p_a) goes from some 110 mmHg (~147 mBar) at the upper neck to some 260 mmHg (347 mBar) at its hoof [43]. Special anatomical adaptations such as increased thickness of the tunica adventitia and tunica media, relatively more collagen compared to elastin, and innervation of the vasa vasorum in the lower extremities and tight skin layers are made such that the blood and extracellular fluids remain within the animal at such high delta pressures [43, 44] but also the extraordinary high pressure differences, as predicted for the extinct *Brachiosaurus brancai* [45].

Another remarkable adaptation of anatomy with respect to the gravity vector can be seen in snakes. Where aquatic snakes spend most of their lives in a (with respect to gravity) horizontal orientation, arboreal snakes orient themselves alongside or opposite the gravity vector when they climb in or descend from a tree, respectively. One of the main functions of the heart is to provide the brain with sufficient oxygen. For a more energy-sufficient anatomy, you see that in the aquatic snake the heart is placed further away from the brain than in arboreal snakes since, in the latter, it has to cope with changes in hydrostatic pressure all the time. A land-living snake holds the average between the two configurations [46].

An extreme deviation from the norm with respect to cardiovascular anatomy is the case of bats where their hemodynamic system is inverted when they are at rest from feet-ward to head-ward pressure increase. In a pilot study, bats were inverted for a prolonged period of time from 8, 15 to 22 days (Fig. 4). It was reported that the ascending

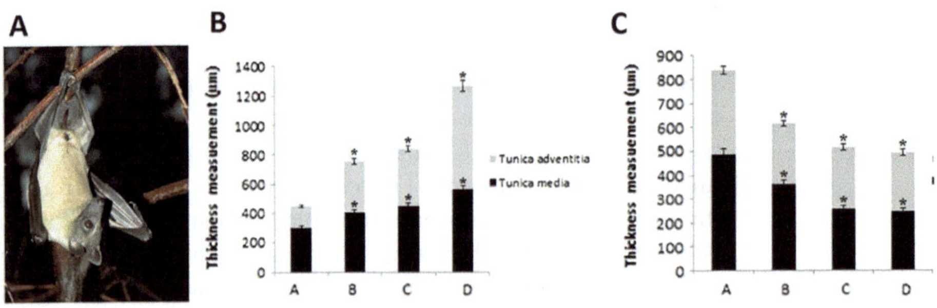

Fig. 4 (**A**) Picture of a fruit bat (*Eidolon helvum helvum*) hanging upside down (picture courtesy of Klaus Rudloff: www.biolib.cz). (**B**) Changes in ascending aorta tunica media and tunica adventitia with prolonged head-down posture. (**C**) Changes in abdominal aorta tunica media and adventitia with prolonged inversion. A is control and B, C, and D are, respectively, 8, 15, and 22 days of inversion (Taken from Olumide Ashaolu et al. [47] with permission)

aorta in the experimental animals exhibited a thickening of the media and adventitia, while the abdominal aorta showed thinning of these same regions, and these changes increased in magnitude with longer periods of inversion. These morphological changes, because of pressure deviations, were reflected in the smooth muscle cells and elastin and collagen content of the vessels [47].

5 Centrifuges

Although some examples of animal hypergravity/centrifugation experiments are addressed earlier, we want to address a few issues that are important with respect to the use of centrifuges.

It can be very illustrative to use centrifuges to explore the impact of gravity or microgravity on living systems, but the disadvantage of centrifuges is that they rotate. This means that the actual linear acceleration of gravity is now not generated by increasing velocity but by changing the direction of the object. This centrifugal force, first described by the Dutch mathematician and physicist Christiaan Huygens in his book *De vi Centrifuga* written around 1659, generates a kind of inertial force given by the equation $F_c = ma_c = m\omega^2 r$, where F_c= the centrifugal force (in g), a_c = centripetal acceleration (m/s^2), ω = angular velocity (rad/s), and r = radius (m).

The other artifact related to centrifugation is the Coriolis force named after the French mathematician Gaspard-Gustave de Coriolis (1792–1843). This pseudo-force is the so-called cross-coupled response, owing to the angular motion in two planes, and contributes to the impurity of gravity generated in a rotating system, so Coriolis forces are only applicable when moving like walking or swimming in a rotating frame. The expression for Coriolis is $a_{Coriolis} = (2\ v \times \omega) / g$, where $a_{Coriolis}$ = Coriolis acceleration expressed in g, v = velocity of the moving object (m/s), ω = angular velocity of rotating system (rad/s), and g = acceleration due to gravity (m/s^2) (see also [48]).

It can be seen from both equations that some artifacts in centrifugation depend to a large extent on the radius of the system, ergo, by the rotations per minute (rpm) the platform has to rotate to generate a certain g level. That is the reason that larger centrifuges are preferred over small radii systems. In a study by Charles Fuller, male Sprague-Dawley rats were used, implanted with a biotelemetry transmitter and exposed to $1.5 \times g$ in centrifuges with a diameter of 1.8, 2.5, or 6.0 m [49]. The circadian rhythms from animals in the larger diameter system were much more regular than the ones with higher rpms. So, although exposed to the same mechanical load of 1.5 g, the rotation of the system affects some physiological parameters.

Other empirical advantages of larger centrifuges are a more homogeneous g field for the samples and a reduction of the inertial shear forces [50]. Finally, as in any moving system, vibrations and noise might be relevant for certain samples at certain g levels.

Animal research, especially research including mammalian species, using hypergravity (HG) started roughly some half a century ago. The initial species investigated were mice [51], rats [52, 53], dogs [54, 55] and chicken [56], turtles [57] or toads/*Xenopus* [58], and rabbits [59], but later snakes [46], crustaceans like hornets [60] and crickets [61], *Drosophilae* [62, 63], and fish [64, 65] were also subjects of studies. Nearly all body systems are investigated and are influenced by an increase in weight. In light of this, it might also be interesting to experimentally revisit Kleiber's law [66].

Most hypergravity studies are performed with rodents, and a general phenomenon reported is the loss of mass when exposed to hypergravity. In a study at the AMC hospital in Amsterdam (NL), a long-duration hyper-g study was conducted in a centrifuge with an arm length of 1.10 m and ventilated free-swinging gondolas (height 0.725 m, width 0.45 m) at the end of the arms. The g vector was always perpendicular to the bottom of the cages (size $0.36 \times 0.44 \times 0.30$) and had a magnitude of either 2.5 or $2.0 \times g$ depending on the type of experiment. The mating and pup gestation all occurred under these HG conditions. However, during the first 10 postnatal days, gravity level was set at $1.8 \times g$ to increase the survival rate of the pups. Thereafter, HG was increased to $2.5 \times g$ again (Fig. 5). Ambient noise level for HG and control rats was approximately the same (53 dBA), and the main source of noise was caused by ventilation. The animals were hypophagic and showed a decreased body weight at the initial phase of centrifugation, and they picked up a normal growth rate after several days [67–70]. While there was a reduction in body mass at HG, most animals were euphagic, indicating an increased metabolic rate [59], and there was a relative decrease in fat mass

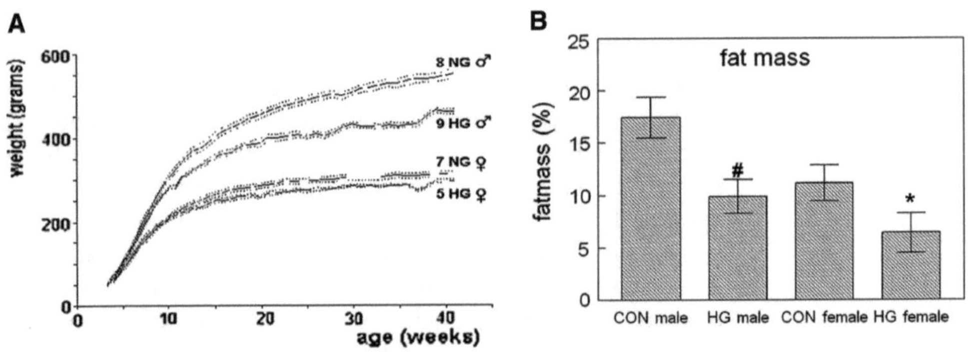

Fig. 5 Long–Evans rats conceived, born, and raised at $2.5 \times g$ for 40 weeks. (**A**): The decreased body weight as noted in this study is in agreement with numerous other chronic hypergravity studies. The reduced weight is more pronounced in males. (**B**): What also is obvious in many HG studies is that animal all show a reduction in fat mass [69]. Also in this study the body fat mass in % was significantly reduced in both male and female HG animals compared to controls. (From (Van Loon et al. [153]))

and a low level of circulating leptin level [71–73, 55, 74]. Based on these chronic animal hypergravity data, one might speculate that with the current worldwide obesity epidemic [75], chronic human centrifugation might be applied for reasons of research and may be treatment of this disease [76].

In the early days of gravitational physiological experimentation with animals, studies by Matthews and colleagues [77] were performed where rats were exposed for three generations to 3 g. Others exposed mice chronically to 7 g for the period of 1 year [78].

Since gravity has an impact on mass, one can make the general statement that the larger (i.e., heavier) the animal is the higher the impact of gravity (see also Fig. 1). A publication from Wunder and colleagues indicated the maximum survival time at various g intensities for a large series of specimens ranging from single cells to larvae, guppies, mice, rats, and humans [78]. What can be deduced from that work is that the impact of gravity also depends greatly on time. Organisms can withstand relatively high g levels but for only a short period, while relatively low gravity levels can be tolerated for chronic exposure. From this work, it is indicated that a mouse will survive a hypergravity level up to 1050 g but for only one minute (which is still about 500–700 heart beats!! [79]), while we have seen that it could survive 7 g for a year [77]. These values are 50 g and 3 g for a rat or 4 g and ~1.3 g for chickens, respectively. In the publication from Chae, the LD50 values for exposure to hypergravity in the $+g_z$ direction in a 70 cm radius centrifuge and at an onset rate of 2.8 g/s were 20.4 g for mice, 10.2 g for rats, and 5.3 g for rabbits, while 27.3 g for finches, 15.9 g for pigeons, and 11.2 g for roosters [80].

Although centrifuges are used to learn about hypergravity as such, they can also be applied to predict what might happen at partial g levels. Basically, there are two methods to explore low gravity levels. (1) The first method is the gravity continuum. In such an experiment design, one performs a series of experiments from higher g levels down to 1 g. The relationship of this data, e.g., linear or exponential, can be extrapolated to lower g levels (Fig. 6, left). This approach can also be used for physical science or technology studies. (2) The second approach is the "Reduced Gravity Paradigm" where a biological system is exposed for a certain period to hypergravity such that it reaches a steady state. When this state is reached, the centrifuge is slowed down or stopped to reach a lower g level. The hypothesis here is that the responses of the system (cells, organoids, plants, animals, etc.) going from a higher to a lower g level are similar (most likely not the same) as in a common spaceflight experiment where you go from Earth 1 g to microgravity (Fig. 6, right). See also [81] as well as [82] and [74] regarding the gravity continuum. The preferred design of study to explore the effect of gravity is to perform an experiment both at micro-g and/or partial g, combined with a hypergravity study. There are not too many examples of that.

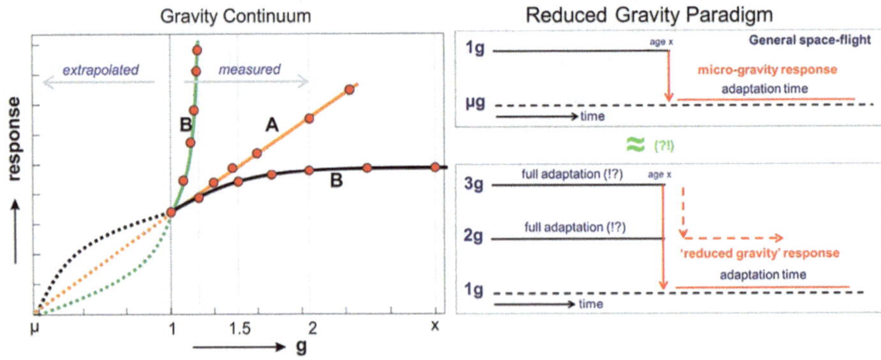

Fig. 6 Left: Cartoon of the Gravity Continuum. The red dots represent measurements at various hyper-g values. Connecting lines based on the measurements (solid) can be extrapolated (dotted lines) to lower g conditions. Right: Graphic representation of the "Reduced Gravity Paradigm," RGP. The directions of the response(s) going from 1 g to μg are (nearly) the same as going from a higher to a lower g level. For more details, see [81]

One is a study done using a mouse facility that was developed by the Italian Space Agency, ASI, called the Mouse Drawer System. In this middeck locker size facility, six mice could be housed. It was a completely automated facility built for a Shuttle where the crew had to provide water and food only after some 15 days [83–85], and the same MDS facility was used to expose mice to 3 g for 15- and 30-day periods in the LDC centrifuge at ESA-ESTEC [86, 87]. An approach that could also be applied for physical sciences studies, e.g., as in studies on liquid waves [88, 89].

Centrifuges have also been applied in-flight, and one of the earliest uses of in-flight centrifuges for animal research was by the Russian series of biocosmos flights (Fig. 7A): Bion-3 (1975), Bion-4 (1977), Bion-5 (1979), Bion-10 (1992), and Bion-11 (1996) [90–93]. On the Kosmos 782 (Bion-3), turtles and fish were exposed to 1 g in flight. The postflight analysis after this 20-day mission showed that the behavior and physiology of 1 g in-flight animals were comparable to ground-based 1 g controls. On the 18.5-day duration Bion-4/Kosmos-936 biosatellite, 10 rats were exposed to 1 g artificial gravity in a 320 mm radius centrifuge (Fig. 7A). Compared to the microgravity animals, the 1 g in-flight rats had, among others, fewer effects on errors during post-flight maze tests; an increase in water intake and a decrease in diuresis; a fall of the resistance of peripheral red blood cells; a decrease of oxygen consumption; a decline in the capacity to keep balance on the rail; or a drop in the static physical endurance [94].

Since the application of centrifuges in Russian missions, other space agencies also started to make use of in-flight centrifuges mainly for on-board controls for space radiation and launch-related effects. For the European entities like DLR and ESA, it was especially the BOTEX, NIZEMI, Biorack, and Biobox facilities [95] that provided

Fig. 7 In-flight centrifuges. (**A**) Rat centrifuge as flown on Kosmos 936/Bion 4 launched in 1977 (image courtesy of Russian Institute for Biomedical Problems (IMBP), Moscow, Russia). (**B**) The Centrifuge-equipped Biological Experiment Facility developed by JAXA accommodating the mouse habitat cage units (HCU) (Image from [102]). (**C**) The middeck locker size Multi-use Variable-gravity Platform (MVP), which can produce artificial gravity in 0.1 g increments, up to a maximum of 2.0 g(Image courtesy of Techshot (now Redwire), Greenville, IN, USA)

this centrifuge capacity. However, due to its limited volume of experiment containers (~2 × 4 × 8 cm), only smaller animals like nematodes [96], fruit flies [97], highly radiation-sturdy rotifers [98], or tadpoles [99, 100] could be housed in such volumes [101].

Currently, the main rodent centrifuge facility on board the International Space Station, ISS, is the Japanese Mouse Habit Unit (MHU) located in the Kibo module [67] (Fig. 7B). Mice housed in the facility exposed to artificial gravity showed that after 35 days the 1 g in-flight mice maintained the same bone density and muscle weight as mice in the ground 1 g control group, unlike the mice that were under microgravity for the same time [102]. The Chinese space station Tiangong accommodates the Varying Gravity Rack (VGR) that is also applied to create on-board artificial gravity [103]. However, facilities like Kubik [104] with a similar experiment volume to that of the above mentioned Biorack or Biobox are also used for smaller animals such as early-stage developments of newts (*Pleurodeles*) [105]. A recent animal centrifuge is the commercial Multi-use Variable-g Platform (MVP) developed by TechShot (Fig. 7C). In its first flight of this 390 mm diameter centrifuge in 2018, it housed *Drosophila*. The facility including numerous animals was returned to Earth for further analyses [106].

Most in-flight centrifuge are or have been used to provide an on-board 1 g reference control. Seeing the advancements made to venture out to the Moon and Mars, these centrifuges are also used to explore these levels of partial gravity, very much a space operational approach. However, if we want to have a broad and basic understanding of what the impact of gravity is on biological (but also physical) systems ranging from single cells to animals and plants, we should also have a more dynamic and segmented research approach. Centrifuges are the perfect instrument to explore this.

One can use an in-flight centrifuge in a classic way as mentioned earlier to compare microgravity with 1 g (Fig. 8A, B). However, with

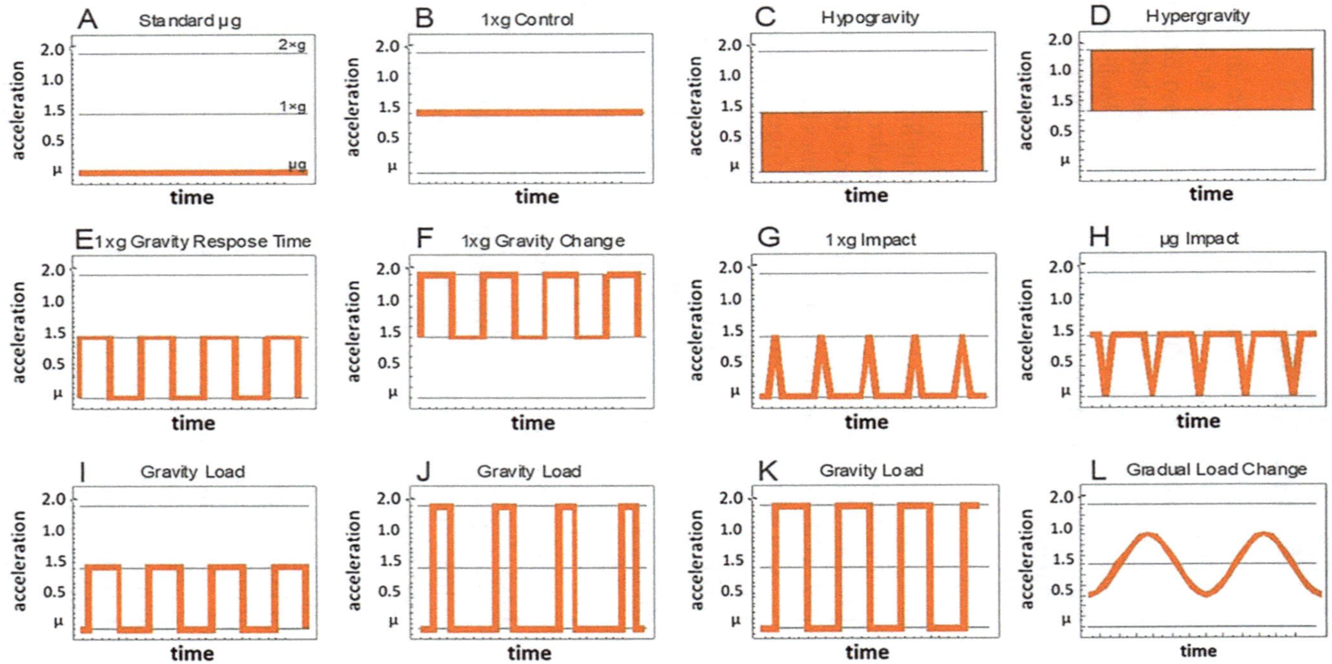

Fig. 8 (**A** and **B**) Standard (classic) gravity profiles. (**C** and **D**) Static hypo- and hypergravity profiles. (**E** to **L**) Dynamic profiles. (**E** and **F**) "Gravity change" profiles. (**G** and **H**) Gravity and microgravity "impact profiles." Profiles (**I**, **J**, **K**, and **L**) Total gravitational loads and gradual changes. (See main text for additional information)

an on-board centrifuge that can run up to 2 g, we can also explore static g exposures to hypogravity (between 1 and μ (Fig. 8C) or hyper-g (Fig. 8D) to evaluate whether the changes found in the hypogravity area continue in the hypergravity domain. To overcome possible effects of microgravity (countermeasures), an exposure for shorter or longer times to 1 × g can be evaluated (Fig. 8E) (see e.g., [10]). In addition, profiles could be applied sequentially to verify whether the exposure to 1 × g, as such, is responsible for a response or whether the transition and change in acceleration levels is capable of producing similar effects (Fig. 8E vs. Fig. 8F). If a microgravity response has already been demonstrated, peak pulses of gravity or microgravity at preset intervals could be applied to define any gravity-sensitive windows [107] within the samples under investigation (Fig. 8G, H) or to define gravity perception/response times, as was very well demonstrated in plant research [108]. To explore whether biological systems are sensitive to various accelerations, the impact of the total sum of loads given by the gravity level multiplied with exposure time could be analyzed. Although the total loads are similar, it could be interesting to investigate how a biological system will respond to such different profiles (Fig. 8I vs. Fig. 8J) and compare these data to an average of 1 × g as per Fig. 8K. Also, gravity perception could be different when a system is suddenly triggered (as in most previous profiles) or when the stimulus is gradually changed (Fig. 8L). It may be clear from these theoretical profile plots that any other gravity profile could be applied by programmable centrifuges. This would enable the science community to explore gravity-related biological phenomena and receptor responses in more detail if sufficient flight opportunities were available. Also, similar profiles could be explored using ground-based centrifuges.

A very illustrating effect of the application of hypergravity and the possible remedial effects with respect to the immune system was demonstrated by Taylor and colleagues showing that *Drosophilae* that were exposed to a fungal infection survived better when exposed to a hypergravity of 4 g, a process involving the *Yuri* gene [109]. Similar positive hyper-g effects were shown in a mouse model of allergic asthma and rhinitis where animals exposed to 5 g for 4 weeks showed an improved immune response [110]. In recent years, more publications have emerged where hypergravity is applied not only to treat various human syndromes and diseases like Raynaud's phenomenon, lymphedema, and peripheral and coronary artery diseases [111] but also for improved mobility in a stroke patient or preventing progressive disability in a case of secondary progressive multiple sclerosis [112, 113].

In-flight animal microgravity experiments combined with the use of centrifuges have shown that artificial gravity is suited to counteract the pathological effects of chronic near weightlessness. Based on this and the numerous on-ground animal and human hypergravity studies using short-arm human centifuges (SAHC), it is to be expected that space agencies or commercial parties will use, for long space missions such as to Mars, artificial gravity and especially long-arm rotation to better protect the crew from various pathologies, morbidities, and co-morbidities related to chronic exposures to microgravity. Besides the many technical advantages of rotating spacecraft, also the huge costs for the current, not fully functional, countermeasure stategies can be saved by the application of a spinning spacecraft [114, 115].

6 In-Flight Animal Studies

Possibly the best known animal with respect to spaceflight is the dog Laika (Fig. 9A). The animal was onboard the second artificial Earth satellite (AES, i.e., Sputnik 2) and launched on November 3, 1957 [116]. Laika was equipped with a potentiometric breath sensor and a carotid collar with a piezoelectric sensor for registering vascular oscillations to measure arterial tension and implanted with a one-lead silver electrode for ECG recording. During the flight, the various physiological parameters were downlinked to ground tracking stations. The vital signs could be monitored for about four orbits, but the temperature had started to rise to some 41°C, which was most likely the cause of Laika's death. In this time period, the Russians favored canines [117] as a larger animal system for physiological experiments, while in the USA, they worked more with non-human primates (Fig. 9B). It was then also not surprising that the first orbital flight of such an animal was by chimpanzee Enos

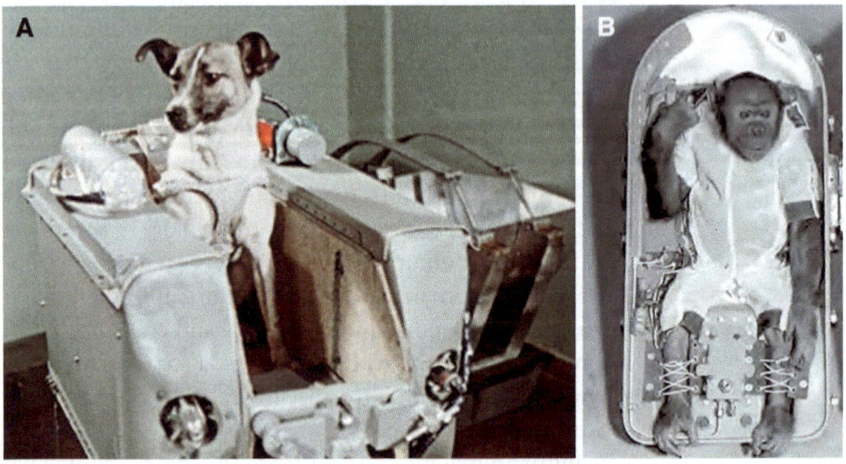

Fig. 9 (**A**) The dog Laika in a model of his enclosure in the capsule that took him into orbiton November 3, 1957, in the Sputnik-2 spacecraft (image courtesy of the Russian Academy of Science (RAS)/Institute of Biomedical Problems (IMBP)). (**B**) The 4-year-old chimpanzee Enos was launched into an orbital flight on November 29, 1961, in a Mercury spacecraft on top of an Atlas rocket(image courtesy of NASA)

in 1961. However, there were many earlier studies where not only animals, fruit flies, spiders, mice, rabbits, rats, and sea urchins but also dogs were launched and exposed to a relatively short duration of free fall (see Burgess & Dubbs for a comprehensive history on this in these USA test flights [118] and [119] for a more graphical representation of animals and spaceflight).

In later studies, it was mainly rodents but also aquatic animals that were exposed to microgravity. The Research Animal Holding Facility was used by the American space agency NASA during Space Shuttle missions [120] and later the Animal Enclosure Module (AEM) for ISS [121], while the Japanese space agency JAXA focused more on aquatic species such as medaka in Shuttle/SpaceLab [122] and a later aquatic habitat facility on board ISS [123]. However, JAXA also developed the Mouse Habit Unit (MHU) as mentioned earlier. Various other modules and facilities have been developed for microgravity studies, most of them without an on-board 1 g control but more and more are including in-flight centrifugation to also explore the impact of partial gravities like on the neuronal development in *Drosophilae* in Moon g [106].

7 Ground Simulations

Although, also because of commercialization of rocketry, spaceflight is getting more and more affordable, there are still quite some limitations in performing in-flight studies. In support of human spaceflight as well as to understand the basic physiology regarding the impact of gravity on living systems, ground-based animal research paradigms have been developed. For human studies, dry immersion [124] and especially 6° head-down tilt bed rest (HDBR [125], Fig. 10C) are used to simulate real spaceflight conditions. Animal analogues are also used, such as 10° head-down tilt for non-human primate rhesus monkey (*Macaca mulatta*) (Fig. 10B) or tail suspension for rodents

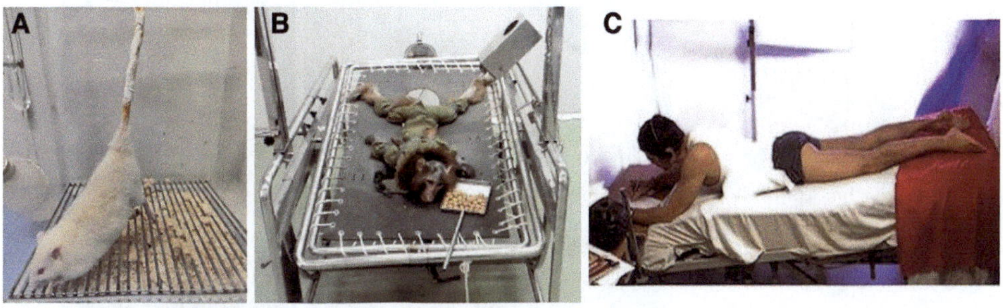

Fig. 10 Ground-based simulation technologies. (**A**) Mice in a tail suspension set-up. (**B**) Head-down tilt configuration of a non-human primate at China Astronaut Research & Training Center (Image courtesy of Dr. Zhongquan & Prof. Li, State Key Laboratory of Space Medicine Fundamentals and Application, Beijing, China). (**C**) Similar set-up to (**B**) but now with a human primate(Image courtesy of European Space Agency, ESA)

(Fig. 10A) [126], a technology that can even be adjusted to simulate Moon or Mars g [127–129]. As the tail suspension method is developed for microgravity simulation of muscle and especially bone [130], the set-up is also used for other systems like behavioral tests [131] and to explore neurophysiological changes [132], although it seems that the simulation for other organ systems is quite debatable [133] and still needs detailed study to be validated for other organ systems.

However, also the Random Positioning Machine (RPM) [134, 135] (Fig. 11A), although nearly exclusively used for cell biological and plant research [136], has also been applied in in vivo animal studies [137–139], although more research is needed to verify how this technology compares to real microgravity studies. The RPM has also been used for studies on *Drosophilae*, and it seems that it is, surprisingly, quite a sufficient simulation of real microgravity, although the flies are free to move in the volume available and not necessarily exposed to a continuously changing gravity vector. Herranz and colleagues showed that 79% of the 6332 differentially expressed genes (DEGs) detected in the ISS are also detected in the RPM experiment in comparisons of the genome-scale correlations of microarray expression values for a 4-day exposure to either real micro-g or simulated μg [140].

Other devices that are used regularly, again mainly in tissue culture studies, are the NASA-developed Rotating Wall Vessel (RWV, Fig. 11B) and the fast-rotating 2D clinostat as developed by Briegleb at the German Space Agency, DLR [141] (Fig. 11C). Research using the RWV showed effects of this device on otolith growth of larval zebrafish [142], although earlier studies comparing the RWV and the 2D clinostat showed no effect on otolith mineralization in the RWV as compared to the 2D clinostat [143]. The RWV was also applied to explore the symbiosis between squid (*Euprymna scolopes*) and their luminescent bacterium *Vibrio liscberi* where it seems that the host immune system is impacted using this system [144]. Although the

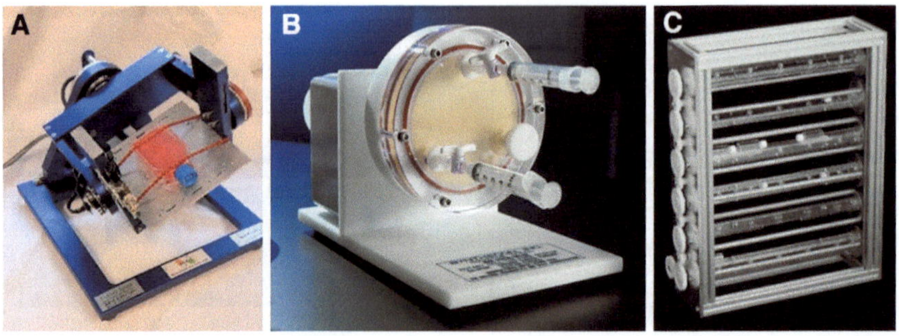

Fig. 11 (**A**) The desktop version of Random Positioning Machine (RPM) (Image courtesy of DESC, The Netherlands). (**B**) The Rotating Wall Vessel system (Image courtesy of American National Aeronautics and Space Administration (NASA, USA). (**C**) Example of a 2D fast-rotating clinostat(Image courtesy of German Space Agency, DLR, Germany)

RWV is used significantly in tissue constructs and stem cell research (see e.g., [145]), others pointed out that this device is not actually simulating microgravity but is basically a terminal velocity device [146].

8 Conclusion

This compact scope overview of gravitational research involving animals provides only a very limited outline of the type of research that has been performed in relation to various species and the instruments or protocols used in spaceflight-related research. It shows the wealth of data on the gravitational physiology that has been produced and used not only as basic science but also as translational research to better understand the effect of short-term or more chronic exposure to altered gravity by humans. The use of species like non-human primates, rabbits, or dogs is not common and even not necessary anymore, surely not in light of reducing the use of animal models via the three R's: Reduction, Refinement, and Replacement [147]. Some of the in vivo studies could and should be replaced by technologies such as tissue spheroids and organoids [148] and related organ-on-chip initiatives [149] in combination with the systems biology multi-omics approach [150], the application of which is growing over the years [151]. However, some animal experiments using more contemporary systems like zebrafish (*Dario dario*), fruit flies (*Drosophilae*), or nematodes (*Caenorhabditis elegans*), but also mice and rats, might still be necessary, although maybe fewer for more metabolic studies where previously mentioned technologies can be applied. However, for studies related to, for example, movement, anatomical adaptations, behavior or social interactions, or a full systemic approach, animal studies are still needed.

To better understand the role of (micro-)gravity in either animal or in vitro (or in silico) studies, the research should explore the full gravity spectrum and not "just" compare microgravity with Earth 1 g, which is more of an operational approach. We should also make more use of in-flight centrifugation to explore the partial gravity domain.

Acknowledgments

Financial support for this work was partially provided by the European Space Agency, ESA-TEC contract #4000136280/21/NL/KML/rk.

References

1. Seckbach J, Walsh M (2008) From fossils to astrobiology: records of life on Earth and the search for extraterrestrial biosignatures, Springer Science & Business Media. Dordrecht
2. Went F (1968) The size of man. Am Sci 56:400–413
3. Albrecht-Buehler G (1991) Possible mechanisms of indirect gravity sensing by cells. ASGSB Bull 4:25–34
4. Pollard EC (1965) Theoretical studies on living systems in the absence of mechanical stress. J Theor Biol 8:113–123
5. Petryshyn V, Corsetti F (2011) Analysis of growth directions of columnar stromatolites from Walker Lake, western Nevada. Geobiology 9:425–435
6. Wager HWT (1910) The effect of gravity upon the movements and aggregation of euglena viridis, ehrb., and other micro-organisms. Proc R Soc Lond B 83:94–96
7. Fenchel T, Finlay B (1984) Geotaxis in the ciliated protozoon Loxodes. J Exp Biol 110:17–33
8. Urban JE (2000) Adverse effects of microgravity on the magnetotactic bacterium Magnetospirillum magnetotacticum. Acta Astronaut 47:775–780
9. van Loon JJWA, Bervoets DJ, Burger EH, Dieudonne SC, Hagen JW, Semeins CM, Doulabi BZ, Veldhuijzen JP (1995) Decreased mineralization and increased calcium release in isolated fetal mouse long bones under near weightlessness. J Bone Miner Res 10:550–557
10. van Loon JJ, Berezovska OP, Bervoets TJ, Montufar-Solis D, Semeins CM, Zandieh-Doulabi B, Rodionova PNV, Duke J, Veldhuijzen JP (2024) Growth and mineralization of fetal mouse long bones under microgravity and daily 1 g gravity exposure. NPJ Microgravity 10:80
11. Vogel SM (2003–2004) Series living in a physical world. Springer, New York
12. Summers AP, Long JRJH (2005) Skin and bones, sinew and gristle: the mechanical behavior of fish skeletal tissues. Fish Physiol 23:141–177
13. Daeschler EB, Shubin NH, Jenkins FAJR (2006) A Devonian tetrapod-like fish and the evolution of the tetrapod body plan. Nature 440:757–763
14. Young JZ (1962) The life of vertebrates, 2nd edn. University Press, New York
15. Sekulic S, Jovanovic A, Zivanovic Z, Simic S, Kesic S, Petkovic B, Capo I, van Loon JJ (2022) Which precocial rodent species is more suitable as the experimental model of microgravity influence on prenatal musculosketal development on international space station?. Life Sci Space Res 33:48–57
16. Bradley NS (1997) Reduction in buoyancy alters parameters of motility in E9 chick embryos. Physiol Behav 62:591–595
17. Ubbels GA, Berendsen W, Narraway J (1989) Fertilization of frog eggs on a sounding rocket in space. Adv Space Res 9:187–197
18. Souza KA, Black SD, Wassersug RJ (1995) Amphibian development in the virtual absence of gravity. Proc Natl Acad Sci U S A 92:1975–1978
19. Ubbels G, Reijnen M, Meijerink J, Narraway J (1995) The role of gravity in the establishment of the dorso-ventral axis in the amphibian embryo (05-NL EGGS). Biorack on Spacelab IML 1:175
20. Houston DW (2016) Vertebrate axial patterning: from egg to asymmetry. In: Vertebrate development: maternal to zygotic control. Springer, pp 209–306. New York
21. Turko AJ, Kultz D, Fudge D, Croll RP, Smith FM, Stoyek MR, Wright PA (2017) Skeletal stiffening in an amphibious fish out of water is a response to increased body weight. J Exp Biol 220:3621–3631
22. Cartwright P, Halgedahl SL, Hendricks JR, Jarrard RD, Marques AC, Collins AG, Lieberman BS (2007) Exceptionally preserved jellyfishes from the Middle Cambrian. PLoS One 2:e1121
23. Ross MD, Pote KG (1984) Some properties of otoconia. Philos Trans R Soc Lond B Biol Sci 304:445–452
24. Popper AN, Hoxter B (1981) The fine structure 0f the sacculus and lagena of a teleost fish. Hear Res 5:245–263
25. Chen W-L, Ko H, Chuang H-S, Raizen DM, Bau HH (2021) Caenorhabditis elegans exhibits positive gravitaxis. BMC Biol 19:1–16
26. van Loon JJ, Krausse J, Cunha H, Goncalves J, Almeida H, Schiller, P (2008) The large diameter centrifuge, LDC, for life and physical sciences and technology. In: Ouwehand L (edn) Life in space for life on Earth, December 2008, ESA Communication Production Office, Anger, France
27. Aceto J, Nourizadeh-Lillabadi R, Maree R, Dardenne N, Jeanray N, Wehenkel L, Alestrom P, van Loon JJ, Muller M (2015) Zebrafish bone and general physiology are differently affected by hormones or changes in gravity. PLoS One 10:e0126928
28. Boyle R, Varelas J (2021) Otoconia structure after short-and long-duration exposure to altered gravity. J Assoc Res Otolaryngol 22:509–525

29. Pedrozo HA, Wiederhold ML (1994) Effects of hypergravity on statocyst development in embryonic Aplysia californica. Hear Res 79:137–146
30. Pedrozo H, Schwartz Z, Luther M, Dean D, Boyan B, Wiederhold M (1996) A mechanism of adaptation to hypergravity in the statocyst of Aplysia californica. Hear Res 102:51–62
31. von Baumgarten RJ, Baldrighi G, Shillinger GLJR (1972) Vestibular behavior of fish during diminished G-force and weightlessness. Aerosp Med 43:626–632
32. Anken R, Forster A, Baur U, Feucht I, Hilbig R (2006) Otolith asymmetry and kinetotic behaviour of fish at high-quality microgravity: a drop-tower experiment. Adv Space Res 38:1032–1036
33. von Baumgarten R, Simmonds R, Boyd J, Garriott O (1975) Effects of prolonged weightlessness on the swimming pattern of fish aboard Skylab 3. Aviat Space Environ Med 46:902–906
34. Hilbig R, Anken RH, Sonntag G, Hohne S, Henneberg J, Kretschmer N, Rahmann H (2002) Effects of altered gravity on the swimming behaviour of fish. Adv Space Res 30:835–841
35. Fuller PM, Jones TA, Jones SM, Fuller CA (2002) Neurovestibular modulation of circadian and homeostatic regulation: vestibulohypothalamic connection? Proc Natl Acad Sci U S A 99:15723–15728
36. Xue B, Skala K, Jones TA, Hay M (2004) Diminished baroreflex control of heart rate responses in otoconia-deficient C57BL/6JEi head tilt mice. Am J Physiol Heart Circ Physiol 287:H741–747
37. Kawao N, Morita H, Obata K, Tamura Y, Okumoto K, Kaji H (2016) The vestibular system is critical for the changes in muscle and bone induced by hypergravity in mice. Physiol Rep 4(19):e12979
38. Lundberg YW, Xu Y, Thiessen KD, Kramer KL (2015) Mechanisms of otoconia and otolith development. Dev Dyn 244:239–253
39. Trudel G, Shahin N, Ramsay T, Laneuville O, Louati H (2022) Hemolysis contributes to anemia during long-duration space flight. Nat Med 28:59–62
40. Michel EL, Johnston RS, Dietlein LF (1976) Biomedical results of the Skylab Program. Life Sci Space Res 14:3–18
41. Smith AH (1975) Principles of gravitational biology. In: Calvin M, Gazenko OG (eds) Foundations of space biology and medicine: ecological and physiological bases of space biology and medicine. 2 v. National Aeronautics and Space Administration, USA/ Academy of Sciences of the USSR
42. Burton R, Smith A (1969) Hematological findings associated with chronic acceleration. Space Life Sci 1:501–513
43. Hargens AR, Millard RW, Pettersson K, Johansen K (1987) Gravitational haemodynamics and oedema prevention in the giraffe. Nature 329:59–60
44. Østergaard KH, Bertelsen MF, Brøndum ET, Aalkjær C, Hasenkam JM, Smerup M, Wang T, Nyengaard JR, Baandrup U (2011) Pressure profile and morphology of the arteries along the giraffe limb. J CompPhysiol B 181:691–698
45. Gunga H-C, Kirsch K, Baartz F, Röcker L, Heinrich W-D, Lisowski W, Wiedemann A, Albertz J (1995) New data on the dimensions of Brachiosaurus brancai and their physiological implications. Naturwissenschaften 82:190–192
46. Lillywhite HB (1996) Gravity, blood circulation, and the adaptation of form and function in lower vertebrates. J Exp Zool 275:217–225
47. Olumide Ashaolu J, Ufuoma Enaibe B, Salihu Ajao M (2015) Prolonged head-down posture of bats induces remodeling of the aorta. Gravit Space Biol 3:29–38
48. van Loon J (2007) The gravity environment in space experiments. In: Brinckmann E (ed) Biology in space and life on Earth. Effects of spaceflight on biological systems. John Wiley and Sons Ltd, Weinheim
49. Fuller C (1997) Report NAG2-795 Centrifuge stopping and diameter: effects of centrifuge diameter & operation on rodent adaptation to chronic centrifugation. Univ. Davis
50. van Loon JJ, Folgering EH, Bouten CV, Veldhuijzen JP, Smit TH (2003) Inertial shear forces and the use of centrifuges in gravity research. What is the proper control? J Biomech Eng 125:342–346
51. Walters GR, Wunder CC, Smith L (1960) Multifield centrifuge for life-long exposure of small mammals. J Appl Physiol 15:307–308
52. Ishay J, Barr-Nea L (1977) Effects of hypergravity on rat fertility, pregnancy, parturition and survival. Experientia 33:244–246
53. Oyama J, Zeitman B (1967) Tissue composition of rats exposed to chronic centrifugation. Am J Physiol 213:1305–1310
54. Mccoy DF, Jankovich JP (1972) Effects of continuous exposure to high gravity on gravity preference in rats. J Comp Physiol Psychol 78:305–310
55. Oyama J (1974) Response and adaptation of beagle dogs to hypergravity. In: Sneath PHA (Ed.) Life science and space research. De Gruyter, São Paulo, SP, Brazil

56. Negulesco J, Clark D (1976) Hypergravity effects on normal and avulsed developing avian radii. Aviat Space Environ Med 47:821–825
57. Wunder C, Dodge C, Cook K, Moressi W (1977) Food and oxygen requirements for growing mice and turtles after hypergravitational development. Life Sci Space Res 15:245–249
58. Todd JR RE (1940) The cellular components of centrifuged frog eggs. J Exp Zool 83:359–377
59. Pace N, Smith A (1983) Scaling of metabolic rate on body mass in small mammals at 2.0 g. Physiologist 26:S125–126
60. Ishay J, Sadeh D (1975) Direction finding by hornets under gravitational and centrifugal forces. Science 190:802–804
61. Horn E, Sebastian C (1996) A hypergravity related sensitive period during the development of the roll induced vestibuloocular reflex in an amphibian (*Xenopus laevis*). Neurosci Lett 216:25–28
62. Serrano P, van Loon JJWA, Medina FJ, Herranz R (2012) Relation between motility, accelerated aging and gene expression in selected drosophila strains under hypergravity conditions. Microgravity Sci Technol 25:67–72
63. Le Bourg É, Minois N, Bullens P, Bearet P (2000) A mild stress due to hypergravity exposure at youngage increases longevity in Drosophila melanogaster males. Biogerontology 1:145–155
64. Lawrence EA, Aggleton JA, van Loon JJ, Godivier J, Harniman RL, Pei J, Nowlan N, Hammond CL (2021) Exposure to hypergravity during zebrafish development alters cartilage material properties and strain distribution. Bone Joint Res 10:137–148
65. Rahmann H, Hilbig R, Flemming J, Slenzka K (1996) Influence of long-term altered gravity on the swimming performance of developing cichlid fish: including results from the 2nd German Spacelab Mission D-2. Adv Space Res 17:121–124
66. Kleiber M (1932) Body size and metabolism. ENE 1
67. Shimbo M, Kudo T, Hamada M, Jeon H, Imamura Y, Asano K, Okada R, Tsunakawa Y, Mizuno S, Yagami K, Ishikawa C, Li H, Shiga T, Ishida J, Hamada J, Murata K, Ishimaru T, Hashimoto M, Fukamizu A, Yamane M, Ikawa M, Morita H, Shinohara M, Asahara H, Akiyama T, Akiyama N, Sasanuma H, Yoshida N, Zhou R, Wang YY, Ito T, Kokubu Y, Noguchi TA, Ishimine H, Kurisaki A, Shiba D, Mizuno H, Shirakawa M, Ito N, Takeda S, Takahashi S (2016) Ground-based assessment of JAXA mouse habitat cage unit by mouse phenotypic studies. Exp Anim 65:175–187
68. Wade CE, Harper JS, Daunton NG, Corcoran ML, Morey-Holton E (1997) Body mass change during altered gravity: spaceflight, centrifugation, and return to 1 G. J Gravit Physiol 4:43–48
69. Katovich MJ, Smith AH (1978) Body mass, composition, and food intake in rabbits during altered acceleration fields. J Appl Physiol Respir Environ Exerc Physiol 45:51–55
70. Pitts G (1977) Effects on body size and composition of chronic exposure to altered gravity. [centrifuging stress in mammals]. NASA Research Center
71. Warren LE, Hoban-Higgins TM, Hamilton JS, Horwitz BA, Fuller CA (2000) Effects of 2G exposure on lean and genetically obese Zucker rats. J Gravit Physiol 7:61–69
72. Wunder CC (1961) Food consumption of mice during continual centrifugation. In: Proceedings of the Iowa Academy of Science, pp 616–624
73. Fuller PM, Warden CH, Barry SJ, Fuller CA (2000) Effects of 2-G exposure on temperature regulation, circadian rhythms, and adiposity in UCP2/3 transgenic mice. J Appl Physiol (1985) 89:1491–1498
74. Pace N, Smith A, Rahlmann D (1985) Skeletal mass change as a function of gravitational loading. Physiologist 28:S17–20
75. Koliaki C, Dalamaga M, Liatis S (2023) Update on the obesity epidemic: after the sudden rise, is the upward trajectory beginning to flatten. Curr Obes Rep 12:514–527
76. van Loon JJWA, Baeyens JP, Berte J, Blanc S, Braak L, Bok K, Bos J, Boyle R, Bravenoer N, Eekhoff M, Chouker A, Clement G, Cras P, Cross E, Cusaud MA, de Angelis MDEDC, Delavaux T, Delfos R, Poelma C, Denise P, Felsenberg D, Fong K, Fuller C, Grillner S, Groen E, Harlaar J, Heer M, Heglund N, Hinghofer-Szalkay H, Goswami N, Hughes-Fulford M, Iwase S, Jkaremaker M, Langdahl B, Linnarsson D, Lüthen C, Monici M, Mulder E, Narici M, Norsk P, Paloski W, Prisk GK, Rutten M, Singer P, Stegeman D, Stephan A, Stienen GJM, Suedfeld P, Tesch P, Ullrich O, van den Berg R, van de Heyning P, Delahaye A, Veyt J, 2010 Vico L, Woodward E, Young L, Wuyts FL (2012) A large human centrifuge for exploration and exploitation research. Ann Kinesiol 3:2010–2015
77. Matthews BH (1953) Adaptation to centirifugal acceleration. J Physiol 122(supplement):31p
78. Wunder C, Lutherer LO (1964) Influence of chronic exposure to increased gravity upon growth and form of animals. In: Felts WJL, Harrison RJ (Eds) International review of general and experimental zoology. Academic Press, New York

79. Janssen PM, Biesiadecki BJ, Ziolo MT, Davis JP (2016) The need for speed: mice, men, and myocardial kinetic reserve. Circ Res 119:418–421
80. Chae EU (1975) Tolerance of small animals to acceleration. Aviat Space Environ Med 46:703–708
81. van Loon JJWA (2016) Centrifuges for microgravity simulation. The reduced gravity paradigm. Front Astron Space Sci 3:21
82. Wade CE (2005) Responses across the gravity continuum: hypergravity to microgravity. Adv Space Biol Med 10:225–245
83. Cancedda R, Liu Y, Ruggiu A, Tavella S, Biticchi R, SantuccI D, Schwartz S, Ciparelli P, Falcetti G, Tenconi C, Cotronei V, Pignataro S (2012) The Mice Drawer System (MDS) experiment and the space endurance record-breaking mice. PLoS One 7:e32243
84. Liu Y, Biticchi R, Cilli M, Piccardi F, Ruggiu A, Cancedda R (2009) Mouse Drawer system (MDS): an automated payload for supporting rodent research on the international space station. Basic Appl Myology 19:87–95
85. Rizzo AM, Corsetto PA, Montorfano G, Milani S, Zava S, Tavella S, Cancedda R, Berra B (2012) Effects of long-term space flight on erythrocytes and oxidative stress of rodents. PLoS One 7:e32361
86. Rizzo AM, Murgia G, Lentini A, Zava S, Ferranti F, Tavella S, Santucci D, van Loon JJ, Colombo I, Corsetto PA (2025) Hypergravity influences mouse erythrocyte membrane lipid composition and antioxidant potential. Acta Astronautica 234:99–105
87. Bonetto V, Magnelli V, Sabbatini M, Caprì F, van Loon JJ, Tavella S, Masini MA (2023) The importance of gravity vector on adult mammalian organisms: effects of hypergravity on mouse testis. PLoS One 18:e0282625
88. Cazaubiel A, Mawet S, Darras A, Grosjean G, van Loon J, Dorbolo S, Falcon E (2019) Wave turbulence on the surface of a fluid in a high-gravity environment. Phys Rev Lett 123:244501
89. Berhanu M, Falcon E, Michel G, Gissinger C, Fauve S (2020) Capillary wave turbulence experiments in microgravity. EPL (Europhys Lett) 128:34001
90. Schmidt MA, Goodwin TJ, Pelligra R (2016) Incorporation of omics analyses into artificial gravity research for space exploration countermeasure development. Metabolomics 12:36
91. Ilyin EA (2000) Historical overview of the Bion project. J Gravit Physiol 7:S1–8
92. Kotovskaya A (2018) Human tolerance of G-loads in space missions and artificial gravity. Hum Physiol 44:779–793
93. Larina I, Nichiporuk I, Veselova O, Vasilieva GY, Popova I (2015) Shifts in metabolism and its regulation under the effect of spaceflight factors. Hum Physiol 41:689–698
94. Gurovsky NN, Gazenko OG, Adamovich BA, Ilyin EA, Genin AM, Korolkov VI, Shipov AA, Kotovskaya AR, Kondratyeva VA, Serova LV, Kondratyev Yu I (1980) Study of physiological effects of weightlessness and artificial gravity in the flight of the biosatellite Cosmos-936. Acta Astronaut 7:113–121
95. Kübler U, Brinckmann E, Grinberg A (2021) Space research centrifuges. Handbook of bioastronautics. Springer, Cham
96. Johnson TE, Nelson GA (1991) Caenorhabditis elegans: a model system for space biology studies. Exp Geront 26:299–309
97. Vernos I, Gonzalez-Jurado J, Calleja M, Marco R (1989) Microgravity effects on the oogenesis and development of embryos of Drosophila melanogaster laid in the Spaceshuttle during the Biorack experiment (ESA). Int J Dev Biol 33:213–226
98. Rebecchi L, Altiero T, Guidetti R, Cesari M, Bertolani R, Negroni M, Rizzo AM (2009) Tardigrade resistance to space effects: first results of experiments on the LIFE-TARSE mission on FOTON-M3 (September 2007). Astrobiology 9:581–591
99. Horn E (2006a) Animal development in microgravity. In: Fundamentals of space biology. Springer, pp 171–226. New York
100. Horn ER (2006b) Microgravity-induced modifications of the vestibuloocular reflex in Xenopus laevis tadpoles are related to development and the occurrence of tail lordosis. J Exp Biol 209:2847–2858
101. Brinckmann E (2012) Centrifuges and their application for biological experiments in space. Microgravity Sci Technol 24:365–372
102. Shiba D, Mizuno H, Yumoto A, Shimomura M, Kobayashi H, Morita H, Shimbo M, Hamada M, Kudo T, Shinohara M, Asahara H, Shirakawa M, Takahashi S (2017) Development of new experimental platform 'MARS'—multiple artificial-gravity research system—to elucidate the impacts of micro/partial gravity on mice. Sci Rep 7:10837
103. Wang S, Wang K, Zhou Y, Yan B, Li X, Zhang Y, Wu W, Wang A (2019) Development of the varying gravity rack (VGR) for the Chinese space station. Microgravity Sci Technol 31:95–107
104. Leandro LJ, Szewczyk NJ, Benguria A, Herranz R, Lavan D, Medina FJ, Gasset G, Loon JV, Conley CA, Marco R (2007) Comparative analysis of Drosophila melanogaster and Caenorhabditis elegans gene expression experiments in the European Soyuz flights to the International Space Station. Adv Space Res 40:506–512

105. Huin-Schohn C, Guéguinou N, Schenten V, Bascove M, Koch GG, Baatout S, Tschirhart E, Frippiat JP (2013) Gravity changes during animal development affect IgM heavy-chain transcription and probably lymphopoiesis. FASEB J 27:333–341
106. Mhatre SD, Iyer J, Petereit J, Dolling-Boreham RM, Tyryshkina A, Paul AM, Gilbert R, Jensen M, Woolsey RJ, Anand S (2022) Artificial gravity partially protects space-induced neurological deficits in Drosophila melanogaster. Cell Rep 40(10):111279
107. Mesland DA (1992) Possible actions of gravity on the cellular machinery. Adv Space Res 12:15–25
108. Bizet F, Pereda-Loth V, Chauvet H, Gérard J, Eche B, Girousse C, Courtade M, Perbal G, Legué V (2018) Both gravistimulation onset and removal trigger an increase of cytoplasmic free calcium in statocytes of roots grown in microgravity. Sci Rep 8:11442
109. Taylor K, Kleinhesselink K, George MD, Morgan R, Smallwood T, Hammonds AS, Fuller PM, Saelao P, Alley J, Gibbs AG, Hoshizaki DK, von Kalm L, Fuller CA, Beckingham KM, Kimbrell DA (2014) Toll mediated infection response is altered by gravity and spaceflight in Drosophila. PLoS One 9:e86485
110. Jang TY, Jung AY, Kim YH (2016) Hormetic effect of chronic hypergravity in a mouse model of allergic asthma and rhinitis. Sci Rep 6:27260
111. Isasi EE, Isasi ME, van Loon JJ (2022) The application of artificial gravity in medicine and space. Front Physiol 13:1627
112. Kourtidou-Papadeli C, Frantzidis C, Machairas I, Giantsios C, Dermitzakis E, Kantouris N, Konstantinids E, Bamidis P, Vernikos J (2023) Rehabilitation assisted by Space technology—a SAHC approach in immobilized patients—a case of stroke. Front Physiol 13:1024389
113. Kourtidou-Papadeli C, Frantzidis CA, Bakirtzis C, Petridou A, Gilou S, Karkala A, Machairas I, Kantouris N, Nday CM, Dermitzakis EV, Bakas E, Mougios V, Bamidis PD, Vernikos J (2022) Therapeutic benefits of short-arm human centrifugation in multiple sclerosis-a new approach. Front Neurol 12:746832
114. van Loon JJWA, Cras P, Roozendaal W, Bouwens W, Vernikos J (2020) Gravity deprivation: is it ethical for optimal physiology?. Front Physiol 11:470
115. van Loon JJWA, Lobascio C, Boscheri G, Goujon C, Voglino S, Zeminiani E, González-Cinca R, Ewald R (2024) Benefits of a rotating—partial gravity—spacecraft. Acta Astronaut 220:403–415
116. Grigor'ev A, Il'in E (2007) Animals in space: on the 50th anniversary of space biology. Her Russ Acad Sci 77:541–549
117. Cambiaghi M, Sacchetti B (2015) Ivan Petrovich Pavlov (1849–1936). J Neurol 262:1599–1600
118. Burgess C, Dubbs C (2007) Animals in space: from research rockets to the space shuttle. Springer Science & Business Media, New York
119. Maggiacomo T, Stegmaier A (2021) Animals in space. National Geographic
120. Martin TP, Edgerton VR, Grindeland RE (1988) Influence of spaceflight on rat skeletal muscle. J Appl Physiol 65:2318–2325
121. Niederwieser T, Gerren R, Koenig P, Tozer S, Stodieck L, Rieger S Hoehn A (2014) AEM-E–A small life support system for the transport of rodents to the ISS. In: 44th International conference on environmental systems
122. Ijiri K (1994) A preliminary report on IML-2 medaka experiment mating behavior of the fish medaka and development of their eggs in space. Biol Sci Space 8:231–233
123. ChatanI M, Mantoku A, Takeyama K, Abduweli D, Sugamori Y, Aoki K, Ohya K, Suzuki H, Uchida S, Sakimura T, Kono Y, Tanigaki F, Shirakawa M, Takano Y, Kudo A (2015) Microgravity promotes osteoclast activity in medaka fish reared at the international space station. Sci Rep 5:14172
124. Tomilovskaya E, Shigueva T, Sayenko D, Rukavishnikov I, Kozlovskaya I (2019) Dry immersion as a ground-based model of microgravity physiological effects. Front Physiol 10:284
125. Pavy-Le Traon A, Heer M, Narici MV, Rittweger J, Vernikos J (2007) From space to Earth: advances in human physiology from 20 years of bed rest studies (1986-2006). Eur J Appl Physiol 101:143–194
126. Mu Y, Wei D, Yao L, Xu X, Li S, Cao R, Chen T, Zhang Z (2023) Choroidal circulation disturbance is an initial factor in outer retinal degeneration in rats under simulated weightlessness. Front Physiol 14:1198862
127. Wagner EB, Granzella NP, Saito H, Newman DJ, Young LR, Bouxsein ML (2010) Partial weight suspension: a novel murine model for investigating adaptation to reduced musculoskeletal loading. J Appl Physiol (1985) 109:350–357
128. Zhang S, Adachi T, Zhang S, Yoshida Y, Takahashi A (2022) A new type of simulated partial gravity apparatus for rats based on a pully-spring system. Front Cell Dev Biol 10:965656
129. Gnyubkin V, Vico L (2022) Animals: unloading, casting. In: Beysens D, van Loon JJWA (eds) Generation and applications of

extra-terrestrial environments on earth. River Publishers, Aalborg, Denmark
130. Morey ER (1979) Spaceflight and bone turnover: correlation with a new rat model of weightlessness. BioScience 29:168–172
131. Bellone JA, Gifford PS, Nishiyama NC, Hartman RE, Mao XW (2016) Long-term effects of simulated microgravity and/or chronic exposure to low-dose gamma radiation on behavior and blood–brain barrier integrity. NPJ Microgravity 2:16019
132. Nday CM, Frantzidis C, Jackson G, Bamidis P, Kourtidou-Papadeli C (2019) Neurophysiological changes in simulated microgravity: an animal model. Neurol India 67:221
133. Kulikova E, Kulikov V, Sinyakova N, Kulikov A, Popova N (2017) The effect of long-term hindlimb unloading on the expression of risk neurogenes encoding elements of serotonin-, dopaminergic systems and apoptosis; comparison with the effect of actual spaceflight on mouse brain. Neurosci Lett 640:88–92
134. van Loon JJWA (2007) Some history and use of the random positioning machine, RPM, in gravity related research. Adv Space Res 39:1161–1165
135. Borst AG, van Loon JJWA (2008) Technology and developments for the random positioning machine, RPM. Microgravity Sci Technol 21:287–292
136. Kiss JZ, Wolverton C, Wyatt SE, Hasenstein KH, van Loon J (2019) Comparison of microgravity analogs to spaceflight in studies of plant growth and development. Front Plant Sci 10:1577
137. Saba A, Pani G, Galleri G, Meloni M, Pippia P (2009) "In vivo" physiological experiments in simulated microgravity conditions on rat bon marrow cells mineralization. J Biol Res-Boll Soc Ital Biol Sper 82:19–21
138. Song C, Kang T, Gao K, Shi X, Zhang M, Zhao L, Zhou L, Guo J (2023) Preparation for mice spaceflight: indications for training C57BL/6J mice to adapt to microgravity effect with three-dimensional clinostat on the ground. Heliyon 9:e19355.
139. Yotov VV, Marovska J, Turiyski V, Ivanov SI (2022) A new random positioning machine modification applied for microgravity simulation in laboratory experiments with rats. Inventions 7:85
140. Herranz R, Benguria A, Lavan DA, Lopez-Vidriero I, Gasset G, Javier Medina F, van Loon JJ, Marco R (2010) Spaceflight-related suboptimal conditions can accentuate the altered gravity response of Drosophila transcriptome. Mol Ecol 19:4255–4264
141. Briegleb W (1992) Some qualitative and quantitative aspects of the fast-rotating clinostat as a research tool. ASGSB Bull 5:23–30
142. Li X, Anken R, Liu L, Wang G, Liu Y (2017) Effects of simulated microgravity on otolith growth of larval zebrafish using a rotating-wall vessel: appropriate rotation speed and fish developmental stage. Microgravity Sci Technol 29:1–8
143. Brungs S, Hauslage J, Hilbig R, Hemmersbach R, Anken R (2011) Effects of simulated weightlessness on fish otolith growth: clinostat versus rotating-wall vessel. Adv Space Res 48:792–798
144. Foster JS, Khodadad CL, Ahrendt SR, Parrish ML (2013) Impact of simulated microgravity on the normal developmental time line of an animal-bacteria symbiosis. Sci Rep 3:1340
145. van den Nieuwenhof DW, Moroni L, Chou J, Hinkelbein J (2024) Cellular response in three-dimensional spheroids and tissues exposed to real and simulated microgravity: a narrative review. NPJ Microgravity 10:102
146. Hasenstein KH, van Loon JJ (2015) Clinostats and other rotating systems—design, function, and limitations. In: Loon BV (ed) Generation and applications of extra-terrestrial environments on Earth. Rivers Publishers, Aalborg, Denmark
147. Balls M (2022) Alternatives to laboratory animals: trends in replacement and the three rs. ATLA: Altern Lab Anim 50:10–26
148. Kim W, Gwon Y, Park S, Kim H, Kim J (2023) Therapeutic strategies of three-dimensional stem cell spheroids and organoids for tissue repair and regeneration. Bioact Mater 19:50–74
149. Leung CM, De Haan P, Ronaldson-Bouchard K, Kim G-A, Ko J, Rho HS, Chen Z, Habibovic P, Jeon NL, Takayama S (2022) A guide to the organ-on-a-chip. Nat Rev Methods Primers 2:33
150. Gutierrez Reyes CD, Alejo-Jacuinde G, Perez Sanchez B, Chavez Reyes J, Onigbinde S, Mogut D, Hernandez-Jasso I, Calderon-Vallejo D, Quintanar JL, Mechref Y (2024) Multi omics applications in biological systems. Curr Issues Mol Biol 46:5777–5793
151. Taylor K, Modi S, Bailey J (2024) An analysis of trends in the use of animal and non-animal methods in biomedical research and toxicology publications. Front Lab Chip Technol 3:1426895
152. Shubin NH, Daeschler EB, Jenkins FA,JR (2006) The pectoral fin of Tiktaalik roseae and the origin of the tetrapod limb. Nature 440:764–771
153. van Loon J, Tanck E, van Nieuwenhoven F, Snoeckx L, de Jong H, Wubbels R (2005) A brief overview of animal hypergravity studies. J Gravit Physiol 12:5–10

Chapter 3

Assessing the Effects of Microgravity on Eukaryotic Cells Using a Statistical Thermodynamic Approach

Aleksandr S. Nikitiuk, Yuri V. Bayandin and Oleg B. Naimark

Abstract

This chapter has the aim to propose a theoretical framework for describing the qualitative changes in the deformation behavior of isolated human breast cancer cells under varying gravitational conditions. We describe interdisciplinary principles of statistical thermodynamics and nonlinear dynamics of eukaryotic cells. The framework aims to capture how microgravity-induced alterations in cytoskeletal organization lead to distinct mechanobiological phenotypes.

Keywords Mechanobiology, Cancer cell, Microgravity, Phenotype, Statistical thermodynamic, Internal variables

1 Introduction

The advancement of astronautics and the preparation for long-duration manned space missions have brought to the forefront the fundamental question of how microgravity influences cellular phenotype transformation. It is well-documented that biological systems—ranging from microorganisms and bacteria to cells and tissues—undergo significant structural and functional alterations under reduced gravitational forces [1–5]. These changes are particularly evident in various human cell lines, such as epithelial gland cells, osteoblasts, and lung cancer cells, which exhibit marked morphological variations under both real and simulated microgravity conditions [6–8].

For instance, studies have shown that 24 h of microgravity exposure disrupts the uniformity of cell shape, leading to the emergence of distinct morphological phenotypes [6, 7]. In breast cancer cells, this results in the formation of two distinct populations: one comprising flattened, spindle-shaped adherent cells (AD), and the other consisting of small, rounded cells that aggregate into 3D organoid-like clusters (CLUMP). These phenotypic changes are not transient; the cells remain viable and continue to proliferate, maintaining their

Mariano Bizzarri (ed.), *Biomedicine and Biotechnology in Space*, Methods in Molecular Biology, vol. 3000,
https://doi.org/10.1007/978-1-0716-5174-2_3,

distinct morphologies for extended periods (over 7 days) in microgravity. Remarkably, both phenotypes revert to their original forms within 6–24 h upon re-exposure to normal gravity, suggesting that these transformations are not driven by permanent genetic alterations but rather by reversible, self-organizing processes influenced by environmental factors [9, 10].

The role of physical forces—such as gravity, adhesion, and surface tension—in shaping cellular behavior cannot be overlooked [11, 12]. Unlike molecular motors, these forces operate on a global scale, influencing tissue morphology, and state across large spatial dimensions. They contribute to the establishment of a "morphogenetic field," integrating large-scale correlations between chemical and physical cues. In this context, gravity emerges as a critical physical factor governing biological organization principles.

Biological structures, as out-of-equilibrium thermodynamic systems, exhibit nonlinear dynamics and can transition into new self-organized states characterized by symmetry breaking [10]. This phenomenon is observed in cells exposed to microgravity, where the subsequent application of a conventional gravitational field causes two distinct stable states to merge into a single phenotypic configuration [13].

Previous hypotheses suggested that alterations in environmental physical parameters (e.g., changes in gravity or shear stress) could explain gravitational effects and promote cell aggregation, leading to the formation of multicellular aggregates and organoids [14, 15]. However, many cellular responses to gravity—such as cytoskeletal reorganization, proteomic shifts, and gene expression changes—remain only partially understood, as they cannot be fully explained by discrete fluctuations in biophysical parameters [16]. The notion of cellular "gravity sensors" has also been proposed, but conclusive evidence for such mechanisms is lacking [17].

The assumption that isolated cells are insensitive to weak external influences holds for quasiequilibrium thermodynamic systems, where the energy of such influences (e.g., electromagnetic or gravitational) is negligible compared to thermal energy. However, for out-of-equilibrium dissipative systems, the presence of structural transitions alter this paradigm. Interactions between structural elements, combined with external fields, can overcome thermal fluctuations, rendering intracellular structures sensitive to gravitational changes [18].

The biochemical mechanisms underlying cellular behavior are inherently dissipative and nonlinear, often resembling phase transitions that give rise to new self-organized configurations [19]. Specialized cytoskeletal structures, for example, exhibit bifurcation transitions that lead to macroscopic self-organized states [20]. In the absence of external constraints, out-of-equilibrium systems can undergo symmetry breaking, increasing their degrees of freedom [21]. When physical constraints are present—whether external or internal—the system selects specific configurations, implying that gravity directs biological systems toward deterministic forms of self-organization. Under microgravity, the system approaches a bifurcation point, where dynamic fluctuations in state parameters amplify, making self-organization processes highly sensitive to weak external or internal stimuli [22].

Experimental studies have shown that cellular evolution is guided by monotonic changes in bifurcation parameters, which alter the landscape of "attractors" through bifurcations. These observations underscore the importance of microgravity as a tool for studying phenotype dynamics, particularly in the context of cytoskeletal reorganization. The hypothesis that phenotypic transitions are linked to qualitative changes in cytoskeletal dynamics offers a framework for understanding the reversibility of these transitions, with potential implications for reversing pathological processes such as tumor progression.

Mathematical models of eukaryotic cells, incorporating statistical-thermodynamic principles, have been developed to describe cytoskeletal mechanics [23–25]. These models account for viscoelasticity, power-law stress relaxation, and fluidization under load, providing insights into how cytoskeletal dynamics respond to external conditions. Critical parameter values in these models can trigger irreversible changes in cytoskeletal behavior, mirroring phenomena observed in oncogenesis.

The objective of this chapter is to propose a theoretical framework for describing the qualitative changes in the deformation behavior of isolated human breast cancer cells under varying gravitational conditions. By integrating principles of statistical thermodynamics and nonlinear dynamics, the framework aims to capture how microgravity-induced alterations in cytoskeletal organization lead to distinct mechanobiological phenotypes. This approach seeks to bridge the gap between observed cellular responses (e.g., viscoelasticity, power-law stress relaxation, and fluidization) and the underlying physical mechanisms, providing a predictive model for phenotype transitions. The proposed framework not only advances fundamental understanding of cell mechanobiology in space environments but also offers potential applications in targeted cancer therapy and tissue engineering.

2 Main Principles of the Statistical-Thermodynamic Model of the Eukaryotic Cell

To describe the qualitative changes in the deformation behavior of isolated human breast cancer cells under varying gravitational conditions, a model capable of reflecting these changes and accounting for the influence of a weak physical signal are justified. The mathematical model of the eukaryotic cell proposed in the works of [23–28]. Its main points are summarized below.

2.1 Microscopic and Macroscopic Internal Variables of the Cytoskeletal Filament Ensemble

The mechanical behavior of eukaryotic cells is modeled by considering the orientation properties of cytoskeletal filaments, primarily actin, intermediate filaments, and microtubules. In describing the mechanics of the cytoskeleton, it is necessary to determine the form of micro- and macroscopic parameters that take into account local symmetry changes due to the orientation properties of the filaments (actin, intermediate, microtubules), as well as the form of the dependence of the thermodynamic potential on these structural variables. The internal variable related to filaments was introduced in [23] by analogy with the structural variables used in the theory of liquid crystals and in the description of deformation properties of solid polymers. Orientation polymerization and depolymerization of an ensemble of filaments are determined by a traceless second order tensor $\boldsymbol{\eta}$, which is the average of the "microscopic" value

$$\mathbf{s} = \frac{3}{2}\left(\mathbf{pp} - \frac{1}{3}\mathbf{I}\right),$$

where $\mathbf{p}$ is a unit vector along the segment axis, $\mathbf{I}$ is the unit tensor. The tensor $\mathbf{s}$ characterizes the orientation of the filament segment. "Macroscopic" order parameter $\boldsymbol{\eta}$ is obtained by averaging the tensor s over a small representative volume (RV) containing a large number of filaments. The scale of the representative volume is much smaller than the size of the cell.

Then, the macroscopic order parameter is defined by the relation

$$\boldsymbol{\eta} = \frac{3}{2}\eta\left(\mathbf{dd} - \frac{1}{3}\mathbf{I}\right), \tag{1}$$

where $\mathbf{d}$ is the director (indicates the direction of loading), η is the anisotropy parameter, which varies from −0.5 (the case of planar isotropy with director $\mathbf{d}$ orthogonal to the loading plane) to 1 (achieved when all segments are fully orientated in the loading direction $\mathbf{d}$), where the value 0 corresponds to an isotropic distribution of filaments. The norm in relation (1) is chosen so that in the coordinate system where $\mathbf{d} = (0, 0, 1)$, we have $\eta_{33} = \eta$. It is known that in polymer systems the process of fiber orientation leads to material deformation. Then let one introduce a coefficient γ, characterizing the "orientational susceptibility" of the filament segments with respect to the external stimulus, and the concentration of segments in a representative volume N, with the help of which one can put the deformation $\boldsymbol{\varepsilon}_o$ in correspondence with the order parameter. Formulate the relationship between strain due to orientation effects of eukaryotic actin filaments and the order parameter $\boldsymbol{\eta}$ as

$$\boldsymbol{\varepsilon}_o = N\gamma\boldsymbol{\eta}. \tag{2}$$

Note that the coefficient γ has the dimension of volume. Let one assume that the value of γ is determined by the volume of an elementary structural unit of the system under consideration. From relation (2) and the assumption of small strain, it follows that $N \sim 0.01\ \gamma^{-1}$. Then, the strain $\boldsymbol{\varepsilon}_o$ varies in the range of values from −0.005 to 0.01, where a negative value corresponds to the RV compression scenario and a positive value corresponds to the stretching scenario. The same applies to the meaning of the plus/minus signs for stress values.

2.2 Effective Field Method

Let the orienting influence experienced by a single ordering "microscopic" element be determined by the deviator of the stress tensor $\boldsymbol{\sigma}$ and the already achieved degree of orientation $\boldsymbol{\eta}$, which is determined by the biochemistry of the process [29]. The effective field acting on a filament segment accounts for both mechanical and biochemical interactions, expressed as:

$$\mathbf{H} = \sigma_o + \lambda \varepsilon_o. \tag{3}$$

Here, $\boldsymbol{\sigma}_o = \boldsymbol{\sigma}_{o,active} + \boldsymbol{\sigma}_{o,passive}$, with $\boldsymbol{\sigma}_{o,passive}$ representing stresses from external loads and $\boldsymbol{\sigma}_{o,active}$ arising from motor protein activity (e.g., myosin–actin interactions) and $\lambda = \lambda_o + \lambda_c$. The parameters λ_o and λ_c denote the effective volumetric energies of filament cross-links and biochemical signals, respectively, with c representing the concentration of signaling molecules (e.g., calcium ions). The inclusion of $\boldsymbol{\sigma}_{o,active}$ and $\boldsymbol{\sigma}_{o,passive}$ accounts for active cellular processes, such as contractility and signaling-driven remodeling, which are critical for capturing the dynamic nature of the cytoskeleton.

The orientation part of the filament segment energy E is determined by the relation

$$E = -\mathbf{H} : \gamma \mathbf{s}_o.$$

The probability of different orientations of segments in a representative volume of the cell cytoskeleton corresponds to the Boltzmann–Gibbs distribution in the presence of an effective field and is given in the form of

$$W = Z^{-1} \exp\left(-\frac{E}{\Theta}\right), \qquad Z = \int_{\Omega} \exp\left(-\frac{E}{\Theta}\right) d\Omega, \tag{4}$$

where Θ is the effective temperature factor, Ω is the space of possible orientations of filament segments, and relation (4) defines the norm of the distribution. Thus, the out-of-equilibrium state of the system is characterized not only by external parameters and temperature (or energy), but also by an internal variable in the presence of some (additional) effective field. This means that the thermodynamic potentials and entropy of the out-of-equilibrium state correspond to the potentials and entropy of the equilibrium state characterized by the values of the state variables of the system, but in the presence of an additional force field.

2.3 Free Energy

Using the self-consistent field method [30], one can introduce the self-consistency equation of microscopic and macroscopic order parameters

$$\varepsilon_o = N \int_\Omega \gamma s W d\Omega. \tag{5}$$

Equation (5) when considering the case of uniaxial deformation of the cage is numerically solved in the following form

$$\xi = \frac{3\gamma}{2\Theta}(\sigma_o + \lambda\varepsilon_o) = -\left(\frac{c_1}{c_4 + \varepsilon_o} + \frac{c_2}{c_5 + \varepsilon_o} + \frac{c_3}{c_6 + \varepsilon_o}\right). \tag{6}$$

Hence, the phenomenological representation of the orientation part of the Helmholtz free energy of the cell F_o is defined in the form

$$F = -\frac{2\Theta}{3\gamma}(c_1 \ln(-c_4 - \varepsilon_o) + c_2 \ln(-c_5 - \varepsilon_o) + c_3 \ln(-c_6 - \varepsilon_o)) - \frac{\lambda\varepsilon_o^{\,2}}{2}, \tag{7}$$

where c_1, c_2, c_3, c_4, c_5, and c_6 are the coefficients of the approximation function of Eq. (5). Table 1 summarizes their values.

We introduce a mechanical analogue of the representative volume of the cell, corresponding to a structural–mechanical scheme in the form of parallel connected elements of Maxwell-type and purely elastic type. The choice of this scheme is due to the fact that the cytoskeleton, as a representative of the class of polymeric materials, experiences reversible deformations of entropic character. A Maxwell-type element is defined as elastic and orientation elements connected in series. The elastic element describes reversible deformations, while the orientational element is associated with the corresponding inelastic deformation mechanism of the filament segments ordering, whose free energy is given by the relation (7).

Table 1
Coefficients of the approximation function of Eq. (5)

c_1	c_2	c_3	c_4	c_5	c_6
1.6×10^5	1.6	1.8	-9×10^4	−1	0.6

Reversible deformations of the other cell organelles are responsible, are described by a purely elastic element coupled in parallel to a Maxwell-type element.

It follows from the second law of thermodynamics that energy dissipation can be represented as

$$T\dot{S} = \left(\boldsymbol{\sigma}_o - \frac{\partial F}{\partial \boldsymbol{\varepsilon}_o}\right) : \dot{\varepsilon}_o \geq 0, \tag{8}$$

where T is the temperature, S is the entropy.

According to the hypothesis of linear relationship of thermodynamic forces and fluxes, as well as Onzager's principle, the following constitutive relation for the representative volume of the cell follows from Eq. (8)

$$\frac{\nu_o}{E_e}\dot{\sigma} + \sigma = \frac{\nu_o(E_r + E_e)}{E_e}\dot{\varepsilon} + E_r\varepsilon + w(\varepsilon, \sigma, E_r, E_e), \tag{9}$$

$$w(\varepsilon_o) \equiv \frac{\partial F}{\partial \varepsilon_o} = \frac{2\Theta}{3\gamma}\xi(\varepsilon_o) - \lambda\varepsilon_o, \tag{10}$$

$$\varepsilon_o = \varepsilon - \frac{\sigma - E_r\varepsilon}{E_e}, \tag{11}$$

where E_r, E_e are the elastic moduli of the cell and the cytoskeleton, respectively, υ_o is the viscosity. The viscosity in the general form for cells depends on the state parameters and can be represented as

$$\nu_o = E_e\tau_o, \tag{12}$$

where τ_o is the characteristic time scale of the response of actin filaments. This time at a known volumetric free energy density can be determined by activation-type relation

$$\tau_o = \tau_0 \exp\left(\frac{\gamma\Delta\Psi}{\Theta}\right), \tag{13}$$

where τ_0 is the period of natural oscillations of microfilament segments (values inversely proportional to natural frequency), $\Delta\Psi$ is the potential barrier that the system needs to overcome in order to all segments orientation of the representative volume in one particular direction.

Note that the energy barrier is affected by the applied stress, which follows from the definition of the thermodynamic potential $\Psi = F(\varepsilon_o) - \sigma_o\varepsilon_o$. The higher its value, the lower the barrier.

Also influential is the effective temperature factor decrease in temperature leads to a decrease in the barrier value. If the values of $\Delta\Psi$ are much larger than Θ, this leads to an increase in relaxation times; if the potential barrier becomes zero, τ_0 ceases to depend on the stress. Relationship (9) in its structure resembles the constitutive relation for the standard linear body model, except for the last summand, which is responsible for accounting for temperature and orientation effects.

2.4 Phenomenology of Eukaryotic Cells

As can be seen from the above basic statements of the statistical-thermodynamic model of the eukaryotic cell, it takes into account its viscoelastic properties.

The nature of viscoelasticity is due to the specific features of the structural relaxation mechanisms of cells. Isolated cells over a wide range of time scales exhibit at first a short-term stress relaxation obeying an exponential law, which is subsequently changed to a power-law dependence. A combined type of models is usually used to describe this effect. In particular, a poroelastic model is applied to describe the initial reaction [31], and for modeling the second type of cell response, models with a power-law of stress relaxation are used. The last class of models either include fractal (or fractional) elements instead of viscous elements, or are representatives of the class of generalized Maxwell models. Equation (13) allows one to introduce into the model a continuous spectrum of relaxation times that depends on the current structure of the modeling object, and to ensure that this feature is described in one unified approach. In this case, the range of time scales on which the time dependence of stresses is observed is 1–2 orders of magnitude. As shown by numerous experimental data, this is not enough, since the step dependence of stresses on time in some cases reaches four orders of magnitude.

The power-law of stress relaxation can be introduced either as is done in models using fractional calculus or generalized Maxwell models, or similarly to the approach taken in the soft glassy rheology model. This model was proposed by Sollich [32, 33] for soft disordered materials such as foams, gels and suspensions and is related to the molecular or mesoscopic consideration of the object of study. The theory suggests that the observed invariant behavior of the system is a natural consequence of the metastability of the internal structure of the material, which consists of many disordered elements that are held together by weak interaction forces in potential minima. This statement correlates with the ideas of the statistical-thermodynamic model described above in the context of the generalized Ginzburg–Landau free energy describing the power-law of stress relaxation over a wide range of time scales due to the metastability "decomposition."

3 Numerical Modeling of Cell Reactions to Gravitational Effects

Let us obtain a mathematical formulation of the problem of modeling the reactions of the cell to gravitational effects of different magnitude. For simplicity, consider a representative volume of the cell. The setting for compression of unit volume simulates the effect of gravity. Then, relation (9) has the form

$$\dot{\varepsilon}(t) = \frac{1}{\tau_0 \exp(\frac{\gamma\Delta\Psi}{\Theta})(E_r + E_e)}(\sigma - E_r\varepsilon(t) - w(\varepsilon(t), \sigma, E_r, E_e)), \quad (14)$$

where σ is the stress value that corresponds to the gravitational field of a particular magnitude. Supplementing Eq. (14) with the initial condition $\varepsilon(t_0) = \varepsilon_0$, we can find the dependence $\varepsilon(t)$ corresponding to a given value of the effective temperature factor Θ. In case the effective temperature factor is defined by an exponential distribution in the form $\Theta = \Theta_0 + \mathrm{Exp}(\Theta_\lambda)$, the time dependence of the strain can be calculated as an average over realizations

$$\langle \varepsilon(t) \rangle_{\Theta=\Theta_0+\mathrm{Exp}(\Theta_\lambda)} \quad (15)$$

To estimate the stress acting on a representative cell volume and corresponding to the acceleration of free fall, it is necessary to know the area of force application and the magnitude of the force itself. To determine the first, let us estimate RV of the cytoskeleton. The filament segment length ranges from 72 to 10,000 nm. With the known microfilament diameter, 8–25 nm, the segment volume γ can be estimated as $10^3 \div 10^6$ nm^3. Then, the concentration of N segments in a representative volume ranges from 10^{-3} to 10^{-6} nm^{-3} or from 10^6 to 10^3 mkm^{-3}. Thus, the representative volume of the cytoskeleton is 1 mkm^3.

The force is estimated as the mass of RV multiplied by the acceleration of free fall. One 10^3 nm segment contains about 10^2 units. Then with an effective mass of segment equal to 100 kDa or 10^{-23} kg and a concentration of 10^5 mkm^{-3}, the force is the order of 10^{-15} N. Hence, for an area value of 1 mkm^2 or 10^{-12} m^2 of stress, the corresponding free-fall acceleration is the order of 10^{-3} Pa. Based on this, the microgravity effect corresponds to 10^{-9} Pa. As this can be seen from the estimates obtained, the magnitude of the stress is very small.

The model parameters are given in Table 2. Young's modulus E_r and E_e were identified and verified for the statistical-thermodynamic model in [25]. The average value of the effective temperature factor Θ_0 is obtained from the constants defined in the paper and the estimation of the filament segment volume, and it significantly exceeds $k_B T$ corresponding to the temperature $T = 310$ K (k_B s the Boltzmann constant). The intensity of the distribution Θ_λ s set so that the free

Table 2
Parameters of statistical-thermodynamic model

E_r, Pa	E_e, Pa	τ_0, s	λ, Pa	γ, nm³	N, nm⁻³	Θ_0, J	Θ_λ, J	m, kg
10^3	10^6	2×10^{-3}	10^6	$10^3\div10^6$	$10^{-6}\div10^{-3}$	$1\div2.3\times10^{-16}$	10^{-17}	10^{-16}

energy profiles vary slightly in their shape. The parameter λ is assumed to be proportional to the Young's modulus E_e.

In order to simplify the analysis of the model and identify the parameters, we introduce the following dimensionless values

$$s'=\gamma Ns, \varepsilon_o'=\frac{\varepsilon_o}{\gamma N}, \sigma'=\frac{\sigma}{\Theta N}, \chi=\frac{\Theta}{\lambda\gamma}, E'=\frac{E}{\Theta}, F'=\frac{F}{\Theta\gamma N^2}$$
$$E_r'=\frac{\gamma}{\Theta}E_r, E_e'=\frac{\gamma}{\Theta}E_e, t'=t\sqrt{\gamma^{1/3}\lambda/m}, \tau_0'=\tau_0\sqrt{\gamma^{1/3}\lambda/m}, \quad (16)$$

where the dimensionless complex χ plays the role of the bifurcation parameter.

Let one consider the deformation process of a representative volume of a cell, which is in an ordinary and a microgravitational field. As noted in the introduction, in the presence of a physical constraint (occurring outside or inside the cell), the system chooses one among many structural configurations. In terms of the statistical-thermodynamic model of the cytoskeleton, this means that the potential Ψ must have one minimum that corresponds to an equilibrium and steady state. Therefore, for modeling the process of deformation of the cytoskeleton RV, the case when χ is greater than some critical value was considered at the first stage χ_c.

Figure 1a represents the time dependences of strain for $\chi > \chi_c$ (blue color), $\chi \approx \chi_c$ (red color), and $\chi < \chi_c$ (green color). These are all possible (typical) cell reactions. The curves are obtained using a dimensionless formulation. For convenience, we will further omit the strokes, assuming that we are talking about dimensionless quantities. Figure 1b illustrates the dependence of the thermodynamic potential of the system on the values of the orientation strain. In the absence of compressive stress, the minimum of the function $\Psi(\varepsilon_o)$ coincides with the zero value of the orientational strain (blue color). Circles correspond to the values of strain at different moments of time.

The cytoskeletal structure changes when the gravitational influence is weakened. This leads to a change in the effective temperature factor and/or filament segment volume, which can be accounted for by varying the parameter χ. The value of this parameter decreases and becomes less than χ_c. As a result, the form of dependence $\Psi(\varepsilon_o)$ changes and another, less stable, minimum in the negative εo region appears (red and green colors).

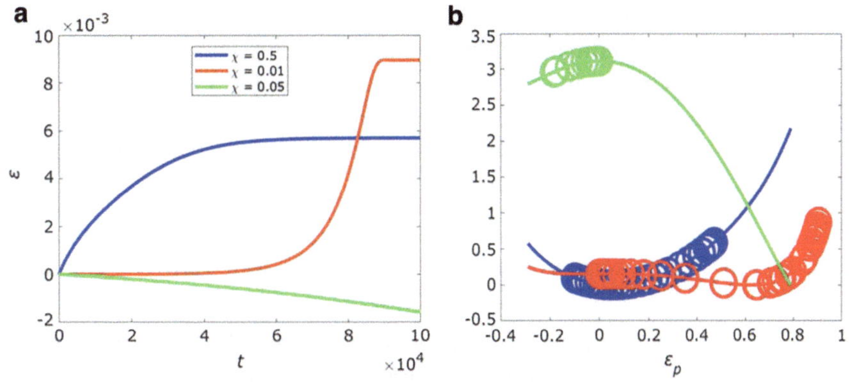

Fig. 1 Dependences of the deformation of the representative cell volume on time (**a**) and of the thermodynamic potential of system on the values of orientational deformation (**b**)

4 Discussion and Conclusion

The proposed theoretical framework provides a comprehensive approach to understanding how gravitational conditions influence the mechanobiological behavior of breast cancer cells through quantitative modeling of cytoskeletal dynamics. Our analysis of the constitutive relations for representative cell volumes revealed the dimensionless parameter χ as a key bifurcation parameter that determines the system's energy landscape and deformation characteristics. This finding offers critical insights into the fundamental mechanisms driving phenotypic transitions under microgravity.

The framework successfully demonstrates that when χ exceeds a critical threshold ($\chi > \chi_c$), cells exhibit a single stable mechanical state, corresponding to their behavior under normal gravity conditions. This is evidenced by the convergence of strain to equilibrium values regardless of initial conditions. The model's ability to reproduce this behavior confirms its validity in describing the deterministic organization of cellular structures under gravitational constraints, supporting previous experimental observations.

More significantly, our framework captures the transition to bistable behavior when χ decreases below χ_c under microgravity conditions. This theoretical prediction aligns remarkably well with experimental reports of two distinct cell populations (AD and CLUMP) emerging in microgravity. The model suggests that this bifurcation occurs due to fundamental changes in the thermodynamic potential landscape, where the effective temperature factor Θ and filament segment volume γ become dominant over gravitational constraints.

The framework's strength lies in its capacity to connect microscopic cytoskeletal reorganization with macroscopic cellular behavior. By introducing the concept of orientational susceptibility γ and incorporating the effective temperature factor Θ, we provide a mechanistic explanation for how minute gravitational changes (10^{-6} g) can induce substantial phenotypic modifications. This addresses a long-standing challenge in cellular mechanobiology—explaining how cells detect and respond to extremely weak physical signals.

Our theoretical approach offers several advantages over previous models:

- It unifies the description of short-term exponential and long-term power-law stress relaxation behaviors within a single thermodynamic framework
- It provides quantitative criteria (through the χ parameter) for predicting phenotype transitions
- It explains the reversibility of phenotypic changes when gravity is restored

The framework's predictions have important implications for cancer research. The demonstrated sensitivity of χ to cytoskeletal parameters suggests potential therapeutic strategies for manipulating cell phenotypes through targeted interventions affecting filament organization or cellular thermodynamics. This could open new avenues for controlling tumor progression by altering the mechanical rather than biochemical properties of cancer cells.

However, several limitations should be noted. The current framework simplifies the complex interactions between different cytoskeletal components (actin, microtubules, intermediate filaments) into effective parameters. Future extensions could incorporate more detailed representations of these subsystems. Additionally, while the model successfully reproduces key experimental observations, direct validation against comprehensive microgravity experiments would further strengthen its predictive power.

The proposed framework also raises important questions for future research:

- How do cell–cell interactions modify the predicted behavior in tissue-like environments?
- What specific biochemical pathways couple with the mechanical changes described by the model?
- Can the χ parameter serve as a measurable biomarker for cancer progression?

These considerations notwithstanding, our theoretical framework establishes a robust foundation for understanding and predicting cellular responses to gravitational variations. By providing quantitative links between physical constraints, thermodynamic parameters, and phenotypic outcomes, it advances both fundamental mechanobiology and potential space medicine applications. The model's ability to describe reversible transitions suggests that mechanical manipulation of cells could become a powerful tool in biomedical research, particularly for developing countermeasures against spaceflight-induced cellular changes or novel cancer therapies.

Acknowledgments

The reported study was made in the framework of the government task, registration number of the theme 124020200116-1.

References

1. Naimark O (2019) Mesoscopic cell dynamics in different environment and problem of cancer. AIP Conf Proc 2167:020237
2. Naimark OB, Bayandin YV, Grishko VV, Nikitiuk AS (2020) Mesoscopic cell mechanobiology and the problem of cancer. Organisms 4:42
3. Bizzarri M, Naimark O, Nieto-Villar J, Fedeli V, Giuliani A (2020) Complexity in biological organization: deconstruction (and subsequent restating) of key concepts. Entropy 22:885
4. Nickerson CA, Ott CM, Wilson JW, Ramamurthy R, Pierson DL (2004) Microbial responses to microgravity and other low-shear environments. Microbiol Mol Biol Rev 68:345
5. Rosenzweig JA, Abogunde O, Thomas K, Lawal A, Nguyen YU, Sodipe A, Jejelowo O (2010) Spaceflight and modeled microgravity effects on microbial growth and virulence. Appl Microbiol Biotechnol 85:885
6. Masiello MG, Cucina A, Proietti S, Palombo A, Coluccia P, D'Anselmi F, Dinicola S, Pasqualato A, Morini V, Bizzarri M (2014) Phenotypic switch induced by simulated microgravity on MDA-MB-231 breast cancer cells. Biomed Res Int 2014:652434
7. Testa F, Palombo A, Dinicola S, D'Anselmi F, Proietti S, Pasqualato A, Grazia Masiello M, Coluccia P, Cucina A, Bizzarri M (2014) Fractal analysis of shape changes in murine osteoblasts cultured under simulated microgravity. Rend Lincei 25:39
8. Pisanu ME et al (2014) Lung cancer stem cell lose their stemness default state after exposure to microgravity. Biomed Res Int 2014:470253
9. Bizzarri M, Pasqualato A, Cucina A, Pasta V (2013) Physical forces and non linear dynamics mould fractal cell shape: quantitative morphological parameters and cell phenotype. Histol Histopathol 28:1
10. Bizzarri M, Masiello MG, Giuliani A, Cucina A (2018) Gravity constraints drive biological systems toward specific organization patterns: commitment of cell specification is constrained by physical cues. BioEssays 40:1700138
11. Bizzarri M, Monici M, Loon JJWAV (2015) How microgravity affects the biology of living systems. Biomed Res Int 2015:863075
12. Heintzelman KF, Phillips HM, Davis GS (1978) Liquid-tissue behavior and differential cohesiveness during chick limb budding. J Embryol Exp Morphol 47:1
13. Pollard EC (1965) Theoretical studies on living systems in the absence of mechanical stress. J Theor Biol 8:113
14. Unsworth BR, Lelkes PI (1998) Growing tissues in microgravity. Nat Med 4(8):901–907
15. Freed LE, Langer R, Martin I, Pellis NR, Vunjak-Novakovic G (1997) Tissue engineering of cartilage in space. Proc Natl Acad Sci U S A 94:13885
16. Bizzarri M, Cucina A, Palombo A, Grazia Masiello M (2014) Gravity sensing by cells: mechanisms and theoretical grounds. Rend Lincei 25:29
17. Vishnyakov GN, Levin GG, Minaev VL, Pickalov VV, Likhachev AV (2004) Tomographic interference microscopy of living cells. Microsc Anal 18:15
18. Moorman SJ, Shorr AZ (2008) The primary cilium as a gravitational force transducer and a regulator of transcriptional noise. Dev Dyn 237:1955
19. Tabony J, Glade N, Papaseit C, Demongeot J (2002) Microtubule self-organisation and its gravity dependence. Adv Space Biol Med 8:19
20. Longo G, Montévil M (2011) From physics to biology by extending criticality and symmetry breakings. Prog Biophys Mol Biol 106(2):340–347
21. Perelson AS (1977) Self-organization in nonequilibrium systems. From dissipative structures to order through fluctuations. In: Nicolis G, Prigogine I (eds) The quarterly review of biology. Wiley, New York, pp 1–100
22. Mojtahedi M, Skupin A, Zhou J, Castaño IG, Leong-Quong RYY, Chang H, Trachana K, Giuliani A, Huang S (2016) Cell fate decision as high-dimensional critical state transition. PLoS Biol 14:e2000640

23. Nikitiuk AS, Koshkina AA, Bayandin YV, Naimark OB (2023) On thermodynamics and relaxation properties of eukaryotic cells. Int J Non Linear Mech 157:104532
24. Nikitiuk AS (2023) Parameter identification of viscoelastic cell models based on force curves and wavelet transform. Comput Res Model 15:1653
25. Nikitiuk AS (2024) Simulation of constant rate loading of eukaryotic cells using statistical thermodynamic methods. Math Biol Bioinforma 19:304
26. Nikitiuk AS, Bayandin YV, Naimark OB (2025) Orientation and shear mechanisms of finite deformation of the cell cytoskeleton. Int J Eng Sci 214:104306
27. Nikitiuk AS, Bayandin YV, Naimark OB (2025) Cell mechanobiology in microgravity conditions. Microgravity Sci Technol 37:1
28. Naimark OB, Nikitiuk AS, Bayandin YuV (2025) Finite deformation of the cell cytoskeleton: orientation and shear mechanisms. Contin Mech Thermodyn 65:0123456789
29. Deshpande VS, McMeeking RM, Evans AG (2006) A bio-chemo-mechanical model for cell contractility. Proc Natl Acad Sci U S A 103:14015
30. Nikitiuk AS, Bayandin YV, Naimark OB (2022) Statistical thermodynamics of DNA with open states. Phys A Stat Mech Its Appl 607:128156
31. Moeendarbary E, Valon L, Fritzsche M, Harris AR, Moulding DA, Thrasher AJ, Stride E, Mahadevan L, Charras GT (2013) The cytoplasm of living cells behaves as a poroelastic material. Nat Mater 12:253
32. Sollich P, Lequeux F, Hébraud P, Cates ME (1997) Rheology of soft glassy materials. Phys Rev Lett 78:2020
33. Sollich P, Cates ME (2012) Thermodynamic interpretation of soft glassy rheology models. Phys Rev E Stat Nonlin Soft Matter Phys 85:031127

Chapter 4

Multiscale Modeling and Systems Biology in Microgravity Investigations

Mariagiovanna Pais, Thomas Cahill, Guillermo H. López-Campos and Gary Hardiman

Abstract

Spaceflight imposes a unique combination of environmental stressors including ionizing radiation, microgravity, and prolonged isolation that disrupt biological systems from the genome to the proteome. These stressors contribute to a range of health effects such as musculoskeletal degradation, neuroimmune dysfunction, and psychological challenges. Advances in countermeasure development, including biomedical devices and therapeutics, have already yielded terrestrial benefits for conditions like osteoporosis and cardiovascular disease. To systematically characterize spaceflight-induced biological alterations, researchers employ a wide array of omics techniques transcriptomics, proteomics, metabolomics alongside silico modeling approaches such as molecular dynamics simulations. Microgravity research is conducted both in true spaceflight environments (e.g., aboard the ISS and suborbital vehicles) and via terrestrial analogs (e.g., clinostats and random positioning machines), using in vitro and in vivo models to capture cellular and organismal responses. Open-access platforms like NASA GeneLab facilitate data sharing and integrative analysis of multiomics datasets generated under spaceflight conditions. In this chapter, we describe computational pipelines for processing space-derived transcriptomic data, with a focus on quality control, alignment, quantification, and downstream functional analysis. These methods support the construction of multiscale models that bridge molecular insights with physiological outcomes, enabling a deeper understanding of space biology and guiding countermeasure development.

Keywords Spaceflight biology, Microgravity, Omics technologies, Transcriptomics analysis, NASA GeneLab, Systems biology, Molecular dynamics simulations, Space medicine, In vitro and in vivo models, Bioinformatics pipelines

1 Introduction

Spaceflight exposes astronauts to unique hazards ionizing radiation, microgravity, and prolonged isolation that disrupt biological systems at every level, from the genome to the proteome [1, 2]. These stressors contribute to a range of health problems including musculoskeletal deterioration (muscle wasting and bone density loss), neurological deficits, and weakened immune defenses [1, 3],

Mariano Bizzarri (ed.), *Biomedicine and Biotechnology in Space*, Methods in Molecular Biology, vol. 3000, https://doi.org/10.1007/978-1-0716-5174-2_4,

as well as psychological challenges such as depression, solitude, and psychosomatic disorders [4]. Efforts to counteract these effects have already yielded benefits on Earth, for example, vibration platforms developed for Soviet cosmonaut training now help treat age-related muscle atrophy. Technologies developed to protect astronaut health such as bone density treatments, cardiovascular monitoring tools, and telemedicine systems have been adapted on Earth to improve care for conditions like osteoporosis, heart disease, and limited access to medical services. As private companies expand human spaceflight capabilities alongside national agencies, it is vital that research findings and multiomics datasets remain openly accessible. Free data sharing accelerates both terrestrial health advances and spaceflight biology, including the creation of robust multiscale biological models [5, 6].

Microgravity investigations leverage both actual spaceflight platforms and terrestrial analogs to dissect gravity-dependent biological processes. On-orbit experiments are conducted aboard the International Space Station (ISS) and, increasingly, on commercial suborbital vehicles (e.g., Virgin Galactic's SpaceShipTwo or Blue Origin's New Shepard), which provide minutes of true microgravity for rapid proof-of-concept studies. Complementary ground-based simulators including clinostats (which rotate samples around a single axis to randomize the gravity vector), random positioning machines (which continuously reorient samples in three dimensions), and rotating wall vessels (which create low-shear, suspension culture conditions) reproduce many aspects of weightlessness for extended-duration cell and tissue experiments. Researchers must also decide between in vitro versus in vivo approaches. In vitro systems (e.g., 2D cell monolayers, 3D spheroids, or organ-on-a-chip platforms) offer precise control over culture conditions, high throughput, and detailed molecular readouts, whereas in vivo models (e.g., rodents flown to the ISS or housed in specialized ground-based clinostat cages) capture systemic, multiorgan responses to microgravity, but involve greater logistical complexity and ethical considerations. Together, these platforms and techniques form an integrated toolkit for unraveling the cellular and organismal adaptations to reduced-gravity environments [7, 8].

At the molecular level, spaceflight studies employ an array of omics and computational techniques to characterize how microgravity alters cellular biochemistry [9]. In proteomics, cells cultured under true or simulated microgravity are lysed and proteins are digested (e.g., trypsin) before mass spectrometry workflows, including liquid chromatography–tandem MS (LC-MS/MS) with data-dependent acquisition, quantify changes in protein abundance, post-translational modifications, and interaction networks [10]. Transcriptomic analyses begin with optimized RNA extraction protocols from limited space-flown samples (often preserved in RNAlater or flash-frozen on orbit), followed by rRNA depletion or poly(A) selection and high-throughput sequencing. Single-cell RNA-Seq adaptations for space utilize microfluidic partitioning (e.g., 10× Genomics Chromium) or plate-based Smart-Seq2 workflows that accommodate low-input, high-background RNA often encountered in microgravity experiments. Metabolomics workflows similarly require careful metabolite quenching (e.g., cold methanol) and extraction from cell lysates, with both nuclear magnetic resonance (NMR) spectroscopy and high-resolution LC–MS methods used to profile small-molecule alterations in central carbon, lipid, and amino acid pathways. Complementing these empirical approaches, molecular dynamics simulations recreate the microgravity environment in silico by modifying force-field parameters, employing explicit solvent models, and introducing pseudogravity constraints (or gravity vector averaging algorithms) to probe how proteins and membranes behave under reduced gravitational forces. Trajectory analyses measuring root-mean-square deviation (RMSD), hydrogen-bond lifetimes, and solvent density reveal microgravity-induced conformational shifts that correlate with experimental omics findings. Together, these methodologies provide a comprehensive, multiscale view of molecular adaptations to spaceflight.

NASA's GeneLab is an open-access, collaborative bioinformatics platform managed by the Space Biosciences Division at NASA's Ames Research Center. It serves as both a data repository and an analysis environment for "omics" experiments conducted in spaceflight and related ground-based simulations. GeneLab enables researchers to upload, download, store, search, share, and analyze genomic, transcriptomic, proteomic, metabolomic, and epigenomic datasets, all accompanied by detailed metadata describing the experimental conditions (e.g., hardware, duration, organism). By aggregating spaceflight and analog experiment data across diverse organisms from rodents and fruit flies to plants and cultured cells, GeneLab enables scientists worldwide to explore how microgravity, radiation, and other space-specific stressors affect biological systems at multiple scales. Its interactive interface and integrated tools lower the barrier for both bioinformatics specialists and domain researchers, fostering hypothesis generation without the need for new experiments. Ultimately, GeneLab accelerates discoveries in space biology, informs countermeasure development for astronaut health, and yields insights with terrestrial applications such as musculoskeletal disease and aging research [11].

In this chapter, we explore multiscale modeling and systems biology approaches to understanding and mitigating the biological challenges of space travel [12]. We will focus on the computational methods involved in preprocessing, transcriptomics data from GeneLab, and further downstream analysis of the raw sequencing reads. We assume the reader had a basic familiarity with the command–line interface. While it may seem daunting at first, many free tutorials are available online, and you can acquire essential skills with only a small-time investment. It also helps to know the common file formats used in sequencing workflows, if you are not yet comfortable with these, we recommend consulting bioinformatics protocols that explain raw data file types in detail [13].

2 Materials and Resources

Genomics, transcriptomics, proteomics, and metabolomics are high-throughput omics technologies that enable large-scale investigation of molecular responses to environmental stressors such as microgravity and, among these, RNA sequencing (RNA-Seq) is a widely used method for the identification of differentially expressed genes, novel transcripts, and regulatory signatures in a variety of physiological conditions [13]. RNA-Seq involves running a standard analytical workflow, which includes quality assessment and preprocessing of raw reads, mapping to a reference genome, gene-level quantification, followed by statistical modeling to identify differential expression [14]. When integrated with functional annotation, pathway enrichment analysis, and network biology, RNA sequencing data underpin systems level investigations of transcriptional responses. This approach is particularly valuable in space biology studies, as it enables the identification of key biochemical pathways mediating adaptation to altered gravity and supports the development of predictive models for long-term physiological changes.

2.1 Computing Resources

Researchers typically rely on high performance computing (HPC) clusters to handle the intensive computational demands of next-generation sequencing data analysis. Many universities offer HPC resources for faculty and students, which can dramatically shorten omics data preprocessing times. Although numerous bioinformatics tools are available for Windows, they often receive more comprehensive support on Linux and macOS. If HPC is not accessible, omics analysis can still be performed on a capable personal computer. Both macOS and Windows systems can run most of the required software. Windows users may need to install a Linux environment (e.g., Windows Subsystem for Linux: https://learn.microsoft.com/en-us/windows/wsl/about). Regardless of platform, ensure the machine meets the minimum requirements for the primary alignment tool: at least five CPU cores and 16 GB of RAM.

2.2 Environment Management

Bioinformatics analyses often rely on various platforms and programming languages, with R and Python being the most used. Both are script languages that are freely available, multiplatform, and contain biomedical development branches Bioconductor (https://bioconductor.org/) for R, and Biopython (https://biopython.org/) for Python. Due to their popularity, it is relatively easy to find online training materials and courses that facilitate the familiarization with and learning of both languages. This chapter will use a combination of software packages that have been developed for either language.

As bioinformatic analysis often involves installing numerous different tools, packages, and dependencies, although not essential, it can be extremely beneficial to use system package managers such as Conda, which allows software to be isolated into separate environments. For example, if a system has Python v3.11.0 installed and one of the tools requires Python v3.9.0 for compatibility, it is unnecessary to remove the existing installation to downgrade. Instead, separate environments can be created with different Python versions installed for distinct tasks. This approach prevents conflicts between programs and allows for seamless switching between environments based on the analysis being performed. Conda is available for Linux, macOS, and Windows (https://anaconda.org/anaconda/conda); it is well supported and adopted, with many free online learning resources.

2.3 Software

2.3.1 Programming Languages

1. **Python** (https://www.python.org/), Linux, macOS, and Windows. Programming language.
2. **R** (https://www.r-project.org/), Linux, macOS, and Windows. Statistical computing environment.

2.3.2 RNA-Seq Preprocessing and Differential Expression Analysis

1. **FastQC** (https://www.bioinformatics.babraham.ac.uk/projects/fastqc/), Linux, macOS, and Windows. Java-based software.
2. **MultiQC** (https://multiqc.info/), Linux, macOS, and Windows. Python tool.
3. **Cutadapt** (https://cutadapt.readthedocs.io/en/stable/), Linux, macOS, and Windows. Python tool.
4. **STAR** (https://github.com/alexdobin/STAR), Linux, macOS, and Windows. RNA-Seq aligner (C++ software).
5. **HTSeq-count**(https://htseq.readthedocs.io/en/release_0.11.1/), Linux, macOS, and Windows. Python tool.
6. **DESeq2** (https://bioconductor.org/packages/release/bioc/html/DESeq2.html), Linux, macOS, and Windows. R Bioconductor package.

2.3.3 Visualization Tools

1. **ggplot2** (https://ggplot2.tidyverse.org/), Linux, macOS, and Windows. R package for data visualization, widely used to generate bar plots, dot plots, PCA plots, and scatterplots in transcriptomic analyses.
2. **pheatmap** (https://cran.r-project.org/web/packages/pheatmap/index.html), Linux, macOS, and Windows. R package.
3. **RColorBrewer** (https://cran.r-project.org/web/packages/RColorBrewer/), Linux, macOS, and Windows. R package.

2.3.4 GO Analysis

1. **org.Xx.eg.db** (https://bioconductor.org/packages/release/data/annotation/html/org.Mm.eg.db.html), Linux, macOS, and Windows. R Bioconductor annotation package.
2. **clusterProfiler**(https://bioconductor.org/packages/release/bioc/html/clusterProfiler.html), Linux, macOS, and Windows. R Bioconductor package.

2.3.5 PPI Network

1. **STRINGdb** (https://bioconductor.org/packages/release/bioc/html/STRINGdb.html), Linux, macOS, and Windows. R Bioconductor package.
2. **visNetwork** (https://cran.r-project.org/web/packages/visNetwork/), Linux, macOS, and Windows. R package.

2.3.6 Machine Learning

Stats (https://stat.ethz.ch/R-manual/R-devel/library/stats/html/00Index.html), Linux, macOS, and Windows. Base R package for statistical methods, including k-means and hierarchical clustering.

3 Methods

3.1 NASA GeneLab Datasets

Transcriptomic analysis of samples collected during spaceflight experiments presents unique challenges due to the physiological complexity induced by microgravity and the scarcity of biological replicates and, for these reasons, a structured pipeline can be implemented, consisting of multiple analytical stages aimed at ensuring data integrity, reproducibility, and biological interpretability. To incorporate NASA's GeneLab datasets into your workflow, first create a GeneLab account and navigate to the Data Repository to identify relevant experiments (e.g., spaceflight vs. ground controls for your organism of interest). For each selected dataset, review the associated metadata, including organism strain, mission duration, tissue type, and sequencing platform, to ensure compatibility with your study design.

Download raw FASTQ files and accompanying metadata tables directly from the GeneLab portal. Next, process transcriptomic data using GeneLab's standardized RNA-Seq pipeline or your own bespoke pipeline: perform quality control (FastQC), trim adaptors (e.g., Trimmomatic or Cutadapt), and map reads to the appropriate reference genome using an aligner (e.g., STAR or HISAT2). Quantify gene expression counts with a tool such as featureCounts or HTSeq, and normalize counts using a method such as DESeq2 or edgeR [15, 16]. For downstream pathway and network analyses, annotate differentially expressed genes with functional databases (e.g., Ensembl, KEGG), and integrate results into network-based tools (e.g., Cytoscape). Throughout the process, document all software versions, parameters, and GeneLab DOI identifiers to ensure reproducibility and compliance with Findability, Accessibility,

Interoperability, and Reuse (FAIR) data standards [17]. RNA-Seq methodology (Fig. 1) consists of three main stages: preprocessing and quality control, differential expression analysis, and downstream interpretation [18, 19]. In space biology studies, it is essential to carefully consider experimental design factors such as sample size, sequencing depth, and biological variability between spaceflight and ground control groups. These elements determine the statistical power to detect differentially expressed genes and the granularity with which functional adaptations can be characterized. Additionally, choosing appropriate analysis tools depends on the biological model (e.g., human, mouse, or rat), the availability of species-specific annotations, and the computational resources at hand—especially since high-throughput sequencing experiments generate large datasets that can be resource-intensive to process [1].

3.2 Case Study: Rodent Research Kidney Transcriptomics

In this chapter, we use kidney transcriptomics datasets to highlight the workflow outlined in Fig. 1. Data for kidney tissue were retrieved from NASA GeneLab (https://genelab.nasa.gov/) using dataset identifiers OSD-102 (Rodent Research-1, RR-1v367) and OSD-163 (Rodent Research-3, RR-3v468). For RR-1, six 16-week-old female C57BL/6J mice were aboard the ISS for 37 days. RR-3 comprised ten 12-week-old female BALB/c mice flown for 39–42 days. After the mission, the crew euthanized the animals onboard. Tissues were stored at –80°C and fully dissected upon return to Earth. Ground control animals were matched by age, sex, strain, housing hardware, and euthanasia protocols to ensure accurate comparative analysis. Further details on transcriptome patterns and biological insights from a comparison of these two different strains of mice flown on the International Space Station, C57BL/6J and BALB/c have been published recently [20]. Kidney stone formation is a significant health risk during long-term space missions, driven by elevated blood calcium levels from bone demineralization under microgravity and space radiation. Spaceflight also induces transcriptomic alterations in various tissues, including the kidney. To investigate these changes, Finch et al., analyzed renal RNA-Seq profiles from two mouse strains (C57BL/6J and BALB/c) flown aboard the International Space Station [20]. The results reveal that spaceflight perturbs lipid metabolism, extracellular matrix remodeling, and TGF-β signaling pathways in the kidney, with a more pronounced transcriptional response in C57BL/6J mice than in BALB/c. This study also uncovered strain-specific differences in hyaluronan metabolism that may mitigate extracellular matrix remodeling by suppressing epithelial–mesenchymal transition. These findings improve our understanding of renal adaptation to spaceflight and may inform novel countermeasures for preventing kidney disease in astronauts and ultimately benefit patient populations on Earth. The objective in this chapter is to exploit this data for demonstrative purposes using systems biology tools. For biological inference, the reader is directed to the paper.

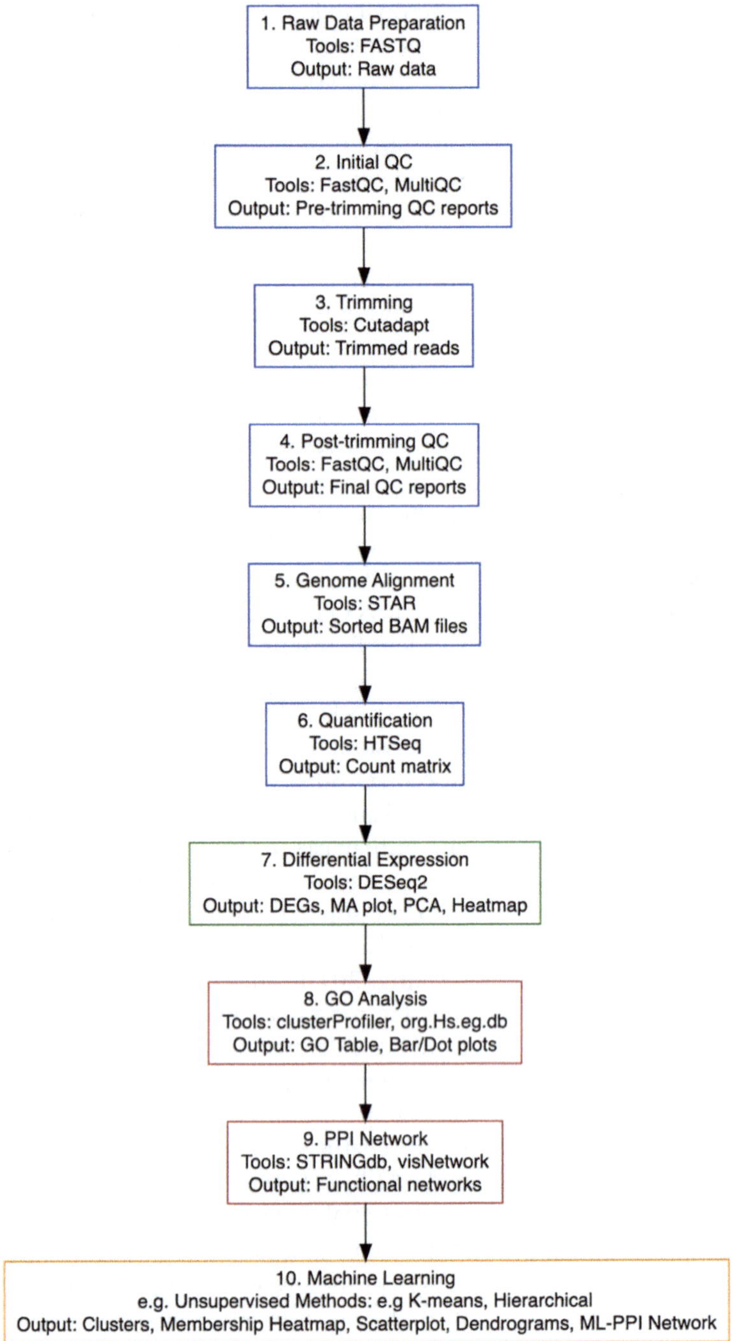

Fig. 1 Flow chart of the RNA-Seq data analysis pipeline. The diagram illustrates the RNA-Seq workflow, starting with the acquisition of raw sequencing data in FASTQ format and concluding with machine learning techniques (e.g., K-means, hierarchical clustering) to identify expression patterns and gene modules

3.3 Preprocessing and Quality Control

To ensure the trustworthiness of downstream analysis, RNA-Seq data is routinely preprocessed and quality controlled before it can be interpreted meaningfully. Preprocessing typically comprises the removal of adapter sequences and trimming of low-quality bases with tools like Cutadapt [21], which helps to reduce alignment errors and enhance mapping efficiency. Quality control tools, such as FastQC [22], are commonly used to assess the quality of raw and processed data. These provide crucial metrics such base quality scores, GC content, sequence length distribution, and duplication rates. MultiQC is an aggregated report prepared with techniques that can help uncover systematic flaws across samples (Fig. 2) [23]. The "General Statistics" table from the MultiQC v1.14 report in Fig. 2 shows read-level metrics for a batch of RNA-Seq samples. Each row corresponds to one FASTQ file (e.g., read pairs identified by sample, lane, and read number). Duplicate rates are moderately high (≈44%–52%), suggesting limited library complexity, likely due to PCR bias or low input, yet these rates are consistent across all samples. Obtaining RNA-Seq data from ISS-flown samples can present several challenges, including potential RNA degradation due to extended storage and return delays, as well as limited sample quantities constrained by payload and crew time. Additionally, environmental contamination, variable microgravity effects during processing, and the complexity of coordinating ground controls under identical hardware conditions can further complicate data consistency and interpretation. GC content remains tightly clustered around 46%–49%, matching the expected transcriptome composition and indicating minimal bias during library preparation. Read lengths consistently average 50–51 bp (with medians at 51 bp), confirming the intended sequencing design and uniform trimming. The percentage of reads failing the sequencer's internal

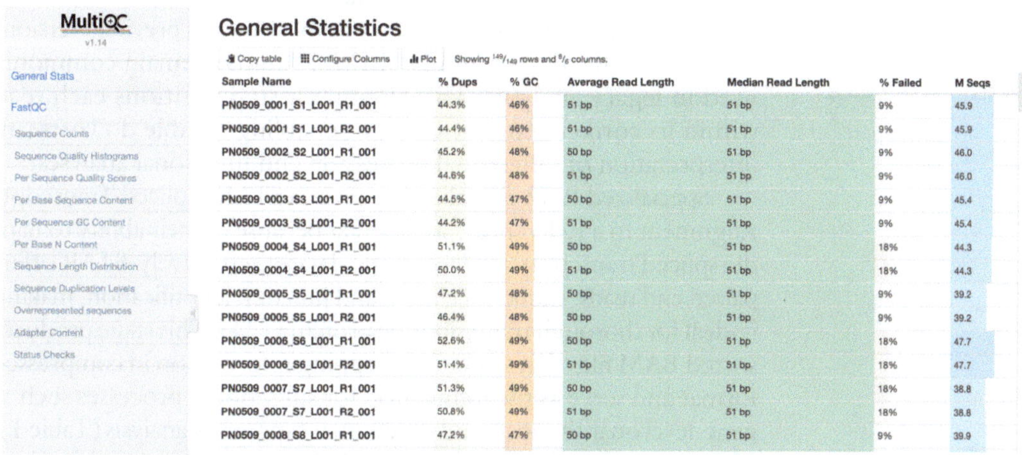

Sample Name	% Dups	% GC	Average Read Length	Median Read Length	% Failed	M Seqs
PN0509_0001_S1_L001_R1_001	44.3%	46%	51 bp	51 bp	9%	45.9
PN0509_0001_S1_L001_R2_001	44.4%	46%	51 bp	51 bp	9%	45.9
PN0509_0002_S2_L001_R1_001	45.2%	48%	50 bp	51 bp	9%	46.0
PN0509_0002_S2_L001_R2_001	44.6%	48%	51 bp	51 bp	9%	46.0
PN0509_0003_S3_L001_R1_001	44.5%	47%	50 bp	51 bp	9%	45.4
PN0509_0003_S3_L001_R2_001	44.2%	47%	51 bp	51 bp	9%	45.4
PN0509_0004_S4_L001_R1_001	51.1%	49%	50 bp	51 bp	18%	44.3
PN0509_0004_S4_L001_R2_001	50.0%	49%	51 bp	51 bp	18%	44.3
PN0509_0005_S5_L001_R1_001	47.2%	48%	50 bp	51 bp	9%	39.2
PN0509_0005_S5_L001_R2_001	46.4%	48%	50 bp	51 bp	9%	39.2
PN0509_0006_S6_L001_R1_001	52.6%	49%	50 bp	51 bp	18%	47.7
PN0509_0006_S6_L001_R2_001	51.4%	49%	51 bp	51 bp	18%	47.7
PN0509_0007_S7_L001_R1_001	51.3%	49%	50 bp	51 bp	18%	38.8
PN0509_0007_S7_L001_R2_001	50.8%	49%	51 bp	51 bp	18%	38.8
PN0509_0008_S8_L001_R1_001	47.2%	47%	50 bp	51 bp	9%	39.9

Fig. 2 Partial view of a typical MultiQC report generated from RNA-Seq preprocessing. The general statistics table aggregates metrics from multiple tools, including FastQC and Cutadapt, providing an overview of sequence quality, duplication levels, adapter content, and per-sample read counts

quality filter ranges from 9% to 18%, higher than ideal but uniform across the batch, implying systematic run-level issues rather than individual library failures. This is explained by the fact that space RNA-Seq experiments face challenges in sample handling, storage, limited replicates, and environmental variability, all of which can complicate RNA preservation and diminish sequencing accuracy. Finally, each sample boasts 39–48 million raw reads, providing robust sequencing depth; even after accounting for duplicates and failed reads, sufficient high-quality data should remain for reliable downstream alignment and quantification. FastQC and MultiQC are run both before and after read trimming to verify data quality, spot technical artifacts, and confirm that only high-quality reads progress to downstream analysis—laying a reliable foundation for accurate gene-expression results. During trimming, tools such as Trimmomatic or Cutadapt detect and remove residual sequencing adapters and low-quality bases, preventing artificial "adapter sequences" from mapping to the genome and inflating expression estimates. These quality-control and trimming steps can be executed on high-performance-computing clusters for large projects or on well-equipped local workstations for smaller datasets.

3.4 Alignment to a Reference Genome

After preprocessing and quality control, sequencing reads are aligned to a species-specific reference genome to accurately quantify gene expression within defined genomic coordinates. It is worth noting that this is one of the most computationally demanding steps, with STAR in particular requiring substantial RAM resources. Proper alignment ensures that transcriptomic analyses reflect true biological signals rather than misassigned or spurious mappings. For example, when studying *Mus musculus*, one would use the mouse reference genome along with corresponding gene annotation files (in GTF or GFF format) that specify the positions of genes, transcripts, and exons [20]. The GRCm39 mouse reference genome is the latest assembly from the Genome Reference Consortium, although previous versions like GRCm38 (mm10) and NCBIM37 (mm9) remain commonly used in legacy analyses. This alignment step positions each read within its correct genomic context, enabling reliable downstream interpretation of differential expression and functional analyses.

Specialized RNA-Seq aligners, such as STAR (Spliced Transcripts Alignment to a Reference), are popular because of their ability to handle spliced transcripts and large-scale datasets effectively. STAR offers paired end reads, strand specificity, and junction identification, making it ideal for thorough transcriptome profiling [24]. This stage produces sorted BAM files that store read mapping information in compressed format and serve as the foundation for subsequent processes such as gene-level quantification and differential expression analysis (Table 1). The alignment statistics for five RNA-Seq samples (that are a subset of the RR mission used as an example) reveal high mapping efficiency across the dataset. Sample 1 yielded approximately 89.2 million total reads, of which about 77.5 million (86.88%) mapped uniquely to the

Table 1

Simplified Example of STAR alignment summary metrics for five RNA-Seq samples taken from GeneLab OSD-102 and OSD-163. The table reports the total number of reads, the number and percentage of uniquely mapped reads, and the number and percentage of multiply mapped reads

Sample ID	Total reads	Uniquely mapped	Percentage of uniquely mapped	Multiply mapped	Percentage of multiply mapped	Sample number
Sample 1	89,194,260	77,493,911	86.88%	6,506,639	7.29%	3
Sample 2	68,991,895	58,262,300	84.45%	4,901,573	7.10%	7
Sample 3	109,455,827	94,319,457	86.17%	8,594,385	7.85%	18
Sample 4	66,260,764	58,136,815	87.74%	4,719,365	7.12%	23
Sample 5	127,703,041	112,205,596	87.86%	8,050,685	6.30%	79

reference genome and 6.5 million (7.29%) mapped to multiple loci. Sample 2 produced roughly 68.9 million reads, with 58.3 million (84.45%) uniquely mapped and 4.9 million (7.10%) multiply mapped. Sample 3 had a read count of 109.5 million, achieving 94.3 million (86.17%) unique alignments and 8.6 million (7.85%) multimapped reads. Overall, unique mapping rates ranged from 84.45% to 87.86%, and multiply mapped reads stayed between 6.30% and 7.85%, indicating consistently high-quality alignments across all samples. In STAR alignments, good-quality RNA-Seq data typically show ≥70–85% uniquely mapped reads, <10–15% unmapped reads, and mismatch rates <0.5% per base. Excessive multimapping, high mismatch rates, or a large fraction of unmapped reads may signal issues with sequencing quality, adapter contamination, or genome annotation mismatches. To ensure reproducibility and efficient software management, RNA-Seq alignment workflows are frequently executed within isolated software environments. This practice is particularly useful when using complex tools like STAR, which may require specific versions of auxiliary libraries or command-line utilities.

3.5 Gene Quantification

After alignment, gene expression levels are quantified by assigning mapped reads to annotated genomic features. Genome annotation plays a critical role in this step, and using different annotation sources can lead to varying results. The most frequently used annotations are those derived from the NCBI, Ensembl, and Encode. In this exercise, Ensembl gene annotations were used. Counting matrices can be generated using tools such as HTSeq [25], which assign reads to gene models based on annotation files; in these matrices, rows correspond to genes and columns to samples (Table 2). The table presents raw RNA-Seq read counts for five genes annotated using Ensembl gene identifiers (ENSG00000000003, ENSG00000000005, ENSG00000000419, ENSG00000000457, and ENSG00000000460) across ten samples (S1–S10).

Table 2

Simplified example of HTSeq count data for RNA-Seq samples showing the number of reads assigned to genes after alignment, providing a gene-level quantification matrix that serves as input for downstream differential expression analysis

Gene ID	S1	S2	S3	S4	S5	S6	S7	S8	S9	S10
ENSG00000000003	609	562	537	650	515	684	616	652	657	726
ENSG00000000005	0	0	0	1	0	0	0	0	0	0
ENSG00000000419	1852	1623	1647	1683	1426	1628	1443	1653	1649	1708
ENSG00000000457	1642	1476	1395	1447	1209	1552	1220	1497	1600	1577
ENSG00000000460	4318	3432	3673	3123	2972	3175	2611	3399	3358	3335

Gene ENSG00000000003 exhibits moderate expression, ranging from 515 to 726 reads per sample, indicating relatively consistent transcription. ENSG00000000005 is effectively unexpressed in nine samples (count = 0) and shows only a single read (count = 1) in S4, suggesting it is largely inactive under these conditions. ENSG00000000419 displays higher expression levels, between 1426 and 1852 reads, reflecting robust transcription across all samples. ENSG00000000457 shows similarly strong expression (range 1209–1642 reads), while ENSG00000000460 is the most highly expressed gene in this subset, with counts from 2611 to 4318 reads. Overall, these data illustrate distinct expression patterns, low to negligible (ENSG00000000005), moderate (ENSG00000000003), and high (ENSG00000000419, ENSG00000000457, and ENSG00000000460), that could be further analyzed for differential expression or functional significance. These data provide a quantitative foundation for downstream statistical modeling of gene expression and for identifying differentially expressed genes across various conditions.

3.6 Differential Expression Analysis

Differential expression analysis is a key step in the RNA-Seq workflow: it identifies genes whose expression levels vary significantly between experimental conditions, thereby revealing regulatory mechanisms underlying biological responses such as cellular adaptation in spaceflight. Several statistical tools have been developed for differential expression analysis; one of the most widely used is DESeq2 [16]. DESeq2 is an R-based package (R Project for Statistical Computing; https://www.r-project.org) specifically designed to accommodate the discrete nature of RNA-Seq count data. It estimates gene-specific dispersion parameters and incorporates normalization for sequencing depth and sample variability within a robust statistical framework, enabling high-confidence identification of differentially expressed genes (DEGs).

Following normalization and model fitting, DEGs are typically selected based on thresholds such as adjusted *p*-values and log2 fold changes (Table 3), which help distinguish biologically relevant transcriptional shifts. The table displays differential expression results (e.g., from DESeq2) for ten mouse genes. Below is an interpretation of each column and the key findings:

Gene ID: Ensembl identifiers for each gene (e.g., ENSMUSG00000000028).

baseMean: The average normalized count across all samples. Higher values indicate that a gene is more abundantly expressed overall.

log2FoldChange: The estimated $\log_2$ ratio of expression between two conditions (e.g., treatment vs. control). A positive value implies upregulation in the treatment, while a negative value indicates downregulation.

lfcSE: (Log_2 Fold Change Standard Error): The standard error associated with the $\log_2$ fold change estimate.

Table 3

Simplified example of DESeq2 differential expression results, which includes key metrics such as $\log_2$ fold change, *p*-value, and adjusted *p*-value (FDR) set at 0.05, which are used to identify genes with significant expression changes between experimental conditions

Gene ID	baseMean	log2FoldChange	lfcSE	stat	pvalue	padj
ENSMUSG00000000028	26.35	0.89	0.37	2.42	0.02	0.07
ENSMUSG00000000049	42,198.68	0.47	0.15	3.19	0.00	0.01
ENSMUSG00000000056	1455.66	−0.92	0.20	−4.60	0.00	0.00
ENSMUSG00000000088	3042.86	−0.22	0.18	−1.25	0.21	0.42
ENSMUSG00000000103	0.00	NA	NA	NA	NA	NA
ENSMUSG00000000120	135.89	−0.66	0.22	−2.97	0.00	0.02
ENSMUSG00000000125	0.23	−1.59	3.53	−0.45	0.65	NA
ENSMUSG00000000126	12.89	0.00	0.50	0.00	1.00	1.00
ENSMUSG00000000131	1022.32	0.26	0.21	1.23	0.22	0.43
ENSMUSG00000000142	176.17	0.74	0.25	3.03	0.00	0.02

stat: The Wald statistic ($\log_2$FC / lfcSE) that is used to test whether the $\log_2$ fold change is significantly different from zero.

pvalue: The raw *p*-value for the Wald test (testing $\log_2\text{FC} \neq 0$).

padj: The Benjamini–Hochberg–adjusted *p*-value to control the false discovery rate (FDR) for multiple hypothesis testing. Genes with padj <0.05 are generally considered statistically significant.

Across the dataset, four genes displayed statistically significant differential expression. ENSMUSG00000000049 (baseMean $\approx 4.2 \times 10^4$) and ENSMUSG00000000142 (baseMean $\approx 1.8 \times 10^2$) were significantly upregulated, with $\log_2$ fold changes of +0.47 (padj = 0.010) and +0.74 (padj = 0.020), respectively. In contrast, ENSMUSG00000000056 (baseMean $\approx 1.5 \times 10^3$) and ENSMUSG00000000120 (baseMean $\approx 1.4 \times 10^2$) were significantly downregulated, exhibiting $\log_2$ fold changes of –0.92 (padj <0.0010) and –0.66 (padj = 0.020), respectively. ENSMUSG00000000028 showed a nonsignificant trend toward upregulation ($\log_2$FC = + 0.89, padj = 0.070), while ENSMUSG00000000088 and ENSMUSG00000000131 exhibited no meaningful change (padj >0.050). ENSMUSG00000000126 remained unchanged between conditions ($\log_2$FC = 0.00, padj = 1.0), ENSMUSG00000000125 had insufficient read counts for reliable inference (high lfcSE, padj = NA), and ENSMUSG00000000103 was not expressed in either condition.

Visualization tools, including ggplot2 (https://ggplot2.tidyverse.org/) and pheatmap (https://cran.r-project.org/web/packages/pheatmap/index.html), can be used to generate

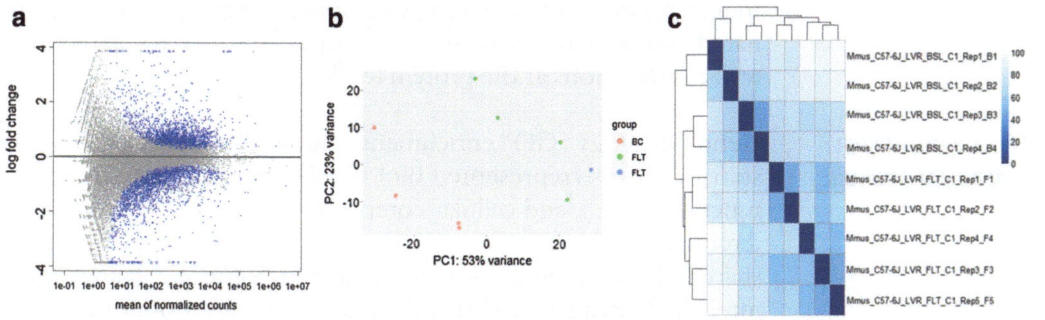

Fig. 3 Overview of transcriptomic changes in murine kidney tissue following spaceflight exposure. (**a**) MA plot shows differentially expressed genes based on log₂ fold change and mean expression levels, (**b**) principal component analysis (PCA) plot depicts clear separation between microgravity exposed (FLT) and ground control samples (BC), and (**c**) Heatmap of the variability of gene expression across all samples, revealing condition-specific expression patterns. All plots are derived from the mouse spaceflight study GeneLab OSD-102 and OSD-163 datasets

exploratory plots such as MA plots, principal component analysis (PCA) plots, and heatmaps (Fig. 3), providing an overview of global expression trends, sample clustering, and variability across conditions [16]. The three panels illustrate quality-control and exploratory analyses of RNA-Seq data, highlighting differential expression, sample clustering by principal components, and intersample correlation.

1. **MA Plot**: Reveals that a sizable subset of genes is significantly up- or downregulated, in ground control vs. space flight particularly among those with moderate to high expression.
2. **PCA Plot**: Confirms that the primary axis of transcriptional variation distinguishes experimental conditions, with replicates clustering tightly, indicating good data quality.
3. **Correlation Heatmap**: Reinforces that biological replicates exhibit strong agreement, while samples from different groups show lower correlation validating that treatment (i.e., space flight and exposure to microgravity) effects dominate over technical noise.

Together, these visualizations provide confidence in data quality (consistent library-prep and sequencing metrics), confirm that biological replicates are reproducible, and demonstrate that experimental conditions produce distinct transcriptomic profiles suitable for downstream differential expression and pathway analyses. This analytical stage, thus, serves as the entry point for downstream functional investigations, such as gene ontology enrichment [26] and network analysis [27].

3.7 Systems Biology and Downstream Analysis

After the identification of differentially expressed genes (DEGs), a variety of downstream analyses can be performed to explore their biological relevance. These analyses are typically used to contextualize transcriptional changes within functional categories, detect coordinated gene programs, identify key molecular pathways, and characterize interactions at the protein level.

3.8 Gene Ontology Enrichment Analysis

Gene Ontology (GO) enrichment analysis can be used to identify statistically overrepresented biological processes (BP), molecular functions (MF), and cellular components (CC) [14] in lists of differentially expressed genes, allowing genes to be grouped based on shared biological functions and large transcriptomic datasets to be interpreted in a structured and hierarchical framework (Table 4).

The table presents Gene Ontology (GO) enrichment results calculated over a background of 13,948 expressed genes, highlighting four immune-related biological processes that are significantly overrepresented among the differentially expressed (DE) genes. Each row denotes a GO term, with the "Gene Ratio" reflecting the proportion of DE genes annotated to that term (e.g., 62 of 13,948 for "Response to interferon-beta"), "Fold Enrichment" indicating how much more common that term is among DE genes versus expectation by chance, and "Adjusted *p*-value" showing statistical significance after correcting for multiple testing. The "DE Genes" column lists example gene identifiers (truncated here with ellipses), and "Count" is the total number of DE genes annotated to that GO term. The R package clusterProfiler (https://bioconductor.org/packages/clusterProfiler/) [16], in combination with organism-specific annotation databases such as org.Mm.eg.db for *Mus musculus* [17],

Table 4

Representative example of GO enrichment results obtained from the spaceflight kidney RNA-Seq dataset and generated in R using clusterProfiler and org.Mm.eg.db. The table reports the most significantly enriched biological processes, showing the number of differentially expressed genes (DE Genes), enrichment metrics, and adjusted p-values. Gene identifiers in the "DE Genes" column refer to Entrez Gene IDs

GO term	Description	Gene ratio	Fold enrichment	Adjusted *p*-value	DE genes	Count
GO:0035456	Response to interferon-beta	62/13,948	6.44	1.27E-07	114,643, 246,729, ...	62
GO:0035458	Cellular response to interferon-beta	54/13,948	5.96	1.58E-06	114,643, 246,729, ...	54
GO:0033141	Pos. regulation of STAT protein phosphorylation	28/13,948	5.48	2.47E-06	19,713, 16,878, ...	28
GO:0002440	Mediator of immune response production	264/13,948	5.87	3.57E-06	50,723, 27,060, ...	264

can be employed to conduct GO enrichment. Results are commonly filtered using adjusted p-values (e.g., FDR-corrected), and the most significant GO terms can be visualized using tools such as ggplot2 (https://ggplot2.tidyverse.org/) in the form of bar plots or dot plots (Fig. 4). The two panels display complementary visualizations of GO enrichment results, highlighting the most significantly over-represented biological processes among a set of differentially expressed genes. Each plot uses the same underlying data, i.e., GO terms, gene counts, gene ratios, and adjusted *p*-values but presents them in different formats to emphasize distinct aspects of the enrichment. Both plots emphasize that immune activation, especially innate, interferon-driven processes is a major driver of the transcriptional changes in this dataset. Secondary processes, such as adaptive immune features (immunoglobulin production) and spermatogenesis gene sets, are also enriched but to a lesser extent. The barplot quantifies absolute gene counts, while the dotplot integrates count, ratio, and significance, allowing one to assess both the relative impact of each GO term and its statistical confidence. These analyses can highlight pathways frequently affected under specific conditions, such as lipid metabolism, oxidative stress response, immune activation, and programmed cell death, providing insights into the systemic impact of space environmental challenges including microgravity.

3.9 Network Biology and Pathway Analysis of Altered Gravity Conditions

Network biology models biological systems as graphs in which nodes represent biomolecules such as genes, proteins, or metabolites, and edges denote functional, regulatory, or physical interactions. By mapping these interactions, researchers can visualize how perturbations at the molecular level propagate throughout the cellular network under specific conditions. For instance, gene regulatory networks elucidate transcriptional control mechanisms, while protein–protein interaction (PPI) networks reveal the assembly and stability of functional complexes. In the context of altered gravity, single-molecule analyses often fail to capture the multifaceted cellular response. A network-based approach uncovers emergent properties by illustrating how interconnected pathways and regulatory modules cooperate to produce systemic effects. Such analyses can identify critical regulatory nodes, construct condition-specific interaction maps, and ultimately support the development of predictive models for cellular adaptation to spaceflight environments [27].

Pathway analysis complements network approaches by functionally categorizing differentially expressed genes into established biological processes, which can be implemented using customized data integration workflows in R [28]. Representative pathways can be selected for focused analysis, such as cellular response to abiotic stimulus. Subnetwork visualizations can then be generated using packages like visNetwork (https://cran.r-project.org/web/packages/visNetwork/) [29], with node attributes representing gene

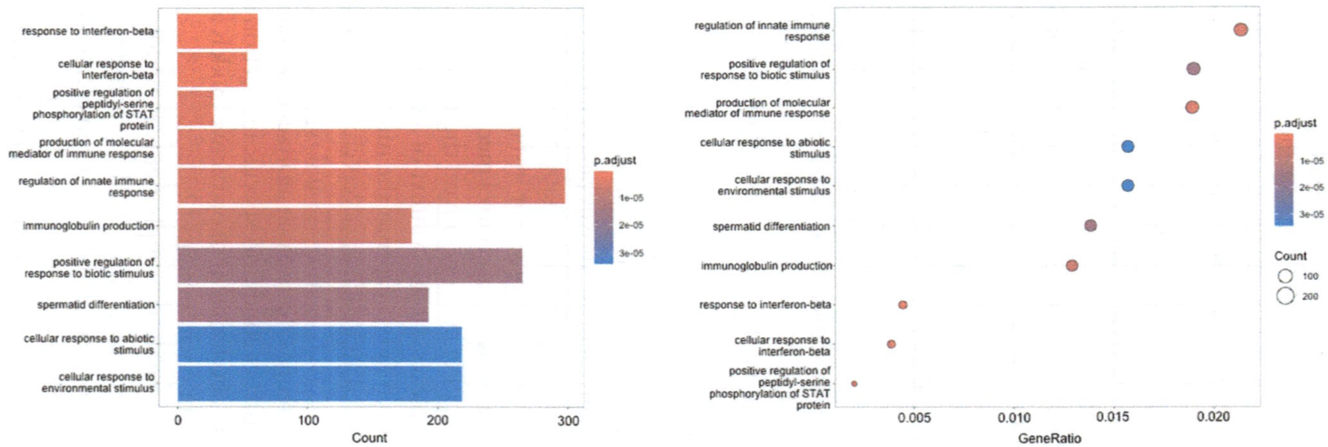

Fig. 4 Bar plot and dot plot generated in R using clusterProfiler and org.Mm.eg.db, based on GO enrichment analysis of differentially expressed genes (DEGs) identified in kidney tissue samples from C57BL/6J mice flown aboard the International Space Station (ISS). The top ten significantly enriched pathways include immune-related responses such as interferon-beta signaling, immunoglobulin production, and regulation of innate immunity, as well as cellular differentiation processes. Deseq2 results used for this analysis derive from Finch et al. (2025)

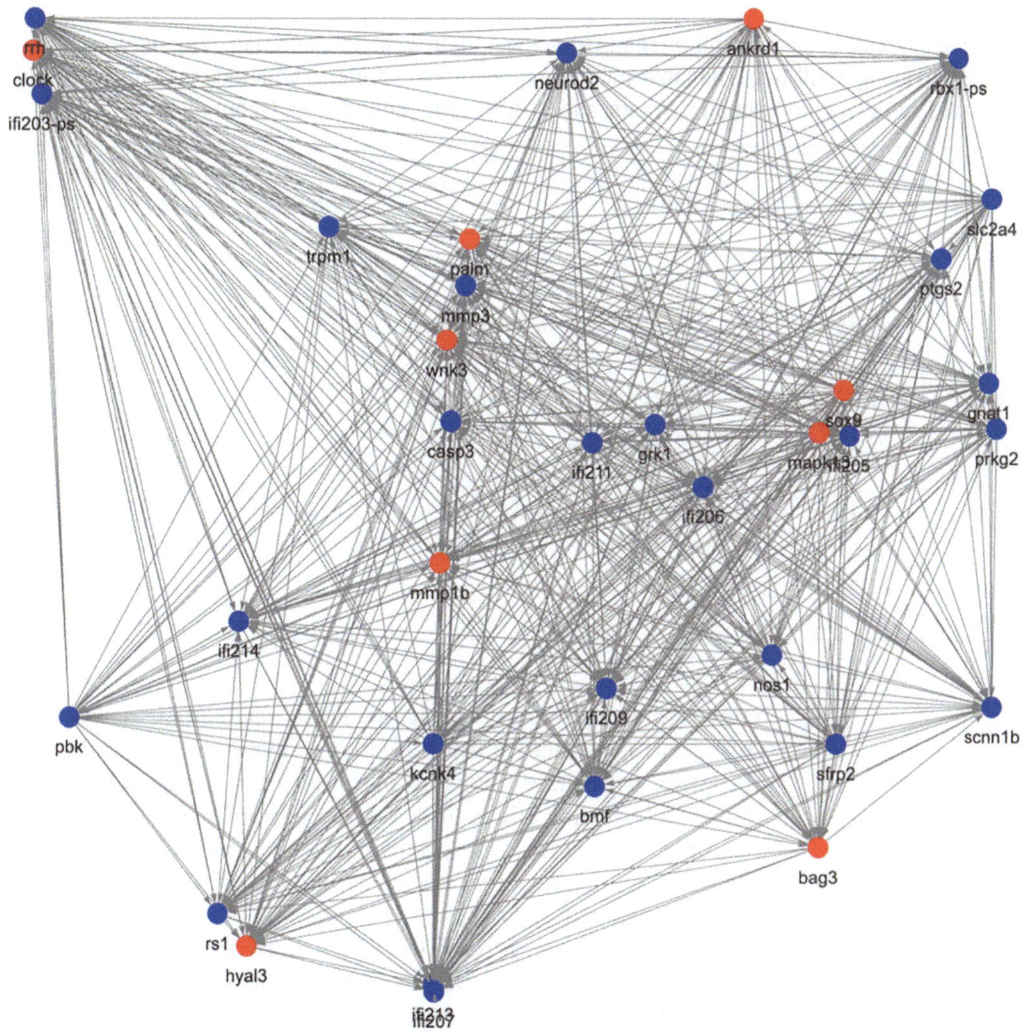

Fig. 5 Example of a gene network in the "Cellular response to abiotic stimulus" pathway in spaceflight-exposed C57BL/6 male mice where nodes represent differentially expressed genes (DEGs), colored by log2FoldChange (red for upregulated, blue for downregulated) and edges indicate comembership within the pathway, visualized using the visNetwork R package to highlight key interactions under altered gravity conditions and RcolorBrewer package for the color

expression direction (e.g., log2FoldChange) and pathway membership and edges may be inferred from shared membership in GO categories, known molecular interactions, or external databases (Fig. 5). This network highlights a core subnetwork of differentially expressed genes in which a small set of "percolator" nodes (red) serve as highly connected regulators, linking to numerous immediate neighbors (blue). At the center lies a tightly clustered interferon-stimulated module (*Ifi209*, *Ifi213*, *Ifi207*), indicating that type I interferon signaling is a primary hub through which many percolator genes

Table 5

Simplified example of selected interactions from the protein–protein interaction (PPI) network constructed using STRINGdb. The table includes representative proteins with their log2 fold changes, associated pathways, interaction partners, and STRING confidence scores (score ≥0.7)

Protein	Log2FoldChange	Pathway	Interactor	Interaction score
Acsl1	+1.14	Cytokine-mediated signaling pathway	Aifl	0.772
Aifl	−1.35	Small GTPase-mediated signal transduction	Aim2	0.801
Aim2	−1.42	Response to interferon-beta	Asns	0.834
Asns	−2.42	Cellular response to external stimulus	Bag3	0.812
Bag3	+1.11	Cellular response to abiotic stimulus	Baiap2	0.788
Baiap2	+1.22	Small GTPase-mediated signal transduction	Acsl1	0.755

converge. Peripheral red nodes such as *Clcf1*, *Bag3*, *Ankrd1*, and *Shox2* form distinct modules related to cytokine signaling, stress response, and kinase-mediated pathways, respectively. Overall, the dense web of gray edges underscores a coordinated, system-level response where central interferon effectors connect multiple regulatory circuits and identify key candidate regulators for further functional validation.

Protein–protein interaction (PPI) networks can also be constructed using databases, such as STRING (https://string-db.org/), filtered by organism (e.g., *Mus musculus*, taxonomy ID: 10090) and confidence thresholds of ≥0.7, which are commonly applied in STRING-based networks to retain only high-confidence protein interactions (Table 5), ensuring a balance between specificity and network coverage [30]. The table lists six proteins with their corresponding $\log_2$ fold-changes (indicating up- or downregulation), the biological pathway each protein is primarily associated with, a known interacting partner (Interactor), and the experimentally or computationally derived interaction score (0–1 scale). Overall, the network of interactions reinforced by high interaction scores (0.755–0.834)—reveals coordinated shifts in inflammatory signaling, interferon responses, metabolic regulation, and stress adaptation. Upregulated nodes (*Acsl1*, *Bag3*, *Baiap2*) appear to drive adaptive processes (cytokine signaling, proteostasis, cytoskeletal rearrangements), whereas downregulated nodes (*Aifl*, *Aim2*, *Asns*) suggest attenuated innate immune activation and amino acid biosynthesis. This interplay highlights how these pathways may be coregulated in response to the experimental condition under study.

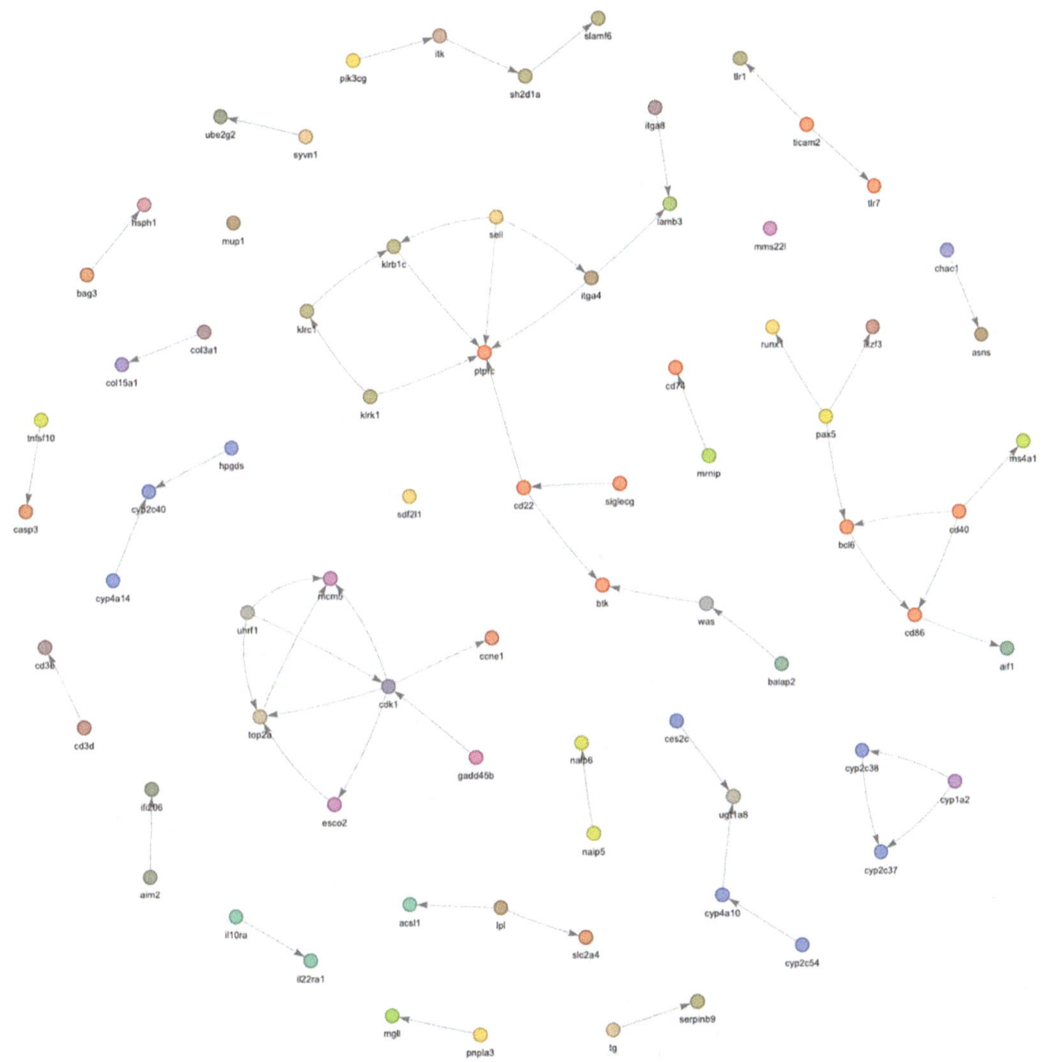

Fig. 6 Visualization of the protein–protein interaction network (PPI) obtained from differentially expressed genes in mice exposed to microgravity relative to ground controls. The interactions were extracted from the STRING database with a confidence score ≥0.7. The nodes, representing proteins, are colored, using RcolorBrewer package, according to the biological pathways to which they belong, while the arcs indicate the interactions between proteins. Nodes that do not have an assigned pathway are colored gray. The visualization was obtained using the VisNetwork package in R

Furthermore, DEGs annotated with enriched GO terms can be mapped onto the PPI network and coloured according to expression profile and pathway association (Fig. 6), using R packages such as RcolorBrewer (https://cran.r-project.org/web/packages/RColorBrewer/) for consistent and visually interpretable color scales. This network diagram depicts multiple interconnected gene clusters, each representing distinct functional modules. At the top-center, a small cluster (*Ifi209*, *Aim2*, *Itga8*, *Lamb5*) highlights interferon and extracellular matrix signaling. To its right, a densely connected hub (*Fos*, *Stat1*, *Stat3*, *Mapk1*, *Jun*, *Kirrel1*, *Kirrel3*) underscores a core transcriptional regulatory network likely driving immune or stress responses. Directly below, a triangle of heat-shock proteins (*Hspa1*, *Hspa4l*, *Bag3*) indicates coordinated chaperone activity, while adjacent, *Mmp9* and *Timp1* form a metalloprotease inhibitor module involved in extracellular matrix remodeling. Combined, these methods enable the identification of transcriptionally coregulated gene modules, the characterization of immune and stress-related interaction clusters, and the prioritization of candidate regulatory hubs. Ultimately, this integrative framework supports investigation into how environmental stressors such as microgravity drive coordinated changes in gene expression and molecular interactions.

In transcriptomic and multiomics analyses, integrative approaches can be used to combine results from different analyses. One common strategy involves the concatenation of outputs from distinct analytical layers, such as functional annotation, molecular connectivity, and expression-based modules, into a unified framework centered on gene-level information, to facilitate the identification of biologically meaningful groupings and enhances the interpretability of complex datasets (Table 6). The table presents eight key mouse genes, *Angptl4*, *Cd36*, *Apob*, *Fads2*, *Hmgcs2*, *Mboat7*, *Pnpla3*, and *Gckr*, each significantly enriched for lipid or energy-metabolism processes (adjusted $p \leq 7.8 \times 10^{-5}$). *Angptl4*, *Apob*, and *Mboat7* are annotated to "regulation of lipid metabolic process," *Cd36* and *Pnpla3* to "energy homeostasis," Hmgcs2 to "cholesterol metabolic process," Fads2 to "oxidoreductase activity," and Gckr to "regulation of small-molecule metabolic process." High-confidence protein–protein interactions (PPI scores ≥ 0.823) link each gene to a functionally relevant partner, e.g., *Angptl4* with *Lpl* (0.988) in triglyceride clearance, *Apob* with *Hadhb* (0.999) in VLDL assembly and fatty acid oxidation, and *Gckr* with *Apoe* (0.999) in glucose-lipid regulatory crosstalk. Collectively, these genes form a tightly interconnected network that underpins coordinated regulation of lipid handling and energy balance.

By integrating these complementary data sources, researchers can uncover emergent patterns that single-layer analyses often miss, facilitating a deeper understanding of the underlying molecular systems and their coordinated transcriptional programs. Moreover, this multilayered framework supports system-level investigations by identifying coregulated gene sets and prioritizing candidate regulatory elements, laying the groundwork for predictive modeling and hypothesis generation across diverse biological contexts [31].

3.10 Machine Learning and Microgravity Systems Biology

Studying biological adaptation to microgravity is challenging due to the complexity of molecular responses and the scarcity of experimental data. Traditional analytical methods often cannot discern meaningful patterns within the high-dimensional, multiomics datasets

Table 6

Representative example of integrated data table for transcriptomic analysis in spaceflight-exposed mice. The table combines genomic features (e.g., gene identifiers, chromosomal coordinates, strand, and biotype), results from gene ontology (GO) enrichment analysis (main GO description, ontology category, adjusted p-values), and protein–protein interaction (PPI) mapping using the STRING database

Genomics information					Pathway enrichment analysis			PPI mapping (String database)		
Ensembl gene ID	**Symbol ID**	**Chr:Start-End**	**Strand**	**Gene biotype**	**Main GO description**	**Ontology**	**p.adjust**	**Ensembl protein ID**	**Main interaction**	**PPI score**
ENSMUSG 00000002289	Angptl4	17:33992724-34000804	−1	Protein_ coding	Regulation of lipid metabolic process	BP	3.15E-07	ENSMUSP 00000002360	Lpl	0.988
ENSMUSG 00000002944	Cd36	5:17986688-18093799	−1	Protein_ coding	Energy homeostasis	BP	2.21E-06	ENSMUSP 00000036044	Mttp	0.823
ENSMUSG 00000020609	Apob	12:8027648-8066835	1	Protein_ coding	Regulation of lipid metabolic process	BP	3.15E-07	ENSMUSP 00000088249	Hadhb	0.999
ENSMUSG 00000024665	Fads2	19:10040129-10079110	−1	Protein_ coding	Oxidoreductaseactivity	MF	1.75E-04	ENSMUSP 00000102665	Fmo4	0.924
ENSMUSG 00000027875	Hmgcs2	3:98187751-98218054	1	Protein_ coding	Cholesterol metabolic process	BP	7.81E-05	ENSMUSP 00000037107	Lclat1	0.99
ENSMUSG 00000035596	Mboat7	7:3680788-3696522	−1	Protein_ coding	Regulation of lipid metabolic process	BP	3.15E-07	ENSMUSP 00000037259	Cdc40	0.939
ENSMUSG 00000041653	Pnpla3	15:84052038-84071437	1	Protein_ coding	Energy homeostasis	BP	2.21E-06	ENSMUSP 00000096847	Cck	0.934
ENSMUSG 00000059434	Gckr	5:31454787-31484658	1	Protein_ coding	Regulation of small molecule metabolic process	BP	6.86E-06	ENSMUSP 00000133008	Apoe	0.999

generated by spaceflight research. In this context, artificial intelligence (AI) and machine learning (ML) have emerged as essential tools, offering data-driven approaches to analyze, model, and predict cellular responses under spaceflight conditions [9, 32, 33]. ML can effectively handle the high-dimensional, noisy, and heterogeneous datasets characteristic of systems biology and enhance extraction of meaningful patterns [34, 35].

Unsupervised learning is frequently used in transcriptome studies when class labels are unavailable, as it uncouples analysis from predefined categories [36]. Clustering algorithms such as k-means and hierarchical clustering can reveal coexpression patterns and latent structures in gene expression data. By grouping genes or samples based on expression similarity, these methods simplify the identification of functionally coherent modules [37, 38]. Determining the optimal number of clusters can be challenging. To address this, researchers often use the Elbow method in R (https://www.r-project.org). This approach calculates the within-cluster sum of squares for various cluster counts and plots these values. The "elbow" point, where the decrease in sum of squares begins to stabilize, indicates the number of clusters beyond which including more clusters yields minimal improvement in the model fit (see Fig. 7) [39].

After choosing the optimal number of clusters, you can visualize clustering results using a membership heatmap or scatterplots in R. For example, the pheatmap package can generate a heatmap where

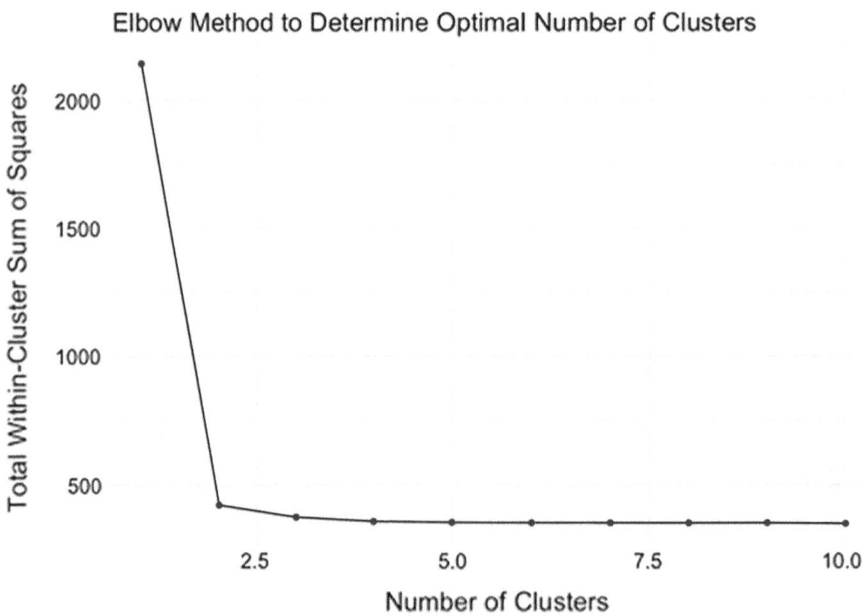

Fig. 7 Elbow method to determine optimal number of clusters (R). Applied to log2FoldChange values from the percolation-filtered subnetwork. The inflection at $k=3$ indicates the optimal number of clusters for K-means analysis in spaceflight-exposed renal tissue

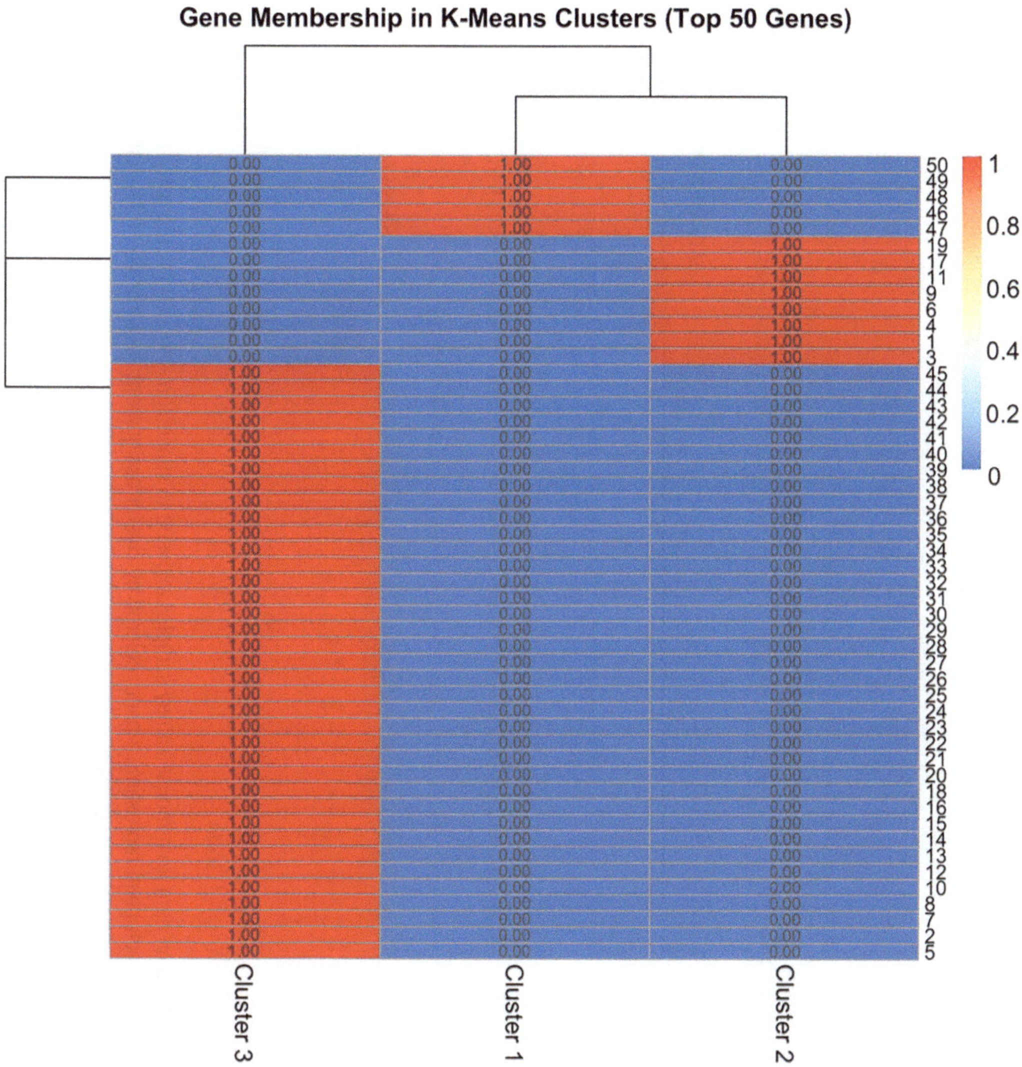

Fig. 8 *K*-means membership heatmap of the top 50 differentially expressed genes (DEGs) ranked by absolute log₂ fold change where rows represent individual genes, while columns indicate cluster membership. Binary values (1 = member, 0 = nonmember) highlight gene assignment across clusters, enabling the identification of coexpressed gene modules. Pheatmap package was used to generate the plot

each row is a gene and each column is a cluster, with binary values (1 = member, 0 = nonmember) indicating cluster assignment (Fig. 8). Alternatively, ggplot2 can create scatterplots showing gene-level $\log_2$ fold changes color-coded by cluster, making it easy to inspect expression trends across groups (Fig. 9).

For improved biological interpretation, clustering outputs can be integrated with network biology frameworks. Clustered genes can be mapped onto protein–protein interactions (PPI) [20] to assess connectivity and functional coherence. Packages such as visNetwork

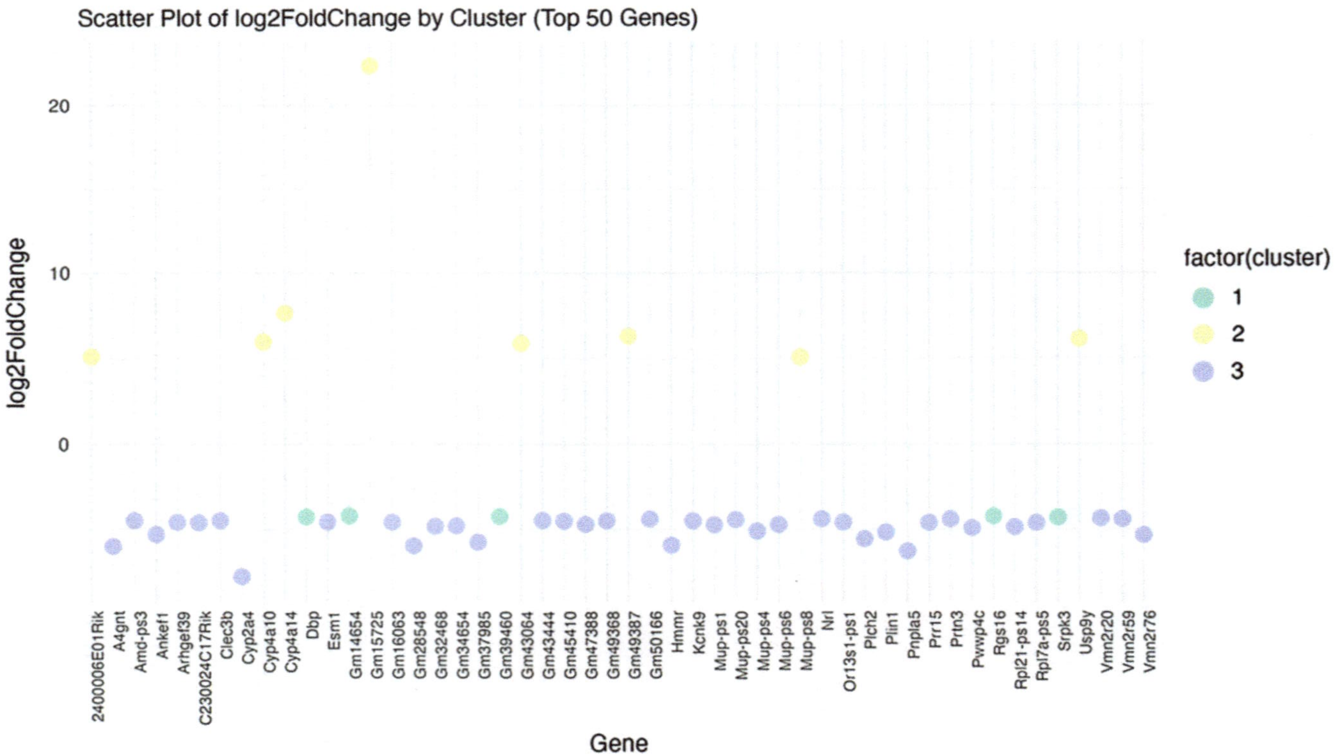

Fig. 9 Scatterplot obtained from K-means clustering performed on data from mouse kidney samples exposed to spaceflight-induced microgravity, using the stats package in R, and the plot was generated with ggplot2 where each point represents a gene, colored by cluster assignment. log₂ fold change values are displayed, highlighting expression variation and grouping across transcriptional profiles

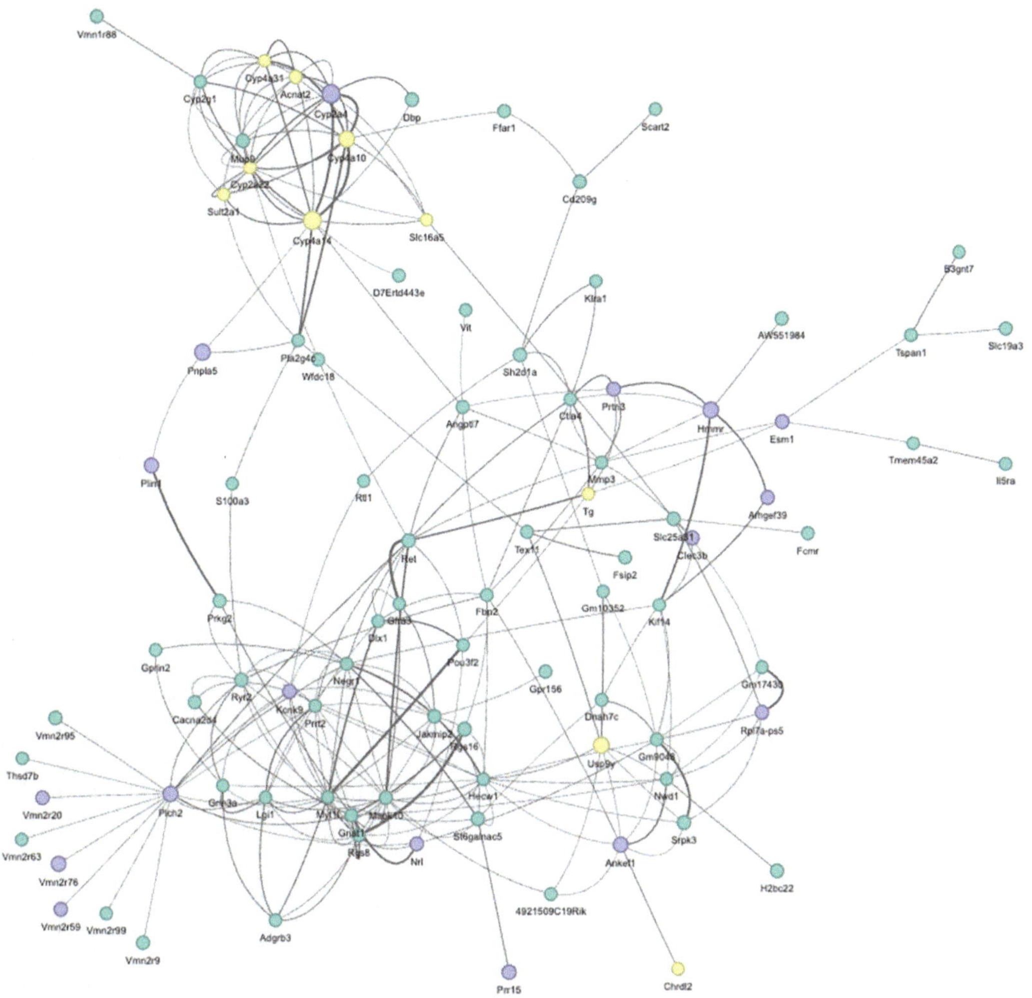

Fig. 10 Integrated PPI network with *K*-means clustering where genes were clustered using the stats package in R after subnetwork extraction by percolation. Protein–protein interactions were added from the STRING database to enhance biological context, where nodes are colored by cluster, and edges represent known or predicted interactions. The visualization was obtained using the VisNetwork package in R and RcolorBrewer package for the color

[19] support the generation and interactive visualization of such networks within R environments, offering an additional layer of insight into the systems level organization of transcriptomic data (Fig. 10).

On the other hand, hierarchical clustering provides a tree-based representation of gene similarity by iteratively grouping genes based on their expression distance, resulting in a dendrogram that reveals multilevel relationships between coexpressed genes (Fig. 11). When applied to transcriptomic data from altered environments like microgravity, these methods help generate data-driven hypotheses and build predictive models of cellular adaptation [9].

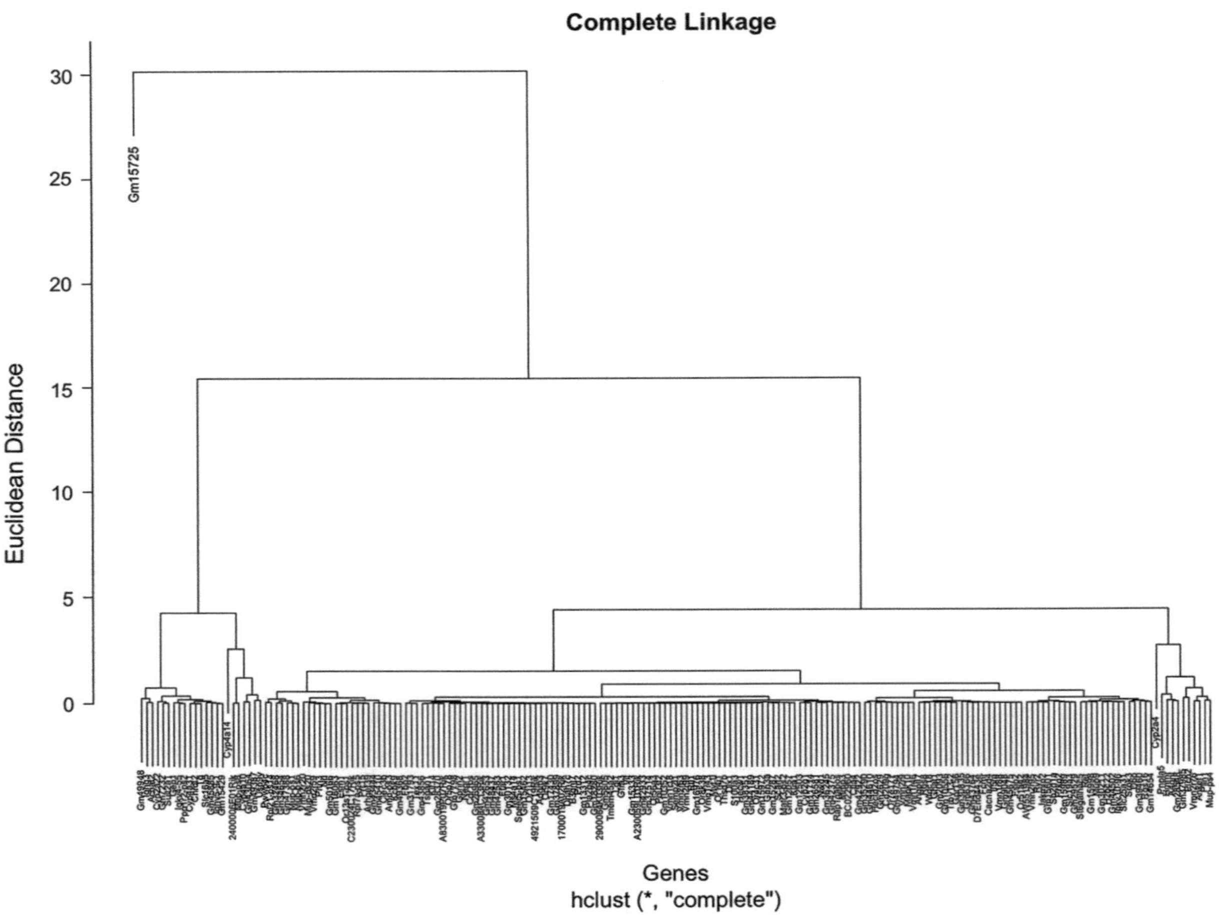

Fig. 11 Hierarchical clustering dendrogram of differentially expressed genes using complete linkage. The dendrogram illustrates the clustering structure of the top differentially expressed genes based on Euclidean distance and complete linkage method. Each branch represents a gene, and their grouping reflects similarity in expression profiles. This unsupervised clustering approach enables the identification of coexpressed gene modules and highlights potential regulatory patterns relevant to microgravity-induced transcriptional responses Stats package was used to generate the plot

3.11 Feature Selection and Dimensionality Reduction for Space Omics Data

Feature selection methods improve biological interpretability in high-dimensional transcriptome analyses by reducing noise and highlighting the most relevant molecular signals. One such method is the percolation technique, a network-based approach that identifies genes that are both strongly regulated and occupy central positions within interaction networks [40].

The percolation approach starts by selecting genes that exhibit large expression changes—often defined by absolute $\log_2$ fold-change thresholds. These selected genes, referred to as "percolator nodes," are then mapped onto a biological interaction network (e.g., a protein–protein interaction network) to identify their immediate neighbors and construct a functionally annotated subnetwork. This procedure reduces dataset dimensionality while preserving biological connectivity and coregulation, since it retains genes with both substantial expression shifts and central network positions. The resulting subnetwork can be further analyzed using methods such as machine learning, clustering algorithms, or pathway enrichment (see Figs. 5, 6, and 10). In fields like spaceflight biology where adaptive responses often depend on coordinated changes across multiple regulatory pathways, these integrated models are especially valuable.

3.12 Multiscale Computational Modeling Approaches

Multiscale computational modeling can be applied to understand how microgravity disrupts biological processes across scales. Biological systems operate across multiple, interdependent organizational levels, from molecular interactions to whole-organism behaviors. Accurate multiscale models must, therefore, capture both upward causation (how individual components influence system-level behavior) and downward causation (how emergent system properties feedback to regulate component dynamics). In practice, achieving this typically requires hybrid frameworks that integrate distinct modeling formalisms (e.g., discrete cellular automata with continuum reaction–diffusion equations) [12]. Modeling strategies generally fall into three categories:

Top-Down Approaches: These begin by characterizing macroscopic system behavior such as overall metabolic flux or population growth trajectories and then infer hypotheses about underlying components. Top-down models are often simpler to construct and parameterize, but because their variables are abstracted from biological detail, they can become decoupled from mechanistic reality and lack the flexibility to incorporate new molecular-level information.

Bottom-Up Approaches: In contrast, bottom-up models explicitly represent individual components (e.g., genes, proteins, or cells) and define their pairwise interactions. System-level behavior emerges from these interactions. Bottom-up frameworks are inherently modular new components or interactions can be added without restructuring the entire model, making them robust and extensible. However, they often demand substantial computational resources, especially when simulating large networks or detailed geometries.

Middle-Out Approaches: These hybrid strategies target an intermediate organizational scale (e.g., a single cell or tissue subunit). Once the mid-level dynamics are established, the model is incrementally expanded both upward (incorporating tissue- or organ-level processes) and downward (capturing subcellular pathways). Middle-out modeling balances computational tractability with biological realism by focusing on a core scale that is most relevant to the question at hand.

All multiscale models require simplifying assumptions such as quasisteady-state kinetics, mean-field approximations, or homogeneous material properties to reduce complexity. These assumptions yield approximate solutions; thus, the validity of a model's predictions depends critically on the appropriateness of its assumptions. The ultimate objective of multiscale modeling is to maximize predictive accuracy within computationally feasible constraints [41–43].

Given the combinatorial complexity of biological networks and physical interactions across scales, multiscale modeling heavily relies on computational methods ranging from numerical partial differential equations (PDE) solvers to stochastic simulation algorithms and high-performance computing clusters to generate and analyze solutions efficiently [44].

Agent-based modeling captures individual cell behaviors under reduced mechanical cues; reaction-diffusion and PDE models quantify how molecular signals propagate through altered fluid environments; finite element analysis predicts tissue deformation and remodeling under zero-g conditions; and hybrid models integrate discrete cellular rules with continuum mechanics to simulate emergent phenomena. By coupling these approaches, researchers can generate predictive insights into spaceflight biology that inform countermeasure design and translate to terrestrial health applications. Through iterative cycles of model calibration, validation against omics and imaging data, and high-performance computing, these methodologies form a robust framework for studying and ultimately mitigating the multifaceted impacts of microgravity on living organisms [12].

Multiscale computational modeling integrates mechanistic descriptions across multiple biological scales molecular, cellular, tissue, and organ to predict how spaceflight conditions perturb complex biological systems. In microgravity research, such models bridge the gap between biochemical reactions and whole-tissue mechanics, allowing investigators to simulate emergent phenomena that arise from interactions at disparate spatiotemporal scales. Four principal approaches are used in multiscale modeling: agent-based modeling, reaction-diffusion and PDE models, finite element analysis, and hybrid models. Each technique provides unique insights into how altered gravity influences cellular behavior, tissue morphology, and organ function [44–46].

3.13 Agent-Based Modeling (ABM)

Agent-based modeling describes a system as a collection of autonomous "agents" (e.g., individual cells or molecules) that follow a defined set of rules. Each agent possesses attributes (state variables) and decision-making algorithms that govern behaviors, such as movement, proliferation, differentiation, and signaling. In a microgravity context, ABMs can simulate how individual cells respond to altered fluid shear, gravitational cues, or modified extracellular matrix stiffness. For example, modeling bone-marrow stromal cells under simulated microgravity involves specifying agents' adhesion dynamics, cytoskeletal remodeling rules, and interactions with diffusing cytokines [47].

3.14 Reaction-Diffusion and PDE Models

Reaction-diffusion (RD) models describe how chemical species (e.g., morphogens, cytokines, metabolites) diffuse through space and react with each other. Partial differential equations (PDEs) extend RD to include additional transport phenomena (convection) and mechanical coupling. Under microgravity, fluid dynamics in extracellular compartments change (e.g., reduced sedimentation, altered cytoplasmic streaming), affect diffusion coefficients and reaction rates. RD/PDE models capture these effects at the tissue and cellular levels [48].

3.15 Finite Element Analysis (FEA)

Finite element analysis extends PDE modeling to include solid mechanics critical for simulating how tissues (e.g., bone, cartilage, vasculature) deform and remodel under microgravity-induced changes in mechanical loading. In microgravity, bones experience near-zero weight-bearing, altering stress-strain distributions and triggering localized bone resorption. FEA captures these phenomena by discretizing tissue geometry into finite elements and solving constitutive equations for stress–strain relationships [49].

3.16 Integration of Transcriptomic Data into Multiscale Modeling Frameworks

MSM provides a framework to investigate how microgravity perturbs biological systems across spatial and temporal scales. While traditionally used to model mechanical and biophysical processes, MSM can be extended to incorporate transcriptomic data, particularly RNA-Seq, to inform cellular behaviors, parameterize biochemical pathways, and validate emergent system-level predictions. This integration is particularly relevant in spaceflight biology, where molecular perturbations often manifest as system-wide phenotypes affecting health and adaptation. Transcriptomic datasets obtained from microgravity-exposed tissues or cells reveal differential expression in pathways related to cytoskeletal regulation, immune function, metabolism, and mechanotransduction. Incorporating this molecular information into MSM enhances biological relevance and allows researchers to simulate how gene-level changes propagate to influence cell behavior, tissue structure, and organ function.

RNA-Seq data can be utilized in MSM in the following ways:
Model Initialization: Define the initial state of biological entities based on gene expression (e.g., immune activation status, differentiation stage).
Parameterization: Use expression values to set kinetic parameters, interaction rates, or behavioral rules (e.g., proliferation rates, migration thresholds).
Model Calibration and Validation: Align model outputs with transcriptomic markers (e.g., *Ki67* for proliferation, *COL1A1* for ECM production) to refine model accuracy.

3.17 Application in Multiscale Modeling Approaches

3.17.1 Agent-Based Models (ABMs)

Transcriptomic data can define agent rules related to adhesion, signaling, differentiation, or migration. For example, ABMs of mesenchymal stem cells may use downregulated focal adhesion genes to simulate reduced substrate adherence under microgravity.

3.17.2 Reaction–Diffusion and PDE Models

These models simulate the transport and interaction of molecular species. Gene expression data for cytokines (e.g., *IL-6*, *TGF-β*) can inform source terms, diffusion coefficients, or degradation rates in microgravity-altered fluid environments.

3.17.3 Finite Element Analysis (FEA)

FEA is used to simulate mechanical stress and strain within tissues. RNA-Seq data indicating altered expression of bone-related genes (e.g., *RANKL*, *OPG*, *SOST*) can modify material properties or remodeling rates in simulations of skeletal adaptation to unloading.

3.17.4 Hybrid Models

These combine discrete and continuous modeling frameworks, allowing the integration of gene regulatory networks with mechanical or spatial processes. RNA-Seq-derived regulatory activity can drive phenotypic transitions within a spatial or mechanical context.

Several computational modeling platforms support the integration of transcriptomic data at different biological scales. These include tools, such as CompuCell3D, Morpheus, and PhysiCell, which enable simulation of gene regulatory networks, spatial patterning, and transcriptomic control of cell behavior in multicellular and agent-based systems (Table 7).

3.17.5 Considerations and Limitations

When incorporating transcriptomic data into multiscale models, it is essential to:

- Validate differential gene expression through robust statistical analysis.
- Choose biologically relevant targets with known functional impact.

Table 7
Modeling platforms supporting transcriptomic integration

Tool	Description
CompuCell3D	Simulates multicellular systems with support for gene regulatory networks.
Morpheus	Integrates spatial models with gene expression to study patterning and behavior.
Virtual Cell	Models biochemical networks and PDE-based transport processes.
PhysiCell	Agent-based platform allowing transcriptomic control of cell phenotypes.
Chaste	Used for modeling cardiac, epithelial, and cancer systems with mechanical links.
BioDynaMo	High-performance platform for large-scale agent-based simulations.

- Ensure that the granularity of the model aligns with the scale of the transcriptomic input (e.g., population-level vs. single-cell).
- Recognize that simplifying assumptions (e.g., steady-state kinetics, homogeneity) are necessary for tractability and should be justified.

Integrating transcriptomic data into multiscale modeling enables a systems level understanding of how molecular changes induced by microgravity lead to emergent biological phenomena. This approach not only enhances model realism but also supports the development of targeted, mechanism-based countermeasures for spaceflight-associated health risks. When used iteratively with experimental validation, transcriptomics-informed MSM becomes a powerful tool for both space and terrestrial biomedical research.

4 Conclusions

Systems biology posits that biological systems cannot be fully understood by examining individual components in isolation. In spaceflight, hazards such as increased radiation exposure and microgravity perturb multiple levels of biological organization from molecular and cellular processes to tissue, organ, and whole-organism function. These stressors manifest as physical pathologies, including muscle atrophy, bone demineralization, neurological deficits, and immune dysfunction, often underpinned by mitochondrial compromise and other cellular alterations. As humanity embarks on longer missions, developing effective countermeasures to protect astronaut health becomes imperative. Advances in multi-omics datasets derived from space-flown cells, model organisms, and astronaut physiological measurements enable comprehensive, systems level investigations.

By integrating mathematical frameworks and computational models, systems biology provides the tools needed to synthesize these diverse data streams into a unified, multiscale portrait of spaceflight's biological impact.

References

1. Afshinnekoo E et al (2020) Fundamental biological features of spaceflight: advancing the field to enable deep-space exploration. Cell 183(5):1162–1184. https://doi.org/10.1016/j.cell.2020.10.050
2. da Silveira WA et al (2020) Comprehensive multi-omics analysis reveals mitochondrial stress as a central biological hub for spaceflight impact. Cell 183(5):1185–1201. https://doi.org/10.1016/j.cell.2020.11.002
3. Cahill T, Cope H, Bass JJ, Overbey EG, Gilbert R, da Silveira WA, Paul AM, Mishra T, Herranz R, Reinsch SS, Costes SV, Hardiman G, Szewczyk NJ, Tahimic CGT (2021) Mammalian and Invertebrate Models as Complementary Tools for Gaining Mechanistic Insight on Muscle Responses to Spaceflight. Int J Mol Sci 22(17):9470. https://doi.org/10.3390/ijms22179470. PMID: 34502375; PMCID: PMC8430797
4. De La Torre GG et al (2012) Future perspectives on space psychology: recommendations on psychosocial and neurobehavioural aspects of human spaceflight. Acta Astronaut 81:587–599. https://doi.org/10.1016/j.actaastro.2012.08.013
5. (2021) Open science in space. Nat Med 27:1485. https://doi.org/10.1038/s41591-021-01508-1
6. Mariana Mazzucato, Douglas K.R. Robinson (2018) Co-creating and directing Innovation Ecosystems? NASA's changing approach to public-private partnerships in low-earth orbit. Technol Forecase Soc 136:166-177. ISSN 0040-1625.https://doi.org/10.1016/j.techfore.2017.03.034
7. Ferranti F, Del Bianco M, Pacelli C (2021) Advantages and limitations of current microgravity platforms for space biology research. Appl Sci 11(1):68
8. Wareing T et al (2024) Flight-scope: microscopy with microfluidics in microgravity
9. Li K et al (2023) Explainable machine learning identifies multi-omics signatures of muscle response to spaceflight in mice. NPJ Microgravity 9(1):90–90. https://doi.org/10.1038/s41526-023-00337-5
10. Kimura Y et al (2024) Identification of gravity-responsive serum proteins in spaceflight mice using a quantitative proteomic approach with data-independent acquisition mass spectrometry. Proteomics 24(9):e2300214. https://doi.org/10.1002/pmic.202300214
11. National A and Space A NASA GeneLab Platform.
12. Millar-Wilson A, Ward Ó, Duffy E, Hardiman G (2022) Multiscale modeling in the framework of biological systems and its potential for spaceflight biology studies. iScience 25(11):105421. https://doi.org/10.1016/j.isci.2022.105421
13. da Silveira WA, Hazard ES, Chung D, Hardiman G (2019) Molecular profiling of RNA tumors using high-throughput RNA sequencing: from raw data to systems level analyses. Methods Mol Biol 1908:185–204. https://doi.org/10.1007/978-1-4939-9004-7_13
14. Mair DB et al (2024) Spaceflight-induced contractile and mitochondrial dysfunction in an automated heart-on-a-chip platform. Proc Natl Acad Sci U S A 121(40):e2404644121. https://doi.org/10.1073/pnas.2404644121
15. da Silveira WA, Hazard ES, Chung D, Hardiman G (2019) Molecular profiling of RNA tumors using high-throughput RNA sequencing: from raw data to systems level analyses. Methods Mol Biol 1908:185–204. https://doi.org/10.1007/978-1-4939-9004-7_13
16. Love MI, Huber W, Anders S (2014) Moderated estimation of fold change and dispersion for RNA-seq data with DESeq2. Genome Biol 15(12):550. https://doi.org/10.1186/s13059-014-0550-8
17. Wilkinson MD et al (2016) The FAIR Guiding Principles for scientific data management and stewardship. Sci Data 3(1):160018. https://doi.org/10.1038/sdata.2016.18
18. Overbey EG et al (2021) NASA GeneLab RNA-seq consensus pipeline: standardized processing of short-read RNA-seq data. iScience 24(4):102361. https://doi.org/10.1016/j.isci.2021.102361
19. Moreno-Villanueva M et al (2024) Transcriptomics analysis reveals potential mechanisms underlying mitochondrial dysfunction and T cell exhaustion in astronauts' blood cells in space. Front Immunol 15:1512578. https://doi.org/10.3389/fimmu.2024.1512578
20. Finch RH et al (2024) Spaceflight causes strain dependent gene expression changes associated with lipid and extracellular matrix dysregulation in the mouse kidney in vivo

21. Martin M (2011) Cutadapt removes adapter sequences from high-throughput sequencing reads. EMBnetjournal 17(1):10. https://doi.org/10.14806/ej.17.1.200
22. Andrews S (2010) FastQC: a quality control tool for high throughput sequence data.
23. Ewels P, Magnusson M, Lundin S, Käller M (2016) MultiQC: summarize analysis results for multiple tools and samples in a single report. Bioinformatics 32(19):3047–3048. https://doi.org/10.1093/bioinformatics/btw354
24. Dobin A et al (2013) STAR: ultrafast universal RNA-seq aligner. Bioinformatics (Oxford, UK) 29(1):15–21. https://doi.org/10.1093/bioinformatics/bts635
25. Anders S, Pyl PT, Huber W (2015) HTSeq—a Python framework to work with high-throughput sequencing data. Bioinformatics (Oxford, UK)31(2):166–169.https://doi.org/10.1093/bioinformatics/btu638
26. Carbon S et al (2021) The Gene Ontology resource: enriching a GOld mine. Nucleic Acids Res 49(D1):D325–D334. https://doi.org/10.1093/nar/gkaa1113
27. Barabási A-L, Oltvai ZN (2004) Network biology: understanding the cell's functional organization. Nat Rev Genet 5(2):101–113. https://doi.org/10.1038/nrg1272
28. Palli R, Palshikar MG, Thakar J (2019) Executable pathway analysis using ensemble discrete-state modeling for large-scale data. PLoS Comput Biol 15(9):e1007317. https://doi.org/10.1371/journal.pcbi.1007317
29. Almende BV, Contributors and Thieurmel B (2015) visNetwork: network visualization using 'vis.js' library. In: CRAN: contributed packages.
30. Szklarczyk D et al (2019) STRING v11: protein–protein association networks with increased coverage, supporting functional discovery in genome-wide experimental datasets. Nucleic Acids Res 47(D1):D607–D613. https://doi.org/10.1093/nar/gky1131
31. Picard M et al (2021) Integration strategies of multi-omics data for machine learning analysis. Comput Struct Biotechnol J 19:3735–3746. https://doi.org/10.1016/j.csbj.2021.06.030
32. Willis CRG et al (2024) How to obtain an integrated picture of the molecular networks involved in adaptation to microgravity in different biological systems. NPJ Microgravity 10(1):50. https://doi.org/10.1038/s41526-024-00395-3
33. NASA (2023) Using AI/ML for space biology research. Nasagov
34. Sidak D, Schwarzerová J, Weckwerth W and Waldherr S (2022) Interpretable machine learning methods for predictions in systems biology from omics data. Front Mol Biosci 9:926623. https://doi.org/10.3389/fmolb.2022.926623
35. Xu C, Jackson SA (2019) Machine learning and complex biological data. Genome Biol 20(1):76. https://doi.org/10.1186/s13059-019-1689-0
36. Naeem S, Ali A, Anam S, Ahmed MM (2023) An unsupervised machine learning algorithms: comprehensive review. Int J Comput Digit Syst 13(1):911–921. https://doi.org/10.12785/ijcds/130172
37. Chen YT, Witten DM (2023) Selective inference for k-means clustering. J Mach Learn Res 24:152. PMID:38264325; PMCID: PMC10805457
38. Kodinariya T, Makwana P (2013) Review on determining of cluster in K-means clustering. Int J Adv Res Comput Sci Manag Stud 1:90–95
39. Langfelder P, Horvath S (2012) Fast R Functions for Robust Correlations and Hierarchical Clustering. J Stat Softw 46(11):i11. PMID: 23050260; PMCID: PMC3465711
40. Torrisi G, Kühn R, Annibale A (2020) Percolation on the gene regulatory network
41. Powathil GG, Swat M, Chaplain MA (2015) Systems oncology: towards patient-specific treatment regimes informed by multiscale mathematical modelling. Elsevier, pp 13–20
42. Castiglione F et al (2014) Modeling biology spanning different scales: an open challenge. BioMed Res Int 2014:902545. https://doi.org/10.1155/2014/902545
43. Arrell DK, Terzic A (2010) Network systems biology for drug discovery. Clin Pharm Ther 88:120–125. https://doi.org/10.1038/clpt.2010.91
44. Dada JO, Mendes P (2011) Multi-scale modelling and simulation in systems biology. Integr Biol (Camb) 3:86–96. https://doi.org/10.1039/c0ib00075b
45. Bardini R, Politano G, Benso A, Di Carlo S (2017) Multi-level and hybrid modelling approaches for systems biology. Comput Struct Biotechnol J 15:396–402. https://doi.org/10.1016/j.csbj.2017.07.005
46. Deisboeck TS, Wang Z, Macklin P, Cristini V (2011) Multiscale cancer modeling. Annu Rev Biomed Eng 13:127–155
47. Railsback, SF, Grimm, V, 2011. Agent-based and individual-based modeling: a practical introduction. Princeton University Press, Princeton, NJ
48. Murray, JD, 2003. Mathematical biology II: spatial models and biomedical applications. Springer, New York, NY
49. Zienkiewicz OC, Taylor RL (2000) The finite element method: volume 1, the basis. Butterworth-Heinemann, Oxford, UK

Part III

Cell and Pathways Studies

Check for updates

Chapter 5

Transcriptomics in Space: A Basic R Pipeline for Analyzing Bulk RNA-Sequencing Data

Gbolaga O. Olanrewaju, Nathaniel J. Szewczyk and Craig R.G. Willis

Abstract

Space life sciences have been revolutionized by the paradigm shift of biology into a "big data" era. Spaceflight experiments are now routinely generating large-scale molecular ("omics") datasets across different tissues and organisms to help better understand the mechanisms of health decline in space. Large-scale study of gene expression ("transcriptomics") has proven to be a particularly useful for producing new insights into the signaling pathways and key molecules that underpin maladaptation to spaceflight. Technologically, bulk RNA-sequencing provides a simple yet powerful platform for studying transcriptome-wide dysregulation in space. This chapter outlines a basic pipeline for analyzing bulk RNA-sequencing data using R, a well-established programming tool for transcriptomics. The pipeline is demonstrated throughout via application to RNA-sequencing data from a real-life spaceflight experiment, and is provided as a single-resource code script (available at: https://github.com/williscrg/MiMB-Space-RNAseq-Pipeline) that the reader can use to run the pipeline from start to finish "as is" or with modification if/as desired or required.

Keywords Spaceflight, Bioinformatics, Omics, Transcriptomics, RNA-seq, R, RStudio

1 Introduction

Spaceflight is an invaluable platform for studying fundamental biological principles in ways that are not viable under terrestrial conditions [1]. Characterized by a unique milieu of stressors, such as microgravity and cosmic radiation, the space environment poses a wide array of cellular and physiological challenges, from DNA damage and mitochondrial dysregulation to cardiovascular dysfunction and musculoskeletal atrophy [2]. Many of the phenotypic changes induced by spaceflight closely resemble hallmarks of disease on Earth (e.g., sarcopenia [3]). Thus, understanding the molecular mechanisms of how life adapts to space is not only beneficial for ensuring astronaut health but also has broader terrestrial implications, such as advancing regenerative medicine and accelerating biotechnological developments [4, 5].

Mariano Bizzarri (ed.), *Biomedicine and Biotechnology in Space*, Methods in Molecular Biology, vol. 3000,
https://doi.org/10.1007/978-1-0716-5174-2_5,

Technological advances mean that today it is possible to quantify tens-of-thousands of molecules simultaneously in a sample. These approaches are denoted with the suffix "-omics" and named based on molecular type, for example, genomics (genes), transcriptomics (transcripts), proteomics (proteins), and metabolomics (metabolites) [6]. By enabling genotype–phenotype relationships to be studied at immense resolution, omics approaches have transformed molecular medicine on Earth and paved the way for breakthrough advances in disease diagnostics and personalized medicine [7]. The use of omics has also been growing exponentially in modern space biology research [8]. Spaceflight experiments leveraging omics have uncovered many novel mechanisms of stress adaptation and physiological deconditioning in outer space [9]. As a result, space agencies worldwide are now coming to recognize omics-driven research as being an essential pillar for developing countermeasures against spaceflight-induced health decline and for advancing the understanding of fundamental biology in extraterrestrial environments [10].

Of the various omics types, transcriptomics has emerged as a particularly useful tool for studying molecular responses to spaceflight. Whereas genomics offers a blueprint of genetic potential, transcriptomics captures dynamic changes in gene expression in response to environmental stimuli, providing a real-time snapshot of cellular activity [11]. Moreover, transcriptomics currently offers a greater detection coverage than proteomics [12], while at the same time encompassing noncoding RNA, which can be just as important as to cellular function as coding RNA [13]. Transcriptomics has been implemented regularly in space biology across different tissues and organisms, shedding light on pathways and key gene expression changes linked to declines in skin, liver, muscle, kidney, and whole-body health during spaceflight [14–20]. Historically, the first type of transcriptomic technology that emerged was the hybridization-based microarray platform, followed by RNA-sequencing (RNA-seq). Compared to microarrays, RNA-seq offers a wider detection coverage across the transcriptome and the ability to identify novel transcripts and single-nucleotide variants [21]. Accordingly, RNA-seq has become the de facto technology for transcriptomic profiling on Earth and in spaceflight experiments. While it is now possible to perform RNA-seq at single-cell resolution (scRNA-seq), bulk RNA-seq arguably remains a more common choice due to being less costly and easier to process.

This chapter outlines a basic pipeline for analyzing bulk RNA-seq data using R, illustrated in the context of space life sciences via application to real-world data from a recent *Caenorhabditis elegans* (*C. elegans*) spaceflight study [22]. *C. elegans* has come to be a primary model organism in space life sciences due to its small size, short life span, ease of culture, and a completely defined genome with strong evolutionary conservation in humans [20]. *C. elegans* also displays many similar phenotypic (e.g., reduced strength) and transcriptional (e.g., impaired metabolism) features as humans during spaceflight [23]. Nevertheless, it is important to emphasize that while the pipeline in this chapter is given in the context of *C. elegans* and spaceflight, it can be easily adapted to be applied to other species and/or biological contexts. The R programming language was chosen due to its being open source, offering a user-friendly integrated development environment (RStudio Desktop), and having a wide range of dedicated extensions for transcriptomics.

2 Materials

2.1 GitHub Repository

An open-source GitHub repository named "MiMB-Space-RNAseq-Pipeline" has been created for this chapter (available at: https://github.com/williscrg/MiMB-Space-RNAseq-Pipeline). This repository includes an RStudio project file called "MiMB-Space-RNAseq-Pipeline.Rproj," the purpose of which is detailed in the Methods section of this chapter. The repository also includes a folder named "Code" that contains an R script called "MiMB-Space-RNAseq-Pipeline-Script.R." This script serves as a single resource that the reader can use to run the pipeline outlined in this chapter from start to finish. Nevertheless, users are welcome to modify the script/pipeline to meet their own needs or requirements.

2.2 Data

As noted above, the pipeline given in this chapter is demonstrated using RNA-seq data from a recent spaceflight experiment in *C. elegans* [22]. The overarching aim of that experiment was to better understand the molecular causes of strength decline during space travel. RNA was extracted from first-generation and second-generation wild-type and *dys-1 C. elegans* strains that were cultured either on Earth or on-board the International Space Station (three biological replicates per strain per generation per condition). Strand-specific (second-strand cDNA synthesis with dUTP) 100 bp paired-end reads were generated per biological replicate. All sequenced reads are publicly available through the NCBI Sequence Read Archive (SRA) (https://www.ncbi.nlm.nih.gov/sra) (BioProject ID PRJNA1026503). For pipeline demonstration, this chapter only utilizes reads from the second-generation wild-type *C. elegans* cultured either on Earth (SRA run IDs for the three biological replicates: SRR26332897, SRR26332902, SRR26332906) or in space (SRA run IDs for the three biological replicates: SRR26332898, SRR26332903, SRR26332907). The approximate amount of storage space required for these sequenced reads is 12 gigabytes. A copy of the *C. elegans* reference transcriptome is also needed. The given pipeline utilizes the *C. elegans* transcriptome from Ensembl (https://www.ensembl.org/index.html) (release 113 at time of writing) [24], for reasons justified in the Methods section of this chapter. The R script provided in the GitHub repository for this chapter includes code for downloading all required data.

2.3 Computational Hardware

Unless otherwise stated, the described pipeline was developed using a 2024 MacBook Pro with the following specifications: Apple M4 Pro chip, 24 GB unified memory, 512 GB internal solid-state drive storage, 12-core central processing unit, 16-core graphics processing unit. In general, all steps of this pipeline should be possible to run locally on a modern computer with standard (8–16 GB) random-access memory, provided there is sufficient internal storage for the data and required computational software (see below). If a user lacks the computational resource to run any of the steps of the given pipeline locally, alternative strategies might be to (for example) run the pipeline via a high-performance computing cluster and/or leverage external hard drive storage, etc.

2.4 Computational Software

The pipeline given in this chapter requires the R software and, for user-friendliness, is set up to also leverage the RStudio Desktop software. When developing this pipeline, R v4.4.3 and RStudio Desktop v2024.12.1+563 were used. This pipeline also relies on several R packages. These packages and the versions of them used during pipeline development are as follows: *ashr* (v2.2-63), *BiocManager* (v1.30.25), *biomaRt* (v2.62.1), *clusterProfiler* (v4.14.6), *DESeq2* (v1.46.0), *EnhancedVolcano* (v1.24.0), *enrichplot* (v1.26.6), *fastqcr* (v0.1.3), *GEOfastq* (v1.14.0), *org.Ce.eg.db* (v3.20.0), *PCAtools* (v2.18.0), *remotes* (v2.5.0), *STRINGdb* (v2.18.0), *tximport* (v1.34.0). At the time of writing, all listed packages are installable through either the Comprehensive R Archive Network (CRAN) or Bioconductor (main repositories for approved R package releases). Other required software and the versions of them used during pipeline development are as follows: *FastQC* (v0.12.1), *kallisto* (v0.50.0). The R script provided in the GitHub repository for this chapter includes code for installing the required R packages (and dependencies) and software (aside for R and RStudio Desktop). These packages and software are contextualized at relevant points throughout the Methods section of this chapter. All required software and R packages are open source. To encourage reproducibility, the script tries to install the same package/software versions as given above (package build tools present: Xcode Command Line Tools [Xcode-select] v2409, and the compilers Apple clang v17.0.0, MacOSX15.4.sdk, and GNU Fortran v12.2.0). Users are welcome to use alternative package/software versions but are advised that outputs may not identically match those generated here. Moreover, some aspects of the script and/or software setup may need to be altered by the user to suit their specific operating system. Some elements of the script also require an internet connection to run.

3 Methods

3.1 Preliminaries

3.1.1 Download GitHub Repository

GitHub supports open science by providing a framework that supports the FAIR principles of Findable, Accessible, Interoperable, and Reusable [25]. The most recent version of the GitHub repository for this chapter can easily be downloaded to a local computer by navigating to its associated webpage (https://github.com/williscrg/MiMB-Space-RNAseq-Pipeline), clicking on the green "Code" button toward the right-hand side of the page, and then selecting "Download Zip" under the "Local" tab (folder may need to be unzipped once downloaded to be able to access its contents). Optionally, the GitHub repository can be "cloned" (downloads repository with complete history) to a local computer via the "Git" command line tool (depending on operating system, users may first need to download, install, and configure Git for command line use) using the command: git clone https://github.com/williscrg/MiMB-Space-RNAseq-Pipeline.git.

3.1.2 Get Setup with R and RStudio Desktop

As elucidated in the Materials section of this chapter, the user needs to download and install the appropriate R and RStudio Desktop software for their operating system. Guidance and links for downloading and installing the R software can be found at: https://www.r-project.org/. Guidance and links for installing the RStudio Desktop software can be found at: https://posit.co/download/rstudio-desktop/. RStudio Desktop should automatically detect the installation of R when in use.

With both R and RStudio installed, the reader can proceed with opening the RStudio project file named "MiMB-Space-RNAseq-Pipeline.Rproj" in their saved copy of the GitHub repository for this chapter. To do this, the reader should first open the RStudio Desktop application, then at the top of the page click "File" and select "Open Project," navigate to the saved copy of the GitHub repository for this chapter, select the file called "MiMB-Space-RNAseq-Pipeline.Rproj" and then click "Open." The top-right corner of the RStudio Desktop application window should then read "MiMB-Space-RNAseq-Pipeline" (may end with "-main" if GitHub repository was downloaded via the green "Code" button), indicating that the RStudio project of the same name was successfully opened. RStudio projects are folders that contain an RStudio project file (denoted by.Rproj file extension). When an RStudio project file is opened in RStudio, the working directory gets automatically set to the RStudio project folder. As such, the user works relative to the location of the RStudio project, helping to minimize problems with fragile paths.

With the RStudio project for this chapter open in RStudio, the user can proceed to open the R script called "MiMB-Space-RNAseq-Pipeline-Script.R," contained in the "Code" directory of the saved copy of the GitHub repository for this chapter. As noted above, the given R script acts as a single resource that can be used to run the entire pipeline outlined in this chapter from start to finish, including installation of all other required software and R packages (and dependencies). To open the R script in RStudio, click "File" at the top of the window, select "Open File," navigate to the folder called "Code" in the saved copy of the GitHub repository for this chapter, select the file called "MiMB-Space-RNAseq-Pipeline-Script.R" and then click "Open." The user should then see the contents of the R script in their RStudio Desktop window. Lines of code can be run by highlighting them (either individually or in chunks) and clicking the "Run" button towards the top-right of the RStudio Desktop window. Readouts of ran code are printed in the console pane in RStudio (located at the bottom of the RStudio Desktop window by default). Note that running code out of order will generate error messages in the console if the code that is ran depends on the output(s) of preceding code. To enhance followability, code in the R script has been (sub)sectioned, with each (sub)section of code numbered to match the (sub)section number of the Methods text it is relevant to.

3.2 Source RNA-seq Reads

3.2.1 Create Metadata Table

Regardless of where RNA-seq reads are obtained from, it is crucial to have an accompanying metadata table that describes the sequenced reads in the context of the experimental design of the study they were derived from. The metadata table serves as a particularly useful tool for differentiating sample groups (e.g., control *versus* disease) in downstream analyses of the RNA-seq data. Since only a small number ($n=6$) of samples are being used to demonstrate the pipeline given in this chapter, a metadata table for the samples is created directly within the script itself (script Sect. 3.2.1), by first initializing a data frame object with a column containing the SRA run IDs of the samples (code line 28) and then adding a column to the data frame that contains "SF" if the sample is a spaceflight sample or else "GC" if the sample is a ground control sample (code line 31). Row names of the data frame are also made to be the SRA run IDs of the samples (code line 34) to facilitate some of the steps later in the pipeline related to downstream analysis. The experimental "condition" of each sample was obtained from its associated SRA webpage. For larger datasets, it may be more efficient to download the metadata table for all samples as a single text-formatted file (most RNA-seq data archives permit this) and read that into RStudio (e.g., using the *read.csv()* R function). Similarly, if a user has generated an "in-house" RNA-seq dataset, they are likely to have a metadata table file that could be read into RStudio.

3.2.2 Obtain RNA-Seq Reads

With the volume of publicly available RNA-seq data growing exponentially, it has become common practice for scientists to leverage existing RNA-seq data to test new research hypotheses or validate reproducibly of their own findings. As noted above, the sequenced reads used herein are publicly available via the NCBI SRA and, by extension, the European Nucleotide Archive (ENA). This data can be downloaded directly from its associated NCBI SRA or ENA webpage(s) or through each's command line tool, with the ENA arguably offering a more straightforward download process. Nevertheless, for the purposes of providing the reader with a single start-to-end resource, the given pipeline leverages the *GEOfastq* R package to download the data from ENA (script Sect. 3.2.2). *GEOfastq* is available through Bioconductor, meaning that to install it, the *BiocManager* R package (which allows users to install and manage packages from Bioconductor) first needs to be installed from CRAN. To install specified versions of R packages from CRAN (as per Materials section), the *remotes* R package can be leveraged. After installing the *remotes* package (and dependencies) (code lines 42–23), the specified version of the *BiocManager* package (and dependencies) can be installed (code line 46) and, in turn, used to install the *GEOfastq* package (and dependencies) from Bioconductor (code line 49). Notably, when installing packages from Bioconductor using the *BiocManager install()* function, the **version** argument is for the version of Bioconductor that the specific package version is derived from (here, Bioconductor v3.20), not the actual package version itself.

Before downloading the sequenced reads, it is also helpful to create a new folder structure in the working directory to store them in (code line 52). Complete URLs for each sequenced read file can then be defined (code lines 56–61) by combining the generic first part of the ENA URL with the sample-specific URL components (obtained using the *GEOfastq* package's *get_dldir()* function), the paired-end designations (1 = R1, 2 = R2), and the file extensions (sequenced reads are normally stored in gzipped fastq file format, denoted by the file extension ".fastq.gz"). These URLs can then be iteratively parsed to the *download.file()* R function to download the sequenced read files into the newly created "Reads" folder, naming each file by its SRA run ID and paired-end designation (code lines 64–66). Users may need to adapt the script accordingly for data housed in repositories other than NCBI SRA or ENA. Similarly, users may have their own dataset(s) already stored locally. In that case, the RNA-seq data download step of this pipeline is redundant, though users are encouraged to deposit their own RNA-seq reads into a public database, if they have not already.

3.3 Quality Control Check of RNA-Seq Reads

A key early step in the handling of generated RNA-seq reads is to undertake quality control checks. One of the most used tools for checking the quality of sequenced reads is *FastQC* (Babraham Bioinformatics: https://www.bioinformatics.babraham.ac.uk/projects/fastqc/). *FastQC* itself can be run as an interactive graphical application (particularly useful for beginners wanting to practice using *FastQC* on individual samples) or noninteractively via the command line. Keeping with the objective of providing readers with a single start-to-end resource, the pipeline given in this chapter leverages the *fastqcr* R package (script Sect. 3.3), which provides a set of helper functions for automatically applying *FastQC* to multiple samples and aggregating the results.

The *fastqcr* R package (and dependencies) can be installed through CRAN (code line 73). Once the package is installed, the entire *FastQC* process can be completed in a mere few lines of code. Since the package directly leverages the *FastQC* tool, the latter first needs to be downloaded and installed. To do this, a new folder in the working directory called "Software" is first created (code line 78) to store *FastQC* (and other downstream software). A URL to the *FastQC* download link is then defined (code line 81) and fed, along with the destination directory of the installation, to the *fastqc_install()* helper function to perform download and installation of the *FastQC* tool (code line 84). If no URL is given to the *fastqc_install()* helper function, it will try to install the latest version of *FastQC* by default. After creating a new folder structure in the working directory to store *FastQC* outputs (code line 89), the *FastQC* tool can then be applied to multiple samples via the *fastqc()* helper function from the *fastqcr* package, specifying the path to the folder where the sequenced reads are (**fq.dir** argument), the path to the directory to save the outputs (**qc.dir** argument) and the path to the *FastQC* executable (**fastqc.path** argument) (code line 92). Running this function will produce, per sequenced read file, an html *FastQC* report and zip file containing the underlying elements of the report. Users are welcome to explore individual outputs separately; however, the helper function *qc_report()* from the *fastqcr* package can also be ran to generate a single user-friendly html report that summarizes all *FastQC* results in a specified folder (code line 95).

FastQC itself reports on the quality of sequenced reads via several metrics ("modules"). Readers are encouraged to explore the range of quality metrics *FastQC* reports on in further detail at: https://www.bioinformatics.babraham.ac.uk/projects/fastqc/. For each metric, *FastQC* provides a user-friendly flag system for interpreting the status of a given reads file for that metric: "pass" (green circle with tick), "warn" (orange circle with exclamation mark), and "fail" (red circle with cross). It is important to note that these flags should be used as guides rather than definitive barriers to progress and interpreted in the context of "typical" behavior anticipated for the sample. For example, a sample from an overexpression study may not pass the "over-represented sequences" *FastQC* module by virtue of the over-expressed sequence naturally being present in a large portion of the sample. Moreover, it is commonplace for RNA-seq reads to fail the "sequence duplication levels" *FastQC* module due to presence of both very highly abundant transcripts and very lowly abundant transcripts. As a minimum, users should be looking for sequenced read files to pass the "per base sequence quality" and "adapter content" *FastQC* modules. The RNA-seq samples being used to demonstrate the pipeline in this chapter were "cleaned" (adaptor sequences, contamination, and low-quality reads all removed) by the sequencing company that generated them (Beijing Genomics Institute, Hong Kong), and thus, meet the desired quality levels to proceed. A snapshot from the summary html report produced by the *qc_report() fastqcr* function is given in Fig. 1 to further illustrate this. For the readers benefit, that cleaning process was carried out using the *SOAPnuke* software [26] to cut adaptors if a sequenced read matches ≥25% of the adaptor sequence, and to discard entire reads if bases with a quality score <20 make up ≥40% of the read and/or "N" content accounts for ≥0.1% of the read. Since the given sequenced reads were precleaned, this pipeline does not attempt to recreate the use of *SOAPnuke* or explicitly demonstrate the utility of any other trimming and/or filtering tools. Nevertheless, if ever required, the user can easily integrate the use of any such external tools into the R script

Summary

```
summary(qc)
```

module <chr>	nb_samples <dbl>	nb_fail <dbl>	nb_pass <dbl>	nb_warn <dbl>
Adapter Content	12	0	12	0
Basic Statistics	12	0	12	0
Overrepresented sequences	12	0	12	0
Per base N content	12	0	12	0
Per base sequence content	12	7	0	5
Per base sequence quality	12	0	12	0
Per sequence GC content	12	0	10	2
Per sequence quality scores	12	0	12	0
Sequence Duplication Levels	12	12	0	0
Sequence Length Distribution	12	0	12	0

1-10 of 10 rows | 1-5 of 7 columns

Column names:

- **module**: fastqc modules
- **nb_samples**: the number of samples tested
- **nb_pass, nb_fail, nb_warn**: the number of samples that passed, failed and warned, respectively.
- **failed, warned**: the name of samples that failed and warned, respectively.

Fig. 1 Snapshot from the summary FastQC report of the samples produced by the *fastqc_report()* function from the *fastqcr* R package. Snapshot illustrates the number of sequenced read files ($n = 12$) that were flagged as a "pass," "warn," or "fail" for each of the *FastQC* modules. In this case, all sequenced read files pass the per base sequence quality and adapter content modules, as desired (see Sect. 3.3 of Methods)

using the same approach that is exampled below in Sect. 3.4.1 of this chapter. If a user is indeed ever required to undertake some form of trimming and/or filtering to remove adaptor sequences and improve read quality (see [27–29] for reviews and direction to other tools that serve these purposes), it is also worthwhile rerunning *FastQC* on the cleaned sequencing reads as a sanity check before proceeding.

3.4 Quantification of Gene Expression from RNA-Seq Reads

3.4.1 Pseudo-Alignment to Quantify Transcript Abundances

As elucidated above, RNA-seq reads are stored in (gzipped) fastq files that consist of sequenced reads and corresponding quality scores. Sequenced reads, therefore, need to be mapped to an appropriate reference sequence to determine gene expression, a process known as "alignment." Conventionally, the alignment process involves first mapping sequenced reads to the reference genome of the organism they were generated from, followed by counting the number of reads that align to a given genomic feature (e.g., exon) of the reference genome. The resultant counts are then used as a measure of the amount of each genes RNA in a sample. Nevertheless, genome alignment can be a computationally expensive process, and recent techniques have attempted to overcome this limitation via a process known as "pseudoalignment." Instead of determining precise locations of sequences on a reference genome, pseudoalignment aims to identify which transcript each read originates from. Transcript abundances are then quantified by based on the number of reads associated with each transcript and used to infer gene counts. As a result, pseudoalignment is more computationally efficient than traditional genome alignment without loss of accuracy [30]. Moreover, transcript-level estimates can improve gene-level inferences in downstream analyses [31]. For these reasons, the pipeline given in this chapter incorporates pseudoalignment, via the popular *kallisto* tool [30] (script Sect. 3.4.1).

At the time of writing, to the authors knowledge, there is not a dedicated CRAN or Bioconductor R package for applying *kallisto* to bulk RNA-seq reads. Nevertheless, the *kallisto* tool can be easily executed via R using the *system()* function, which can be used to invoke a command line prompt directly in an R script. The *kallisto* tool itself (here, v0.50.0 for M1+ Mac operating systems) can be downloaded and installed from the *kallisto* GitHub site (https://github.com/pachterlab/kallisto) (code lines 106–113). The *kallisto* pipeline consists of two main steps: building of a transcriptome index, followed by quantification of transcript abundances. Building a transcriptome index using *kallisto* requires a reference transcriptome. In that case, the developers of *kallisto* recommend using the cDNA sequences fasta file (file ending in.cdna.all.fa.gz) for the relevant species from Ensembl genome database (https://www.ensembl.org/index.html) [24]. After creating a new folder structure in the working directory to store the *C. elegans* Ensembl cDNA fasta file (code line 118), it can be downloaded by parsing its URL to the *download.file()* R function (code lines 120–124). The *kallisto* transcriptome index build process itself utilizes a single function called *index*, and involves specifying the *kallisto* tool executable, *index* function, output location (created in code line 129), and file name for the resultant transcriptome index, and cDNA fasta file to be used to create the transcriptome index (code lines 132–135). Similarly, *kallisto* quantification of transcript abundances uses a single function called *quant*, which as a minimum involves specifying the *kallisto* tool executable, *quant* function, transcriptome index file, output location for the resultant quantification (created in code line 140), and the RNA-seq read files for a given sample. Since only one sample can be processed with the quant function at any one time, the current pipeline runs the process iteratively over the read-pairs for each sample, with the optional argument **--rf-stranded** used to specify that in this exact case the input read-pairs are strand-specific reads with the first read reverse (code lines 142–148). Here, the output for each sample is then a folder named with that sample's SRA run ID, containing the transcript abundances (in both.tsv and.h5 file format) and *quant* run info (.json file format) for that sample. The *quant* run info.json output for SRR26332897 is provided as an example in Fig. 2, and shows that approximately 93% of reads could be pseudoaligned to the reference *C. elegans* transcriptome in this case.

3.4.2 Inferring Gene Expression from Transcript Abundances

Gene counts can be readily generated from transcript abundance output files in R via the dedicated *tximport* R package [31] (script


```
{
  "n_targets": 34689,
  "n_bootstraps": 0,
  "n_processed": 24056417,
  "n_pseudoaligned": 22316975,
  "n_unique": 14557196,
  "p_pseudoaligned": 92.8,
  "p_unique": 60.5,
  "kallisto_version": "0.50.0",
  "index_version": 12,
  "start_time": "Thu May  8 18:39:10 2025",
  "call": "./Software/kallisto/kallisto quant -i
./Outputs/kallisto/Index/WBcel235.cdna.all.transcripts.idx -o
./Outputs/kallisto/Quants/SRR26332897 --rf-stranded ./Data/Reads/SRR26332897_1.fastq.gz
./Data/Reads/SRR26332897_2.fastq.gz"
}
```


Fig. 2 Snapshot run info output produced by the *quant* function from the *kallisto* pseudoalignment tool. Example provided is for SRR26332897. The output information includes the number of transcript targets in the transcriptome index ("n_targets"), the number of reads processed ("n_processed"), the number ("n_pseudoaligned") and percentage ("p_pseudoaligned") of reads that were pseudoaligned, the number ("n_unique") and percentage ("p_unique") of reads that were uniquely pseudoaligned, the version of *kallisto* used ("kallisto_version"), and the specific command line call that was run ("call") (see Sect. 3.4.1 of Methods)

Sect. 3.4.2). The function from the *tximport* R package for inferring gene counts from transcript abundances is itself *tximport()*. The main required inputs to this function are the paths to the transcript abundance files to be used (**files** argument), the software that was used to generate the transcript abundances (**type** argument), and a data frame that maps transcript IDs to the associated gene IDs (**tx2gene** argument).

Several dedicated R packages exist for obtaining transcript-to-gene ID mapping data. Since the cDNA fasta file used to build the *kallisto* transcriptome index in the step above was sourced from the Ensembl genome database, the current pipeline leverages the *biomaRt* R package, which enables users to access Ensembl's database of gene annotations (called "BioMart") directly in R itself [32]. After installing the *biomaRt* R package (and dependencies) from Bioconductor (code line 156), the transcript-gene ID mappings for *C. elegans* can be obtained by first connecting to the appropriate section of the BioMart database for *C. elegans* gene annotations from Ensembl using the *useEnsembl()* biomaRt function (**version** argument set to 113 since cDNA fasta file was obtained from Ensembl release 113) (code line 159) and then pulling the required gene annotation attributes using the *getBM()* function (code line 162). Here, external gene names (i.e., gene symbols) are pulled in addition to Ensembl transcript and gene IDs as they will be helpful later in the pipeline. However, since the *tximport()* function ultimately requires a two-column data frame of transcript IDs and associated gene IDs, a separate copy without the gene symbols is also created (code line 167).

Upon installation of the *tximport* R package (and dependencies) (code line 170), the *tximport()* function can then be successfully run by specifying the paths to the *kallisto* quant outputs generated in script Sect. 3.4.1 (here, the.tsv transcript abundance files are utilized), specifying that *kallisto* was the software used to generate the transcript abundances, and parsing the two-column transcript-gene ID mapping data frame (code line 173). By default, the *tximport()* function expects transcript abundances as input and returns estimated gene counts, with the output a simple list holding the abundances, counts and average transcript lengths for each gene. Moreover, by default the *tximport()* function estimates gene counts from transcript abundances without scaling by library size or average transcript length (relevance is described in the subsequent section).

3.5 Downstream Analysis of RNA-seq Counts

A wealth of downstream methods exists for extracting biologically meaningful information from RNA-seq gene counts. Nevertheless, for brevity, the current pipeline focuses on four core analytical approaches: principal component analysis, differential gene expression analysis, Gene Ontology enrichment analysis, and network analysis (script Sect. 3.5).

3.5.1 Intermediatory Steps

Prior to formal downstream analysis, there are often intermediatory steps that can be taken to optimize data extraction and statistical calculations (script Sect. 3.5.1). In the context of the given pipeline, a helpful first intermediatory step is to initialize a "DESeq dataset" (DDS) object. The DDS object is a special type of data class associated with the *DESeq2* R package [33]. Not only is the *DESeq2* package one of the most widely used tools for performing differential gene expression analysis of RNA-seq gene counts (and thus exploited for this purpose in this pipeline), it also has many inherent functions for handling, processing, and exploring RNA-seq gene count data. At a basic level, a DDS object serves as a single place for storing gene counts, the associated gene annotations, the metadata of each sample, the study design, and calculations and results generated via *DESeq2* functions. Nevertheless, many other transcriptomics-related packages also accept DDS objects as inputs to their own functions, extending the utility of DDS objects beyond just the *DESeq2* pipeline itself. Once the *DESeq2* package and its dependencies have been installed from Bioconductor (code line 185), users can easily create a DDS object from the above-generated object from *tximport()* via the *DESeqDataSetFromTximport() DESeq2* function and specifying the name of the object generated by *tximport()* (**txi** argument), the sample metadata (**colData** argument), and a formula indicating how to model the samples (**design** argument) (code line 188). The latter should recapitulate column names of the sample metadata table. Here, a design formula that simply recapitulates the "Condition" variable in the metadata table is sufficient (i.e., whether a sample is SF or GC). Gene symbols can also be added to the DDS object straightforwardly (stored in the "mcols" DDS object component) by cross-referencing Ensembl IDs between the DDS object and the previously created gene annotation data frame to extract the corresponding gene symbols (code line 191). The *DESeqDataSetFromTximport()* function implicitly handles average transcript lengths for downstream *DESeq2* testing. If deviating from this, the user should consider whether the **countsFromAbundance** setting needs to be altered to explicitly account for transcript lengths when running the *tximport()* function itself [31].

Another useful intermediary step prior to commencing with downstream analyses is to apply prefiltering to remove lowly expressed genes. Doing so is helpful for improving biological and statistical signals, since a gene needs to be expressed at some minimal level before it is likely deemed biologically important, while it is difficult to make reliable statistical judgments on genes with low counts [34]. A consensus on what classes as a lowly expressed gene is lacking, however, for bulk RNA-seq, it is commonplace to define a gene as being expressed in a sample, if it has a read count ≥ 10 in that sample. A standard practice by the authors of this chapter is to define a gene as being expressed in an experiment, if it has a read count ≥ 10 in every sample. This step can be performed in a single line of R code, by keeping each row (genes) of the DDS object for which every column of the associated counts matrix has a value of at least 10 in that row (code line 196).

3.5.2 Principal Component Analysis

A simple yet effective method to apply early in the downstream analysis of RNA-seq gene counts (and large datasets in general) is principal component analysis (PCA) (script Sect. 3.5.2). PCA is an approach to data reduction whereby a large dataset of variables is simplified into a smaller set of uncorrelated variables that capture most of the variation in the larger dataset, called "principal components" [35]. Principal component 1 captures most of the variation, followed by principal component 2, and so on. In the context of RNA-seq gene counts, principal component 1 captures most of the variation in gene expression in the dataset, followed by principal component 2 and so on. Pairs of principal components can then be plotted as a two-dimensional "PCA plot" to visualize how (dis)similar samples are based on their gene expression profiles, offering a snapshot on global shifts in gene expression across conditions. Since principal components 1 and 2 capture most variation in gene expression, they are often the principal components folded into a PCA plot. However, in more complex study designs with multiple experimental factors, relationships among samples can also become apparent by plotting other pairs of principal components.

Though the *DEseq2* R package does have a built-in function for generating PCA plots (the *plotPCA()* function), the ability to modify the visual aesthetics of plots is restricted. In this regard, an alternative option is the *PCAtools* R package available through Bioconductor. The *PCAtools* package offers seamless integration with the *DESeq2* pipeline and DDS objects, while at the same time providing the user with greater flexibility when it comes to changing visual aesthetics. After installing the *PCAtools* package (and dependencies) (code line 202), a first step is to leverage its *pca()* function to extract the PCA data (code line 205). Here, PCA results are generated from variance stabilization–normalized counts using the 25% most variable genes. The *biplot()* function from the *PCAtools* package can then be applied to the output of the *pca()* function to generate an object holding the PCA plot (code line 208). This function offers many different arguments for adjusting the aesthetics of a PCA plot, however, for simplicity, the given example demonstrates how to color points by the "Condition" variable, adjust the point size, reposition the legend to be on the right-hand side of the plot, and modify label sizes of the points. After creating a new folder structure in the current working directory to store the plot (code line 213), the *ggsave()* function from the *ggplot2* R package (installed as a dependency of the *fastqcr* package) can be used to save the object holding the PCA plot in PDF format (code line 216). The generated plot is shown in Fig. 3. Here, the samples differentiate well across the x-axis that itself represents

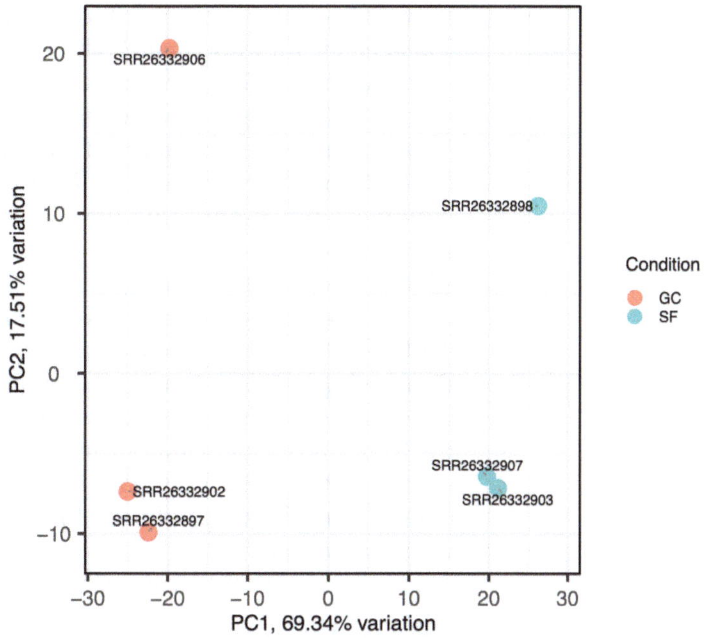

Fig. 3 Principal component analysis plot of spaceflight (SF) and ground control (GC) samples, based on variance stabilization-transformed gene counts of the top 25% most variable genes in the dataset. Principal component 1 is given on the x-axis and principal component 2 is given on the y-axis. The plot was generated using the *PCAtools* R package (see Sect. 3.5.2 of Methods)

principal component 1. Since principal component 1 captures close to 70% of variation in gene expression in the dataset, it, therefore, logically follows that condition (i.e., SF vs. GC) is the main driver for variation in gene expression across samples.

3.5.3 Differential Gene Expression Analysis

Testing for differential gene expression is a cornerstone of transcriptomics analysis. At its core, differential gene expression analysis involves testing each gene in a dataset to determine if its expression is statistically different between groups of samples [29]. In the context of this chapter, differential expression analysis is performed to determine genes that have significantly higher or lower expression in SF samples compared to GC samples (script Sect. 3.5.3).

The *DESeq2* package wraps its differential expression analysis steps into a single helper function, *DESeq()*, which can be applied directly to the DDS object (code line 222). When called, the *DEseq()* function first performs estimation of size factors to recapitulate differences in library size, then performs estimation of gene dispersions to measure the spread in gene count data, before finally fitting a negative binomial model (gene counts typically follow a negative binomial distribution) to each gene that integrates the size factor and dispersion estimates. By default, Wald significance tests are performed to test for significance in the negative binomial models and generate the associated *P*-values for differential gene expression. The updated DDS object can then be fed to the *results()* function of the *DESeq2* package to extract the differential expression results for a given comparison (contrast) of interest (code line 225), including the log2-fold changes, nominal *P* values, and *P* values adjusted for multiple comparisons using the Benjamini–Hochberg procedure. Here, the contrast is set to compare SF samples to GC samples, while Cooks cutoff and independent filtering are not performed so that *P* values are assigned for every gene (users are encouraged to read the *DESeq2* vignette for further details on these arguments). Positive log2-fold changes in the results table then indicate higher gene expression in SF samples compared to GC samples, and negative log2-fold changes indicate lower gene expression in SF samples compared to GC samples. The *DESeq2* developers further suggest undertaking shrinkage of log2-fold change estimates to enhance visualization and ranking of genes. Broadly, shrinkage methods aim to improve log2-fold change estimates by borrowing information across all genes [33]. To achieve this, the "ashr" method [36] is demonstrated. After installing the *ashr* package (and dependencies) from CRAN (code line 230), the *lfcshrink() DESeq2* function can be applied to perform shrinkage of log2 fold-change estimates, specifying the DDS object, the associated results table object, and the "ashr" shrinkage method (code line 233).

After setting up an appropriate folder structure for it in the current working directory (code line 238), the final table of differential expression analysis results can be saved as a text file for later use (code lines 240–244). From a visualization standpoint, a common plot type for illustrating the results of differential gene expression analysis is a volcano plot. In volcano plots, the log2-fold change of each gene is plotted *versus* its statistical significance (normally on the negative log10 scale), with the top genes by log2-fold change and/or statistical significance then often also labeled to emphasize promising candidate targets. The developers of the *PCAtools* R package have also created a dedicated R package for creating volcano plots called *EnhancedVolcano*, which can again be installed (along with its dependencies) via Bioconductor (code line 249). The function from the *EnhancedVolcano* package to produce volcano plots is itself *EnhancedVolcano()* and offers many different argument options for tweaking visual aesthetics of the plot. Here, its utility is demonstrated to produce a volcano plot wherein log2-fold changes are plotted *versus* negative log10 of adjusted *P*-values, with light green used to illustrate significant differentially expressed genes (adjusted $P<0.05$ used for this), with darker green plus labeling used to highlight significant genes that also have |log2-fold change| >3 (arbitrarily chosen for demonstration purposes) (code lines 251–270). As with the PCA plot, the object holding

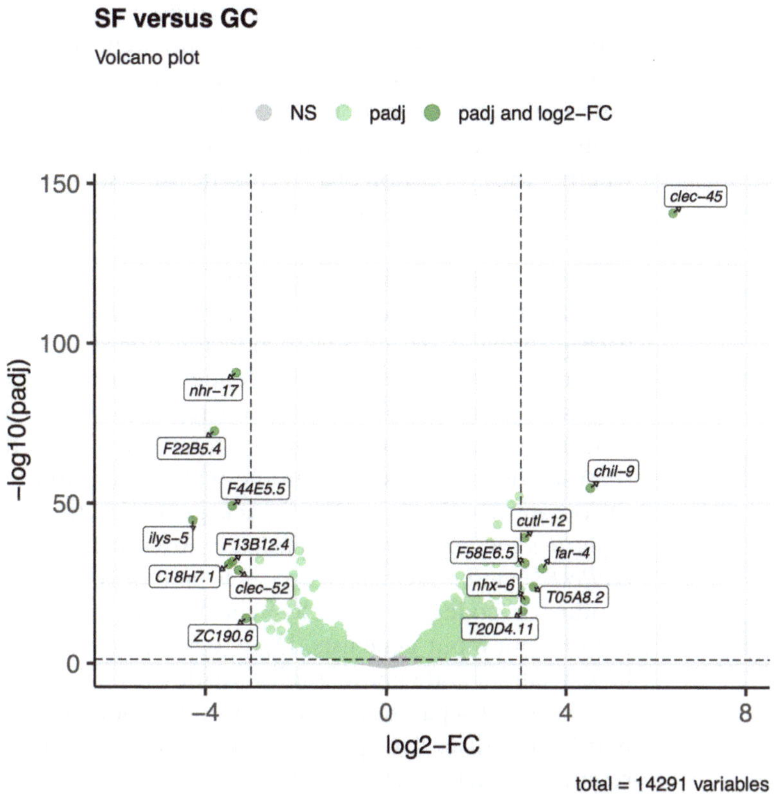

Fig. 4 Volcano plot illustrating the results for differential gene expression analysis of spaceflight (SF) *versus* ground control (GC) samples. Differential expression analysis was performed using the *DESeq2* R package. Light green illustrates significant differentially expressed genes (adjusted $P<0.05$), with darker green plus labeling used to highlight significant genes that also have |log2-fold change| >3 (arbitrarily chosen for demonstration purposes). Volcano plot was generated using the *EnhancedVolcano* R package (see Sect. 3.5.3 of Methods)

the volcano plot can be saved in PDF format using the *ggsave()* function of the *ggplot2* package (code line 273). The saved plot is shown in Fig. 4. Here, the volcano plot illustrates that the gene most significantly upregulated by spaceflight is *clec-45*, which is implicated in immune response [37].

3.5.4 Gene Ontology Enrichment Analysis

Deciphering the biological relevance of observed gene expression changes is an important aspect of downstream transcriptomic analysis. This task can be straightforward to perform manually if there are only a handful of differentially expressed genes. However, with extreme interventions such as spaceflight, there can often be hundreds-to-thousands of differentially expressed genes, lamenting the need for alternative, computational tools. In which case, a widely employed method for elucidating the functional relevance of differentially expressed genes is to test for enrichment of Gene Ontology terms by means of over-representation analysis (script Sect. 3.5.4).

The Gene Ontology represents the largest available source of information on gene functions, organized into defined classes ("terms") based on biological process, cellular compartment, and molecular function (https://geneontology.org/) [38]. Over-representation analysis is itself a method that allows the user to statistically test whether the proportion of a given gene list (e.g., significantly upregulated genes) mapping to a particular term (e.g., "cell adhesion") is more than what would be expected by chance [39].

A valuable tool for performing Gene Ontology enrichment analysis programmatically is the *clusterProfiler* R package [40]. This package provides several user-friendly functions for running over-representation analysis using either defined or customized gene terms. Its basic function for performing over-representation analysis of Gene Ontology terms is the *enrichGO()* function. To run this function, the user also requires the genome-wide annotation R package for the organism of interest. The R package serving that function for *C. elegans* is the *org.Ce.eg.db* R package. After installing both the *clusterProfiler* and *org.Ce.eg.db* packages (and dependencies) from Bioconductor (code lines 280–284), the user can then proceed to define the relevant input gene lists. Here, genes significantly upregulated by spaceflight and genes significantly downregulated by spaceflight are extracted separately to one another (code lines 288–290), to better infer direction-specific gene functional changes. Over-representation analysis also requires an appropriate background list to test enrichment relative to. Best practice is to define the background list as being any gene that could have been a positive hit, which here would be all genes that survived prefiltering and were input into differential expression analysis (code line 293). The *enrichGO()* function can then be run by inputting the list of upregulated (or downregulated) genes (**gene** argument) and the background gene list (**universe** argument), specifying the organism-specific genome-wide annotation package (**OrgDb** argument) and the gene ID system to use (**keyType** argument) (code lines 295–297). By default, the function tests molecular function terms from the Gene Ontology, but for demonstration purposes, biological process terms are tested here (controlled by the **ont** argument). The **readable** argument is also set to TRUE so that gene symbols are returned in the output instead of IDs. Moreover, the function automatically defines statistical significance at the Benjamini–Hochberg corrected $P<0.05$ level, but this can be adjusted in-function if needed (using the **pValueCutoff** and **pAdjustMethod** arguments). Outputs of the *enrichGO()* function are stored as an "enrichResult" R object. All enrichment results (i.e., significantly and nonsignificantly enriched Gene Ontology terms) can be extracted and saved in plain-text file format for later use (code lines 299–301).

A variety of different display forms can be used to effectively plot significant Gene Ontology enrichment results, e.g., heatmaps, bubble plots, or even a simple table. Here, visualization of Gene Ontology enrichment results is demonstrated in the form of an enrichment map. Enrichment maps visualize the significant terms (here, defined at adjusted $P<0.05$ level) as a network with links based on semantic similarity, making for easier interpretation of overlapping gene sets [41]. The developers of the *clusterProfiler* package have also created a dedicated R package for plotting enrichment analysis results including enrichment maps, called *enrichplot*. After installing this package (and dependencies) from Bioconductor (code line 306), a given enrichment map plot object can be produced by applying the *emapplot()* function of the *enrichplot* package to a pairwise semantic similarity matrix for significantly enriched Gene Ontology terms (generated using the *pairwise_termsim() enrichplot* function) (code lines 308–310). Again, objects that hold *enrichplot()* outputs can be saved in PDF format using the *ggsave()* function of the *ggplot2* package (code lines 312–314). The saved plot for significantly downregulated genes is shown in Fig. 5. The plot highlights that spaceflight downregulated genes involved in the insulin response and structure of muscle fibres in this *C. elegans* spaceflight experiment, consistent with previous findings [23].

3.5.5 Network Analysis

While differential gene expression analysis provides important insights into transcriptomic responses, it alone involves screening lots of individual gene expression changes in isolation. However, this overlooks the high complexity of molecular systems, which cannot be represented simply by the sum of their individual components. To overcome this, a useful complementary to differential gene expression analysis that can be performed is network analysis [42]. Interactions in a molecular network can be defined in a knowledge-based or data-driven manner [43]. The former leverages molecular interaction data predefined experimentally and/or computationally in databases, while the latter leverages patterns in the dataset being studied to define interactions. Since data-driven network analysis benefits from large sample sizes than the dataset being used in this chapter [44], the current pipeline provides a demonstration of knowledge-based network analysis (script Sect. 3.5.5).

A comprehensive database to obtain knowledge-based interaction data for transcriptomics (and proteomics) experiments is the online Search Tool for the Retrieval of Interacting Genes/Proteins (STRING) (https://string-db.org/) [45]. STRING facilitates the construction of functional protein association networks by simply providing a list of gene/protein IDs. While STRING provides a user-friendly online webtool, for the purposes of this pipeline, the *STRINGdb* R package is leveraged. This package (and dependencies) can be installed through Bioconductor (code line 320) and offers several commands to interface with the STRING database through R. As a first step, an R instance of the STRING database is created using

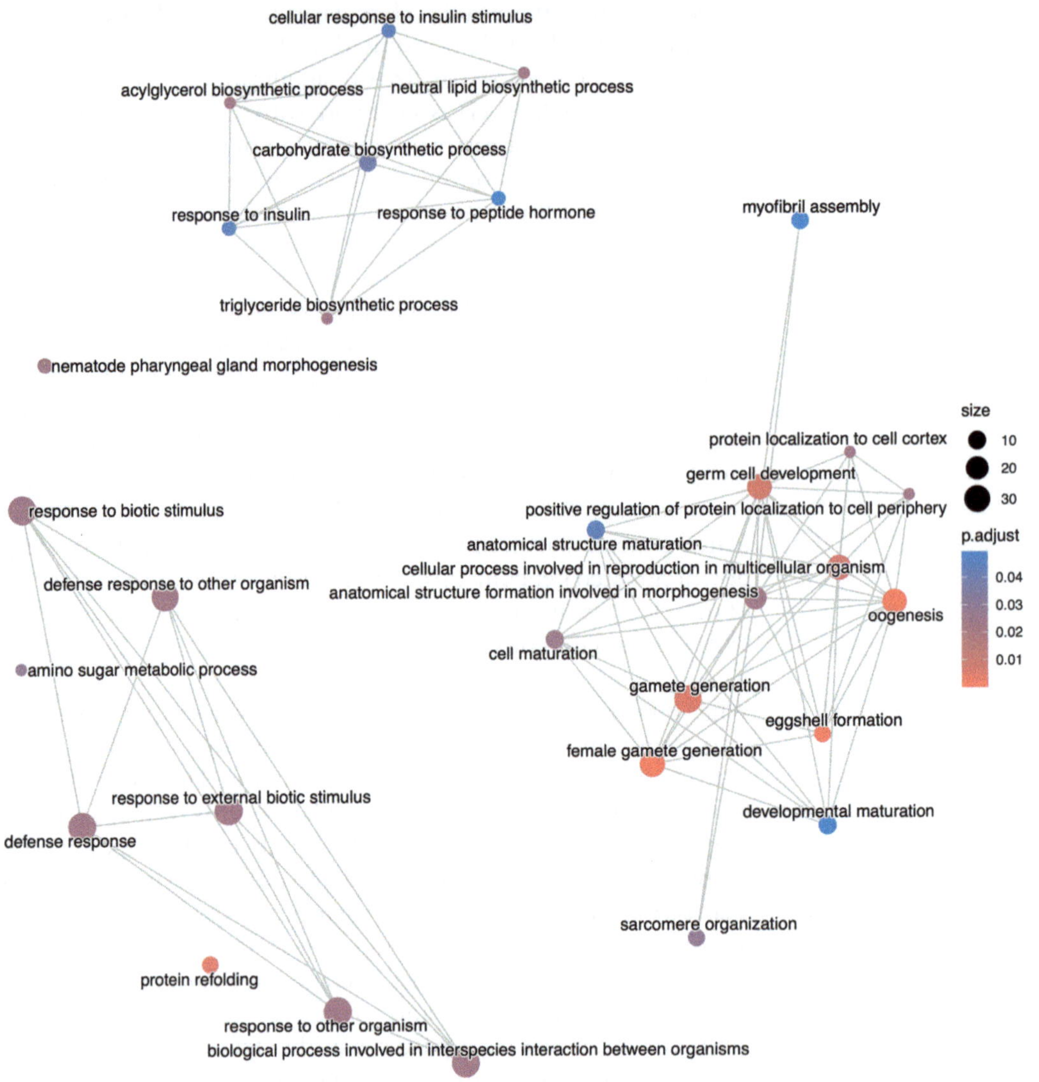

Fig. 5 Enrichment map plot illustrating significantly (adjusted $P < 0.05$) enriched Gene Ontology biological process terms for the genes significantly downregulated by spaceflight in the analyzed dataset. Clusters of connected terms are formed based on pairwise semantic similarity. Enrichment of Gene Ontology biological process terms was performed by means of over-representation analysis using the *clusterProfiler* R package. Nodes sizes are proportional to the number of downregulated genes enriched for the given term, and node colors illustrate strength of statistical significance. Enrichment map plot was generated using the *enrichplot* R package (see Sect. 3.5.4 of Methods)

the *new()* function from the *STRINGdb* package (code line 323). Here, the most recent version of STRING (v12) at the time of writing is specified (**version** argument), as well as the taxonomy ID for *C. elegans* (**species** argument) and the confidence score threshold for defining protein–protein interactions (**score** argument; 0.9 corresponds to highest confidence). After using the *STRINGdb map()* function to map genes to the STRING database and obtain their associated STRING ID's (code line 326), the STRING ID's for significantly upregulated and significantly downregulated genes can be extracted (code lines 328–330) and input into the *STRINGdb get_clusters()* function to identify tightly connected clusters of interacting proteins based on STRING interaction data (code lines 334–340). These cluster assignments can be saved in plain-text format for later use (code lines 342–351). Moreover, a plot for each cluster (or specified clusters) can be saved in PDF format using a coding loop (code lines 355–361). The plot of "cluster 3" for genes significantly downregulated by spaceflight is given in Fig. 6 as an example. In networks, top

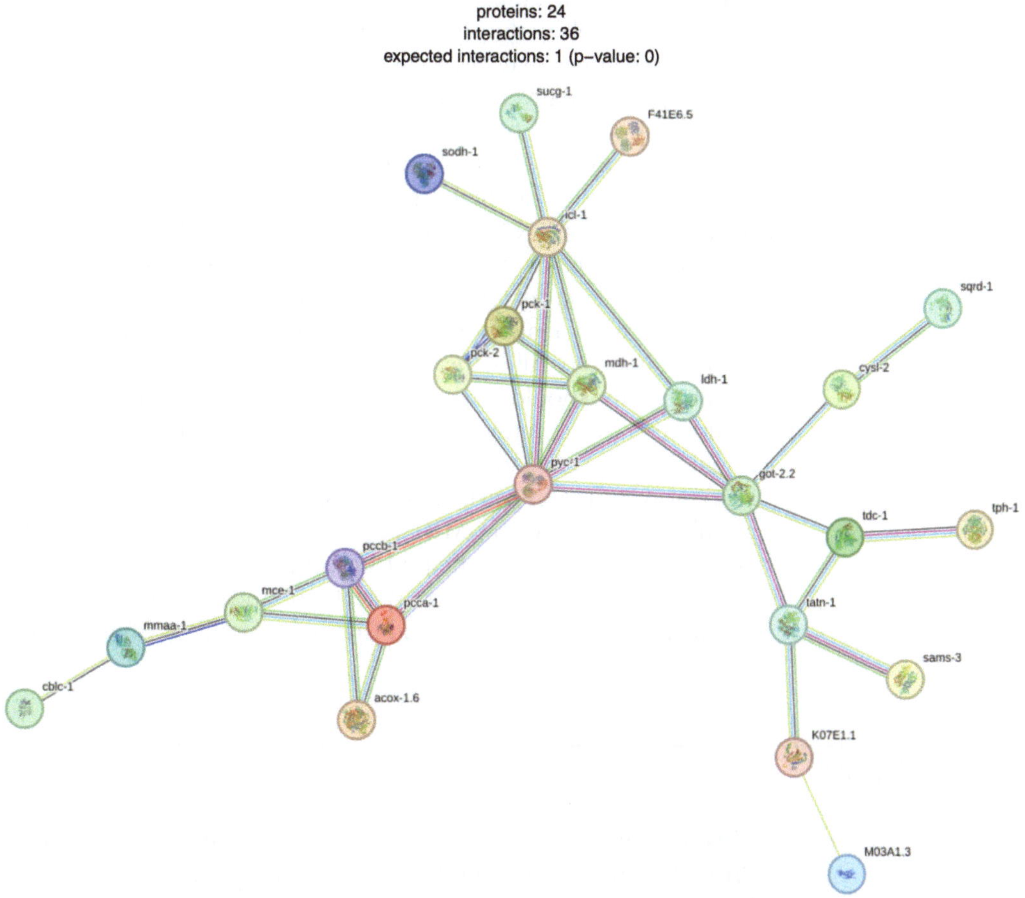

Fig. 6 Network visualization plot of "cluster 3" from cluster analysis of STRING protein–protein interactions for genes significantly downregulated by spaceflight. The number of proteins, the expected number of interactions, and the actual number of interactions for the cluster are given at the top of the plot, along with the *P* value telling if the actual number of interactions is statistically significant. Edge colors illustrate source(s) used to infer interaction (see https://string-db.org/). STRING analysis and network visualization plotting were performed using the *STRINGdb* R package (see Sect. 3.5.5 of Methods)

connected features are referred to as "hubs." and represent the most promising mechanistic candidates [42]. Here, the top hubs by node degree (icl-1 and pyc-1) locate in mitochondrion and are involved in metabolism [37], further consistent with the notion that mitochondrial energy metabolism is dysregulated during spaceflight [9].

4 Conclusion

Bulk RNA-seq offers a powerful strategy to help better understand the signaling pathways and key molecules that underpin health decline caused by exposure to the space environment. The R programming language is a particularly useful tool for analyzing bulk RNA-seq data, owing to the wide range of transcriptomics-centric R packages that exist and the ability to download and execute external transcriptomics tools within R itself. As such, users can develop single R script resources that can analyze RNA-seq data from start to finish. The pipeline given in this chapter provides one such example, leveraging popular, freely available R packages and external tools to demonstrate core tasks in an RNA-seq analysis pipeline, namely: obtaining publicly-archived RNA-seq reads, undertaking quality control checks of RNA-seq reads, quantification of gene expression from RNA-seq reads, and key downstream analyses and visualization of gene expression data (i.e., principal component analysis, differential gene expression analysis, Gene Ontology enrichment analysis, and network analysis). All of the R packages and external tools leveraged in this pipeline should be possible to run locally on standard modern computers, overcoming some of the potential barriers in computational resource. As a reminder, the pipeline is provided as a single R script resource in an open-source GitHub repository (https://github.com/williscrg/MiMB-Space-RNAseq-Pipeline). The authors hope that this serves as a helpful starting resource to the reader for analyzing RNA-seq data from spaceflight and/or terrestrial experiments that they have generated themselves and/or obtained from public databases.

Acknowledgments

This work was funded, in part, through the University of Exeter's Project ADA (Accelerating Data Science and Artificial Intelligence). Nathaniel J. Szewczyk acknowledges the support of the Osteopathic Heritage Foundation through funding for the Osteopathic Heritage Foundation Ralph S. Licklider, D.O., Research Endowment in the Heritage College of Osteopathic Medicine.

References

1. Narayanan SA (2023) Gravity's effect on biology. Front Physiol 14:1199175. https://doi.org/10.3389/fphys.2023.1199175
2. Afshinnekoo E et al (2020) Fundamental biological features of spaceflight: advancing the field to enable deep-space exploration. Cell 183(5):1162–1184. https://doi.org/10.1016/j.cell.2020.10.050
3. Cannavo A et al (2022) Are skeletal muscle changes during prolonged space flights similar to those experienced by frail and sarcopenic older adults? Life (Basel) 12(12):2139. https://doi.org/10.3390/life12122139
4. Scott JM, Stoudemire J, Dolan L, Downs M (2022) Leveraging spaceflight to advance cardiovascular research on Earth. Circ Res 130(6):942–957. https://doi.org/10.1161/CIRCRESAHA.121.319843
5. Mann V, Sundaresan A (2025) Chapter 12—Regenerative medicine and spaceflight. In: Krittanawong C (ed) Precision medicine for long and safe permanence of humans in space. Academic Press, Cambridge, MA, pp 163–177
6. Hasin Y, Seldin M, Lusis A (2017) Multi-omics approaches to disease. Genome Biol 18(1):83. https://doi.org/10.1186/s13059-017-1215-1
7. Karczewski KJ, Snyder MP (2018) Integrative omics for health and disease. Nat Rev Genet 19(5):299–310. https://doi.org/10.1038/nrg.2018.4
8. Overbey EG et al (2024) The Space Omics and Medical Atlas (SOMA) and international astronaut biobank. Nature 632(8027):1145–1154. https://doi.org/10.1038/s41586-024-07639-y
9. da Silveira WA et al (2020) Comprehensive multi-omics analysis reveals mitochondrial stress as a central biological hub for spaceflight impact. Cell 183(5):1185–1201 e1120. https://doi.org/10.1016/j.cell.2020.11.002
10. Rutter LA et al (2024) Astronaut omics and the impact of space on the human body at scale. Nat Commun 15(1):4952. https://doi.org/10.1038/s41467-024-47237-0
11. Dettmer K, Aronov PA, Hammock BD (2007) Mass spectrometry-based metabolomics. Mass Spectrom Rev 26(1):51–78. https://doi.org/10.1002/mas.20108
12. Joanisse S et al (2020) Recent advances in understanding resistance exercise training-induced skeletal muscle hypertrophy in humans. F1000Res 9:F1000 Faculty Rev-141. https://doi.org/10.12688/f1000research.21588.1
13. Zhao Y et al (2016) NONCODE 2016: an informative and valuable data source of long non-coding RNAs. Nucleic Acids Res 44(D1):D203–D208.https://doi.org/10.1093/nar/gkv1252
14. Camera A et al (2024) Aging and putative frailty biomarkers are altered by spaceflight. Sci Rep 14(1):13098. https://doi.org/10.1038/s41598-024-57948-5
15. Li K et al (2023) Explainable machine learning identifies multi-omics signatures of muscle response to spaceflight in mice. NPJ Microgravity 9(1):90. https://doi.org/10.1038/s41526-023-00337-5
16. Cope H et al (2024) Transcriptomics analysis reveals molecular alterations underpinning spaceflight dermatology. Commun Med (Lond) 4(1):106. https://doi.org/10.1038/s43856-024-00532-9
17. Vitry G et al (2022) Muscle atrophy phenotype gene expression during spaceflight is linked to a metabolic crosstalk in both the liver and the muscle in mice. iScience 25(10):105213. https://doi.org/10.1016/j.isci.2022.105213
18. Siew K et al (2024) Cosmic kidney disease: an integrated pan-omic, physiological and morphological study into spaceflight-induced renal dysfunction. Nat Commun 15(1):4923. https://doi.org/10.1038/s41467-024-49212-1
19. Beheshti A et al (2019) Multi-omics analysis of multiple missions to space reveal a theme of lipid dysregulation in mouse liver. Sci Rep 9(1):19195. https://doi.org/10.1038/s41598-019-55869-2
20. Willis CRG et al (2020) Comparative transcriptomics identifies neuronal and metabolic adaptations to hypergravity and microgravity in *Caenorhabditis elegans*. iScience 23(12):101734. https://doi.org/10.1016/j.isci.2020.101734
21. Fernandez-Gonzalo R, Willis CRG, Etheridge T, Deane CS (2022) RNA-sequencing muscle plasticity to resistance exercise training and disuse in youth and older age. Physiologia 2(4):164–179
22. Soni P et al (2023) Spaceflight induces strength decline in *Caenorhabditis elegans*. Cells 12(20):2470. https://doi.org/10.3390/cells12202470
23. Scott A et al (2023) *Caenorhabditis elegans* in microgravity: an omics perspective. iScience 26(7):107189. https://doi.org/10.1016/j.isci.2023.107189
24. Dyer SC et al (2025) Ensembl 2025. Nucleic Acids Res 53(D1):D948–D957. https://doi.org/10.1093/nar/gkae1071
25. Wilkinson MD et al (2016) The FAIR guiding principles for scientific data management and stewardship. Sci Data 3:160018. https://doi.org/10.1038/sdata.2016.18
26. Chen Y et al (2018) SOAPnuke: a MapReduce acceleration-supported software for integrated quality control and preprocessing of

high-throughput sequencing data. Gigascience 7(1):1–6. https://doi.org/10.1093/gigascience/gix120
27. Liao Y, Shi W (2020) Read trimming is not required for mapping and quantification of RNA-seq reads at the gene level. NAR Genom Bioinform 2(3):lqaa068. https://doi.org/10.1093/nargab/lqaa068
28. Corchete LA et al (2020) Systematic comparison and assessment of RNA-seq procedures for gene expression quantitative analysis. Sci Rep 10(1):19737. https://doi.org/10.1038/s41598-020-76881-x
29. Conesa A et al (2016) A survey of best practices for RNA-seq data analysis. Genome Biol 17:13. https://doi.org/10.1186/s13059-016-0881-8
30. Bray NL, Pimentel H, Melsted P, Pachter L (2016) Near-optimal probabilistic RNA-seq quantification. Nat Biotechnol 34(5):525–527. https://doi.org/10.1038/nbt.3519
31. Soneson C, Love MI, Robinson MD (2015) Differential analyses for RNA-seq: transcript-level estimates improve gene-level inferences. F1000Res 4:1521. https://doi.org/10.12688/f1000research.7563.2
32. Durinck S, Spellman PT, Birney E, Huber W (2009) Mapping identifiers for the integration of genomic datasets with the R/Bioconductor package biomaRt. Nat Protoc 4(8):1184–1191. https://doi.org/10.1038/nprot.2009.97
33. Love MI, Huber W, Anders S (2014) Moderated estimation of fold change and dispersion for RNA-seq data with DESeq2. Genome Biol 15(12):550. https://doi.org/10.1186/s13059-014-0550-8
34. Chen Y, Lun AT, Smyth GK (2016) From reads to genes to pathways: differential expression analysis of RNA-Seq experiments using Rsubread and the edgeR quasi-likelihood pipeline. F1000Res 5:1438. https://doi.org/10.12688/f1000research.8987.2
35. Jolliffe IT, Cadima J (2016) Principal component analysis: a review and recent developments. Philos Trans A Math Phys Eng Sci 374(2065):20150202. https://doi.org/10.1098/rsta.2015.0202
36. Stephens M (2017) False discovery rates: a new deal. Biostatistics 18(2):275–294. https://doi.org/10.1093/biostatistics/kxw041
37. Sternberg PW et al (2024) WormBase 2024: status and transitioning to Alliance infrastructure. Genetics 227(1):iyae050. https://doi.org/10.1093/genetics/iyae050
38. Ashburner M et al (2000) Gene ontology: tool for the unification of biology. The Gene Ontology Consortium. Nat Genet 25(1):25–29. https://doi.org/10.1038/75556
39. Soneson C, Delorenzi M (2013) A comparison of methods for differential expression analysis of RNA-seq data. BMC Bioinformatics 14:91. https://doi.org/10.1186/1471-2105-14-91
40. Wu T et al (2021) clusterProfiler 4.0: a universal enrichment tool for interpreting omics data. Innovation (Camb) 2(3):100141. https://doi.org/10.1016/j.xinn.2021.100141
41. Merico D et al (2010) Enrichment map: a network-based method for gene-set enrichment visualization and interpretation. PLoS One 5(11):e13984. https://doi.org/10.1371/journal.pone.0013984
42. van Dam S et al (2018) Gene co-expression analysis for functional classification and gene-disease predictions. Brief Bioinform 19(4):575–592. https://doi.org/10.1093/bib/bbw139
43. Worheide MA, Krumsiek J, Kastenmuller G, Arnold M (2021) Multi-omics integration in biomedical research–a metabolomics-centric review. Anal Chim Acta 1141:144–162. https://doi.org/10.1016/j.aca.2020.10.038
44. Langfelder P, Horvath S (2008) WGCNA: an R package for weighted correlation network analysis. BMC Bioinformatics 9:559. https://doi.org/10.1186/1471-2105-9-559
45. Szklarczyk D et al (2023) The STRING database in 2023: protein-protein association networks and functional enrichment analyses for any sequenced genome of interest. Nucleic Acids Res 51(D1):D638–D646. https://doi.org/10.1093/nar/gkac1000

Chapter 6

A Systems Biology Approach to Discovery and Validation of MicroRNA Signature Health Risks Associated with the Space Environment

Francisco J. Enguita, Ana Lúcia Leitão and Afshin Beheshti

Abstract

When living organisms are exposed to the space environment, numerous changes occur that affect virtually all cellular functions and molecular pathways. One class of understudied molecules in space biomedicine is noncoding RNAs (ncRNAs), particularly microRNAs (miRNAs). Given that spaceflight induces systemic health risks, miRNAs represent natural candidates for key regulators in space, as they can control hundreds to thousands of genes and influence central biological pathways. Transcriptomic data and pathway analyses are employed to identify the key genes driving specific physiological states, which can in turn be used to predict the miRNAs regulating those states. MiRNAs that act as upstream regulators of these driver genes are selected as candidates for a spaceflight-associated miRNA signature. Once these candidate miRNAs are identified, they must be experimentally validated. Depending on the health risk or biological process under investigation, miRNAs are validated either in circulating fluids (e.g., serum, plasma, blood) or in tissue samples using droplet digital polymerase chain reaction (ddPCR). Compared to conventional PCR methods, ddPCR enables absolute quantification of miRNA copy numbers and overcomes several limitations associated with traditional miRNA expression assays. This approach provides an unbiased prediction of candidate miRNAs, followed by precise and specific verification of their expression profiles, and can be broadly applied to identify miRNA signatures associated with various diseases and physiological stressors. Once validated, miRNAs that are overexpressed in response to spaceflight can be targeted for inhibition as potential countermeasures to mitigate health risks. Harnessing the power of miRNAs not only deepens our understanding of space biology but also accelerates the development of targeted interventions, ultimately enabling safe and sustained human exploration of deep space.

Keywords MicroRNA, miRNA, ncRNA, Space biomedicine, Antagomirs, DdPCR

1 Introduction

Prolonged space missions expose the human body to a unique combination of stressors, including microgravity, elevated radiation exposure, and isolation, which significantly increase health risks [1]. These

A.B. provided the first draft, conceptualization, and edits for the chapter. F.J.E. and A.L.L. provided edits and revisions for the chapter.

Mariano Bizzarri (ed.), *Biomedicine and Biotechnology in Space*, Methods in Molecular Biology, vol. 3000, https://doi.org/10.1007/978-1-0716-5174-2_6,

stressors can lead to widespread molecular and physiological disruptions, including altered gene expression, microbiome imbalance, DNA damage, and mitochondrial dysfunction [1–5]. Among the most promising and understudied regulators of these responses are microRNAs (miRNAs), short, noncoding RNAs that post-transcriptionally regulate gene expression decreasing the translational output from coding genes, and that control major biological pathways. The miRNA regulatory action exerts a "buffer" effect that maintains gene expression levels between homeostatic intervals, and any imbalance in miRNA levels will modify the physiological patterns of the genomic output.

Moreover, cells can actively secrete miRNAs to the extracellular environment, and consequently, miRNAs can be detected in all biofluids. Circulating miRNAs have emerged as minimally invasive biomarkers capable of capturing complex physiological changes, including those driven by spaceflight [4–7]. We recently identified a distinct circulating miRNA signature associated with spaceflight, with implications for cardiovascular dysfunction and muscle degeneration [4–7]. These miRNAs not only correlate with physiological changes but may also serve as viable therapeutic targets. Our work and others have demonstrated that modulation of miRNA expression can serve as a countermeasure strategy to mitigate spaceflight-induced health risks [4, 5, 8].

MiRNAs are particularly intriguing due to their ability to regulate mitochondrial function, a central player in spaceflight-induced stress. Besides their regulatory action over genome-encoded mitochondrial genes, several miRNAs can also translocate to mitochondria, modulating mitochondrial activity and pathways, such as calcium homeostasis, oxidative stress response, and energy metabolism [9, 10]. Given the role of mitochondria in aging and disease, and the fact that spaceflight mimics an accelerated aging model [1, 2, 11, 12], miRNAs represent a powerful tool for studying and potentially intervening in the biological consequences of space travel.

The accelerated aging phenotype observed in astronauts, characterized by increased inflammation, dermal thinning, and elevated oxidative stress, mirrors terrestrial aging and is driven in part by mitochondrial and gene regulatory dysfunctions [1, 3, 13]. The accumulation of mitochondrial DNA damage and declining mitochondrial quality further exacerbate the risk of chronic conditions [14–17]. However, miRNAs offer a means of identifying early molecular changes and developing strategies for intervention.

Recent studies have shown that miRNAs play a fundamental role in regulating aging, cellular senescence, DNA repair, and oxidative stress pathways [4–7, 18]. Their importance was recently recognized with the awarding of the 2024 Nobel Prize in Medicine or Physiology for the discovery of miRNAs to Victor Ambros and Gary Ruvkun, highlighting their impact across biology and medicine. In the context of spaceflight, miRNAs mediate key physiological responses and represent promising biomarkers and targets for mitigating health risks [4, 5, 7, 8, 19].

Beyond their role as biomarkers, miRNAs are increasingly recognized as therapeutic targets. One promising approach involves the use of chemically modified antisense oligonucleotides, known as antagomirs, which are designed to specifically inhibit overexpressed miRNAs [4, 5, 7, 20]. Antagomirs, complementary antisense oligonucleotides that block miRNA action, have shown efficacy in silencing miRNAs implicated in various disease states and physiological stress responses [4, 20]. In the context of spaceflight, where certain miRNAs become dysregulated in response to environmental stressors, antagomirs can serve as targeted countermeasures to restore normal gene regulation [4]. For instance, miRNAs that drive oxidative stress, inflammation, or metabolic dysfunction during long-duration missions could be selectively silenced using antagomirs to prevent or reduce tissue damage. Taking advantage of specific nucleic acid carriers, these therapeutics can be delivered systemically or locally, depending on the target tissue and mechanism of action. The adaptability and specificity of antagomir-based interventions make them especially attractive for spaceflight applications, where pharmacological options are limited, and precision in managing biological responses is essential.

This protocol outlines a standardized methodology for predicting spaceflight-relevant miRNA signatures from transcriptomic datasets through specialized bioinformatic tools, followed by validation of these signatures in both rodent and human tissue samples. Researchers should have access to a clean RNA workstation and standard lab equipment, including centrifuges and pipettes, as well as components required for droplet digital PCR (ddPCR): a droplet generator, thermocycler, plate sealer, and plate reader. For the in vivo and in vitro experiments, researchers will also need access to cell culture facilities, incubators, and all the required facilities for animal care and handling. This approach allows for the unbiased prediction and precise quantification of spaceflight-associated miRNAs, facilitating their use as both diagnostic biomarkers and therapeutic targets in space biomedicine. Finally, we finish this chapter with an approach to inhibit miRNAs to help develop effective countermeasures to mitigate the damage caused to the body by the space environment.

2 Materials

2.1 Small RNA Sequencing

1. QIAgen miRNeasy Mini Kit.
2. QIAgen miRNeasy Serum/Plasma Kit.
3. Qubit RNA HS Assay Kit.
4. Bioanalyzer 2100 (Agilent).
5. TruSeq Small RNA Library Prep Kit (Illumina).
6. NEBNext Small RNA Library Prep Kit.
7. NEBNext Multiplex Oligos.

8. Illumina Sequencer (MiSeq, NextSeq, HiSeq, or NovaSeq).
9. Cutadapt (software for trimming of RNAseq reads).
10. FastQC (https://www.bioinformatics.babraham.ac.uk/projects/fastqc) and MultiQC (https://seqera.io/multiqc) (software for quality control of RNAseq reads).
11. miRTrace (software for quality control and genomic trace of small-RNAseq data) [21].
12. miRDeep2 (software for miRNA detection from RNAseq data) [22].
13. Bowtie (general purpose software for RNAseq read alignment) [23].
14. RNAfold (software for RNA secondary structure prediction by energy minimization and base pairing protocols, belonging to the ViennaRNA package) [24].
15. DESeq2, EdgeR, and Limma (software packages for the determination of differential expression patterns from RNAseq data, belonging to the R-Bioconductor package).
16. RBiomirGS (R-package for miRNA gene set analysis by target mapping and gene expression integration) [25].
17. ggplot2, ComplexHeatmap and EnhancedVolcano (R-packages for the customized graphical representation of different datasets).

2.2 Validation and Quantification of Candidate miRNAs in Biological Samples Using ddPCR

1. QIAgen miRNeasy Mini Kit.
2. QIAgen miRNeasy Serum/Plasma Kit.
3. QIAgen miRCURY LNA RT Kit.
4. QIAgen miRCURY LNA miRNA PCR Assay primers (e.g., hsa-miR-24-3p miRCURY LNA miRNA PCR Assay.
5. BioRad QX200™ ddPCR™ EvaGreen Supermix.
6. BioRad Automated Droplet Generation Oil for EvaGreen.
7. BioRad QX200™ Buffer Control for EvaGreen.
8. Nuclease-free water.
9. BioRad QX200 AutoDG Droplet Digital PCR System.
10. BioRad C1000 Touch™ Thermal Cycler with 96-Well Fast Reaction Module.

2.3 Inhibition of miRNAs

1. AUMantagomir™ AUM BioTech (https://www.aumbiotech.com/products/aumantagomir).
2. Complete culture medium suitable for your cell type.
3. Standard tissue culture plasticware (plates, flasks).
4. Optional: scrambled antagomir control.
5. Sterile 1× PBS.
6. Pipettes and sterile filtered tips.

7. Incubator (37°C, 5% CO_2).
8. Insulin syringes or 27–30G needles.
9. Anesthetics and/or restraining equipment for mice (if required).
10. Animal housing compliant with IACUC or equivalent regulatory protocols.

3 Methods

This protocol outlines a comprehensive and standardized workflow for miRNA sequencing (miRNA-seq) using the different Illumina platforms, from sample collection to differential expression and functional pathway analysis (Fig. 1). It integrates current best practices and widely adopted bioinformatics tools for quality control, read processing, alignment, quantification, and interpretation of small RNA datasets. The protocol can be easily adapted to other second or third-generation sequencing systems (e.g., Nanopore or PacBio systems) by integrating the specific requirements from the different manufacturers, such as the library preparation and sequencing reactions. The pipeline is compatible with murine, human, and other biological samples and supports the discovery of both known and novel miRNAs. Key components include adapter trimming with Cutadapt, mapping and quantification with miRDeep2 [22] and Bowtie, normalization and differential analysis with DESeq2 or edgeR, and functional enrichment using RBiomirGS [25]. The description of each step is complemented by alternate protocols that could also be employed, depending on the technical experience of the researcher. The approach is modular, scalable for high-throughput studies, and designed to generate robust, reproducible miRNA signatures relevant to health, disease, and experimental stressors such as spaceflight.

3.1 miRNA Sequencing and Bioinformatics Analysis

3.1.1 Sample Preparation and Sequencing

(A) **Sample Collection and RNA Extraction**

- Collect tissues or biofluids (e.g., plasma, liver, heart, muscle).
- Extract total RNA using a Qiagen miRNeasy Kit for tissues or biofluids.
- Elute in RNase-free water and store at −80°C until use.

(B) **RNA Quality Control**

- Quantify RNA using Qubit RNA HS Assay.
- Assess integrity using Bioanalyzer 2100 or TapeStation (Agilent Technologies). For solid tissue samples, ensure an RNA integrity number (RIN) >7. Since the RIN is calculated by using a

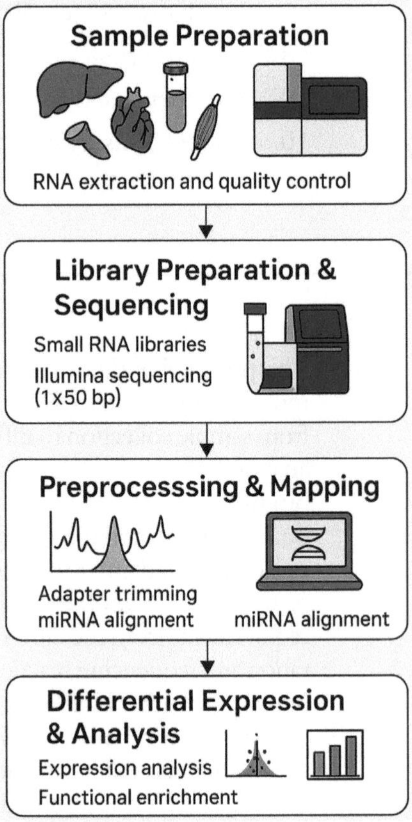

Fig. 1 The standardized miRNA-seq workflow. The miRNA sequencing pipeline begins with sample preparation and total RNA extraction from tissues or biofluids, followed by quality control and small RNA library preparation. Libraries are sequenced using Illumina platforms (single-end, 50 bp), and raw reads undergo adapter trimming and quality filtering. Reads are mapped to known miRNA databases, and novel miRNAs are predicted using secondary structure analysis. Differential expression analysis is performed using statistical models (e.g., DESeq2), followed by downstream functional enrichment of miRNA target genes. The workflow enables high-confidence identification and interpretation of miRNA signatures in experimental models and clinical samples

proprietary algorithm that estimates the ratio between rRNA 28S and 18S, this parameter cannot be calculated for RNA extracted from biofluids. In this case, it would be advisable to include a spike-in synthetic RNA in the samples prior to the RNA extraction to estimate the quality of the extraction protocol.

(C) **Library Preparation**

- Use small RNA library prep kits:

- NEBNext® Small RNA Library Prep Set or Illumina TruSeq Small RNA Kit.

- Recommended input: 50 ng–1 μg total RNA.
- Include unique dual indexes (e.g., NEB #E7600) for multiplexing.
- Recommendable: RNA size selection (140–160 bp) via gel or bead purification to enrich miRNAs.

(D) **Sequencing**

- Performsingle-end50–75 bpsequencingonIlluminaplatforms (e.g., MiSeq, NextSeq, NovaSeq, or HiSeq).
- Recommended depth: 10–20 million reads/sample.
- Convert BCL to FASTQ using bcl2fastq software.

3.1.2 Preprocessing and Quality Control

(A) **Raw Read QC**

- Assess raw FASTQ files using FastQC and summarize with MultiQC. Control the overall quality of the reads per sample and the presence of sequencing adapters.

(B) **Adapter Removal and Length Filtering**

- If required, use Cutadapt to trim 3′ adapters (typically TGGAATTCTCGGGTGCCAAGG).
- Retain reads of 18–26 nt (miRNA size range).
- R command to use: cutadapt -a TGGAATTCTCGGGTGCCAAGG -m 18 -M 26 -o sample_trimmed.fastq sample.fastq.

(C) **Contaminant Filtering (Optional)**

- Use miRTrace software to assess contamination and eliminate rRNA/tRNA/other non-miRNA reads.

3.1.3 miRNA Mapping and Quantification

(A) **Read Alignment and miRNA Quantification**

- Use the standalone software miRDeep2 for alignment, detection, and quantification of miRNAs in the generated dataset. MiRDeep2 uses the real aligner Bowtie against a customized reference database of mature and precursor miRNAs that can be extracted from miRBase or MirGeneDB
  - Typical miRDeep2 commands: mapper.pl sample_trimmed.fastq -e -h -j -m -l 18 -n -o sample_collapsed.fa miRDeep2.pl sample_collapsed.fa genome.fa mature.fa hairpin.fa -t Mouse
- Alternative platforms for miRNA detection under a graphical userinterface,includetheweb-basedsRNAtoolbox(https://arn.ugr.es) [26] and MiRMaster2 (https://ccb-compute.cs.uni-saarland.de/mirmaster2) [27]. Both integrated tools

include pipelines for the detection and quantification of miRNAs and other small ncRNAs from RNAseq data.

(B) **Expression Matrix Generation**

- Output read counts (raw and normalized as RPM or CPM).
- Filter miRNAs with consistently low expression across all samples (e.g., CPM <1 in all samples).

3.1.4 Statistical and Functional Analysis

(A) **Differential Expression Analysis**

- Import count data matrix into R.
- Normalize using DESeq2, EdgeR, or Limma-voom.
- Statistical tests: Wald test, likelihood ratio test (LRT), or ANOVA.
- Adjust for multiple testing with Benjamini–Hochberg FDR.
- R command to use:
  - dds <- DESeqDataSetFromMatrix (countData = counts, colData = coldata, design = ~ condition)
  - dds <- DESeq (dds)
  - res <- results (dds, alpha = 0.05)

(B) **Functional Enrichment Analysis**

- Use RBiomirGS [25] for miRNA-target gene enrichment and functional analysis (GO, Reactome, MSigDB C2 sets).
- Apply a cutoff: FDR <0.25 for pathways.
- Alternatively, use miEAA or miRPathDB for GUI-based pathway tools.

3.1.5 Visualization and Integration

(A) **Visualization**

- Use ggplot2, EnhancedVolcano, and ComplexHeatmap in R for generating Volcano plots, heatmaps, PCA plots and pathway barplots from normalized data and differentially expressed miRNAs.
- Results from the functional enrichment and target analysis can also be represented as interaction networks using software as Cytoscape (https://cytoscape.org) [28] or NAViGaTOR (https://navigator.ophid.utoronto.ca/navigatorwp) [29], to detect relevant nodes. MiRNAs and targets with an increased number of connections will be more important for the biological process studied.

(B) **Integration with Public Datasets**

- Compare identified miRNAs with:
  - NASA GeneLab datasets (e.g., GLDS-334 to GLDS-337)
  - Twin Study miRNA-seq datasets

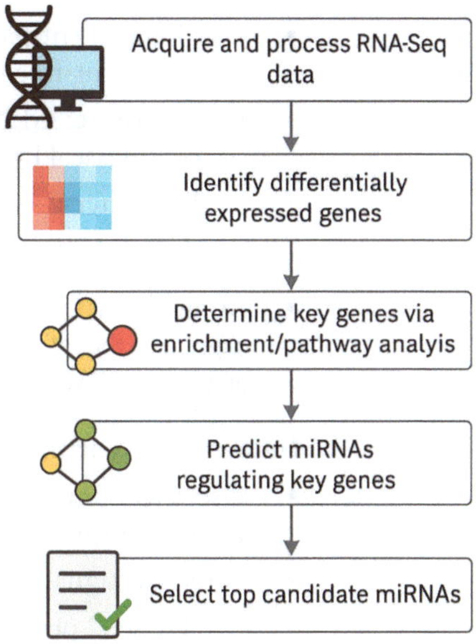

Fig. 2 The prediction of candidate miRNAs from transcriptomic (RNA-seq) data. This workflow outlines the stepwise approach to identifying candidate microRNAs based on differential gene expression analysis from RNA-seq data. Raw transcriptomic datasets are processed to identify significantly altered genes, which are then subjected to pathway and enrichment analyses to determine key driver genes. Upstream regulator analysis is performed to predict miRNAs that control these genes, and the top candidate miRNAs are selected based on functional relevance and regulatory significance for downstream validation

- Use VennDiagram or UpSetR in R to identify shared miRNAs across datasets or conditions.

3.2 Prediction of Candidate miRNAs from Transcriptomic (RNA-Seq) Data

This protocol outlines a computational workflow for predicting candidate miRNAs based on differential gene expression analysis and biological pathway inference from RNA-seq data (Fig. 2). The resulting set of functionally relevant miRNAs can be used for experimental validation and therapeutic targeting.

3.2.1 Acquire and Preprocess Transcriptomic Data

1. Obtain raw RNA-seq data (FASTQ files) from either internal studies or publicly available databases. For spaceflight-related studies, NASA Open Science Data Repository (OSDR) [30] is a recommended source.
2. **Preprocessing and Quality Control**
   - Use FastQC to assess read quality.
   - Use Trim Galore (https://github.com/FelixKrueger/TrimGalore) (which wraps Cutadapt and FastQC) to trim sequencing adapters and low-quality bases.

3. **Transcript Quantification Using Pseudoalignment**
   - Use Kallisto (https://pachterlab.github.io/kallisto) [31] for alignment-free transcript quantification.
     - Reference: GENCODE Release M31 or later (mouse) or Release 41 or later (human)
     - Parameters: --bias --bootstrap-samples=100 --fr-stranded -i [index] -o [output] [fastq1] [fastq2]
   - As an alternative to pseudoaligners, RNAseq reads could be aligned to the reference genomes using new-generation ultrafast aligners as STAR (https://github.com/alexdobin/STAR) [32].
4. **Import and Normalize Quantifications**
   - Use the tximport R package to import transcript-level Kallisto output and summarize to gene level.
   - Filter low signal counts along all the datasets.
   - Use DESeq2 or EdgeR for normalization, statistical testing, and log-fold change estimation.
5. **Data Visualization and Quality Checks**
   - Use ggplot2 or PCAtools to perform PCA and generate diagnostic plots.
   - Use pheatmap or ComplexHeatmap for heatmaps of differentially expressed genes.

3.2.2 Identify Key Differentially Expressed Genes and Pathways

1. **Differential Expression Analysis**
   - Use DESeq2 or limma-voom to identify significantly differentially expressed genes (adjusted p-value ≤ 0.05; log2FC cutoff as needed).
2. **Functional Annotation and Pathway Analysis**
   - Use clusterProfiler or fgsea [33] for gene set enrichment analysis (GSEA), including Hallmark, KEGG, Reactome, and GO terms (C2, C5 sets from MSigDB). Additional functional annotation could also be obtained from web-based tools as Enrichr (https://maayanlab.cloud/Enrichr/) [34].
   - Use Ingenuity Pathway Analysis (IPA) for deeper biological interpretation (if available), focusing on canonical pathways, upstream regulators, and functional networks.
   - Alternatively, open-source options like ReactomePA and Pathview (https://pathview.uncc.edu) can be used.
3. **Identify Central "Driver" Genes**
   - Compile overlapping genes from enriched pathways and upstream regulators.

- Use STRING, Cytoscape, or IPA's Molecule Activity Predictor to determine gene interaction networks and identify central hub genes based on connectivity. Additionally, perform cluster density analysis and distribution within the interaction networks to select the relevant hubs.
- Radial or network plots can be generated using igraph or ggraph in R or using interactive graphical tools such as NAViGaTOR.

3.2.3 Predict Candidate miRNAs Impacting the Gene Network

1. **Prediction Using Upstream Regulator Analysis**
   - In IPA, use the Upstream Regulator tool to identify miRNAs predicted to regulate the central driver genes.
     - Rank miRNAs by adjusted p-value and activation Z-score ($\geq$+2 or $\leq$−2).
     - Select the top miRNAs by statistical significance and regulatory activity. In prior applications, 13 miRNAs were identified, with 4 showing strong activation/inhibition scores.
2. **Additional miRNA-Target Inference Tools** (Optional or Complementary):
   - Use **miRNet** [35, 36] (https://www.mirnet.ca) for miRNA–mRNA interaction network construction and graphical analysis from a user-defined gene list.
   - Use **multiMiR** R package [37] to integrate miRNA-target information from multiple databases comprising predicted and validated targets (TargetScan, miRDB, miRTarBase, etc.) [38] As an alternative only available for human miRNA predicted targets, mirDIP platform can also integrate information from 24 different miRNA target predictors (https://ophid.utoronto.ca/mirDIP/) [39].
   - Use **mirPath v.4** (DIANA Tools) [40] for miRNA pathway enrichment analysis from candidate lists.
3. **Integration and Refinement**
   - Combine results from IPA and open-source tools to refine the list of top candidate miRNAs.
   - Evaluate conservation, tissue expression, and disease relevance using miRbase [41], miRSystem [42], or TargetScan [43].
   - Using multiple datasets from different organs from mice experiencing spaceflight conditions will allow to determine the key overlapping miRNAs that will be truly driving the systemic response.

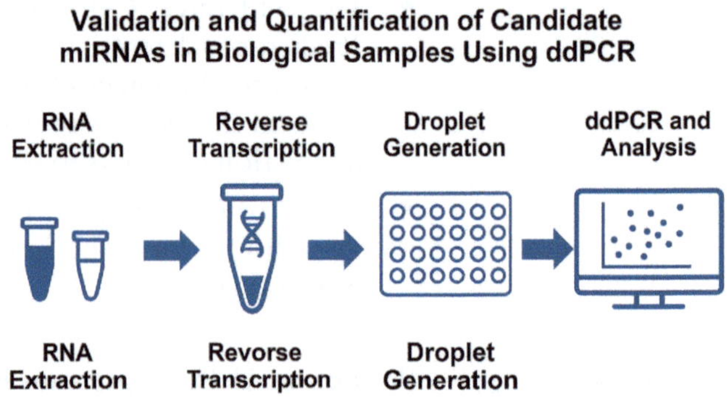

Fig. 3 The validation and quantification of candidate miRNAs using droplet digital PCR (ddPCR). This workflow illustrates the core steps involved in validating predicted miRNA signatures in biological samples. Total RNA is extracted from tissue or biofluid samples, followed by reverse transcription using LNA-based primers. The resulting cDNA is used to generate droplets for ddPCR, which enables absolute quantification of miRNA copy numbers. ddPCR provides high sensitivity and precision for validating differential expression of candidate miRNAs identified through transcriptomic analyses

Output: A curated list of candidate miRNAs that (1) regulate key genes affected by spaceflight or experimental conditions, (2) show significant activation or inhibition predictions, and (3) can be validated experimentally using methods such as ddPCR.

3.3 Validation and Quantification of Candidate miRNAs in Biological Samples Using ddPCR

This protocol describes the process for validating predicted miRNA signatures by quantifying expression levels in tissues or biofluids of interest using droplet digital PCR (ddPCR) (Fig. 3). Digital droplet PCR (ddPCR) offers advantages over real-time PCR (qPCR) including absolute quantification without the need of standard curves or internal references, higher precision, and increased sensitivity, especially for detecting low abundance targets. The goal is to confirm differential expression between experimental and control groups, establishing the biological relevance of candidate miRNAs. The protocol is compatible with serum, plasma, tissue samples, and in vitro tissue models. Validation should be performed on independent sample sets from those used to generate the prediction.

3.3.1 RNA Extraction from Tissues or Biofluids

1. **Sample Preparation**
   - Acquire tissues or biofluids (e.g., serum, plasma) from both experimental and control groups. Maintain samples at −80°C until processing. In the case of plasma samples, avoid the use of heparin as anticoagulant, since it can interfere with the reverse-transcription and amplification steps described below.

2. **RNA Extraction**
   - For serum/plasma: Use the *Qiagen miRNeasy Serum/Plasma Kit* (#217184). Follow the manufacturer's instructions, omitting step 4 (Spike-in Control) if not using an internal standard.
   - For tissues or in vitro models: Use the *Qiagen miRNeasy Mini Kit* (#217004), following standard RNA extraction protocols.
3. **RNA Quantification and Handling**
   - Measure RNA concentration using a NanoDrop spectrophotometer, Qubit RNA HS Assay, or equivalent.
   - Keep all RNA samples on ice during handling. Aliquot and store at −80°C for long-term use.

3.3.2 Reverse Transcription Using LNA miRNA Primers

1. **Prepare cDNA Using the miRCURY LNA RT Kit** (#339340, Qiagen)
   - Order Locked Nucleic Acid (LNA)-enhanced primers for each candidate miRNA of interest.
   - LNA technology is uniquely suited for studying small or highly similar DNA or RNA targets, due to the specific chemical properties of LNAs that result in an increased stability and enhanced annealing specificity. This step can be done in individual PCR tubes or in a 96-well PCR plate.
2. **Reaction Setup (Per Sample):**

Reagent	Volume
5× Reaction buffer	4 μL
Enzyme mix	2 μL
miRNA sample (100 ng)	x μL
Nuclease-free water	to 20 μL
Total volume	20 μL

   - Prepare a master mix of buffer and enzyme for efficiency.
   - For each sample, set up the reaction as shown above with a final miRNA concentration of 5 ng/μL.
   - Dispense RNA and water in tubes or a 96-well plate first, then add 6 μL of the master mix.
   - If you have many samples it is recommended to make a master mix of the 5× Reaction Buffer and Enzyme Mix for all your samples.
     - In each well or tube, add the miRNA samples and Nuclease-free water mix.

- Then, add 6 µL of the master mix to each well or tube.

3. **Thermal Cycling (Reverse Transcription):**

Step	Temperature	Time
RT reaction	42°C	60 min
Enzyme deactivation	95°C	5 min
Hold	4°C	∞

- Store cDNA at 4°C for short-term use or −80°C for long-term storage.

3.3.3 Droplet Generation and ddPCR Setup

1. **Droplet PCR Reaction Setup (Per Well):**

Reagent	Volume
QX200 ddPCR EvaGreen Supermix	12.5 µL
Nuclease-free water	9.9 µL
Primer (LNA miRNA assay)	1.3 µL
Diluted cDNA sample (1:20)	1.3 µL
Total volume	25 µL

- Prepare a master mix of Supermix, water, and primers.
- Add 1.3 µL of each diluted cDNA sample to designated wells in a ddPCR 96-well plate.
- Then, add 23.7 µL of the master mix to each corresponding well.
- For empty wells in a column, add 30 µL of QX200 Buffer Control for EvaGreen.
- It is recommended to make a master mix of the QX200 ddPCR EvaGreen Supermix, Nuclease-free water, and Primer mix.
  - First, add the 1.3 µL cDNA to each well.
  - Then, add the 23.7 µL of the master mix.
- Keep the ddPCR 96-Well PCR Plates on ice while pipetting the above mix to each well.
- For empty wells in a column, add 30 µL of the QX200™ Buffer Control for EvaGreen.

2. **Droplet Generation**
- Keep plates on ice during preparation.
- Load the plate into the BioRad AutoDG Instrument for droplet generation.
- After droplet generation, seal the plate with PCR Plate Heat Seal foil using the PX1™ PCR Plate Sealer with the red horizontal line oriented facing up at the top of the plate.

3.3.4 PCR Amplification of miRNAs in Droplets

Thermal Cycling (PCR):

Step	Temperature	Time	Cycles
Enzyme activation	95°C	5 min	1
Denaturation	95°C	30 s	40
Annealing/extension	58°C	1 min	40
Signal stabilization	4°C	5 min	1
Enzyme deactivation	90°C	5 min	1
Hold	12°C	∞	

- Place the newly sealed plate in the C1000 Touch™ Thermal Cycler with 96–Deep Well Reaction Module with the horizontal red line facing up at the top of the plate.
- **Note:** Annealing temperatures may vary depending on the primer. Use a temperature gradient if amplification is inconsistent. Examples:
  - miR-155-5p: 52°C
  - miR-10b-5p: 54°C

3.3.5 Droplet Reading and Data Analysis

1. **Load Plate into the Droplet Reader**
   - After completion of the above, place the plate in the ddPCR machine.
   - Open QuantaSoft software and configure each well.

Settings:

- **Experiment Type**: ABS (absolute quantification).
- **Supermix**: QX200 EvaGreen Supermix.
- **Target 1**:
  - Name: miRNA primer (e.g., miR-155-5p).
  - Type: Ch1 Unknown.
  - **Target 2**: Leave blank.

2. **Run the Droplet Reading Process**
   - Once complete, export fluorescence and copy number data for downstream analysis (e.g., Excel, R, GraphPad Prism).

Optional: Data Normalization and Statistical Analysis

- Normalize miRNA copy numbers to internal controls. In the case of tissue samples, an endogenous stable miRNA or other small ncRNA like U6 can be used. For biofluids as

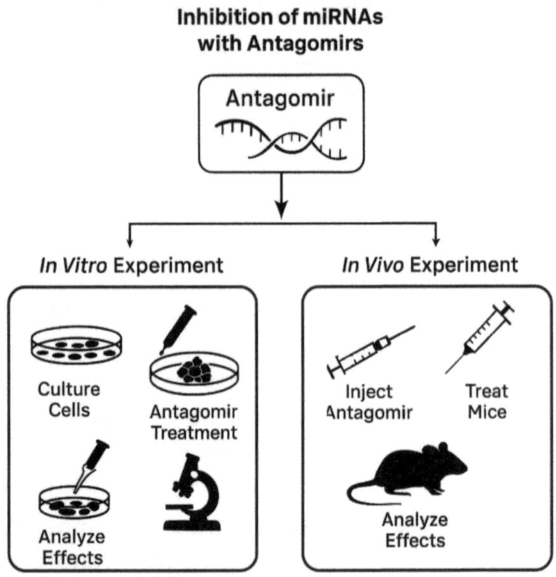

Fig. 4 The inhibition of microRNAs using antagomirs in both in vitro and in vivo experimental models. Antagomirs, chemically stabilized antisense oligonucleotides, are delivered to either cell culture systems (2D or 3D) or live animals (e.g., mice) to inhibit specific miRNAs implicated in disease or experimental stress responses. In vitro, cultured cells are treated with antagomirs and subsequently analyzed for molecular and phenotypic outcomes. In vivo, antagomirs are administered via intraperitoneal injection in animal models, followed by monitoring and endpoint analyses. This dual-platform approach enables robust validation of miRNA function and therapeutic targeting in both cellular and systemic contexts

plasma, a spiked-in RNA or a stably expressed miRNA like miR-16 could be suitable internal normalizers.

- Use R (ggplot2, dplyr, DESeq2) or GraphPad Prism to visualize expression differences and perform statistical comparisons between control and experimental groups.

3.4 Inhibition of miRNAs with Antagomirs In Vitro and In Vivo Experiments

This protocol describes the use of antagomirs, chemically stabilized antisense oligonucleotides, to inhibit specific microRNAs in in vitro cell culture models (2D or 3D) and in vivo mouse models (Fig. 4). The protocol is designed for validating miRNA function as biomarkers or therapeutic targets in disease or experimental conditions, such as spaceflight stressors, radiation, or aging models.

3.4.1 In Vitro Antagomir Treatment (2D or 3D Culture Models)

1. **Cell Culture Preparation**
   - Culture your cells (any type) in either 2D (monolayer) or 3D (spheroids, organoids, or matrix-embedded) format under standard growth conditions.

- Ensure cells are in the exponential growth phase and at an appropriate confluence or developmental stage for antagomir treatment. For 2D monolayers, a 70%–80% confluence is recommended.

2. **Antagomir Dilution**
   - Prepare antagomirs in sterile 1× PBS or nuclease-free water per manufacturer's instructions.
   - Final working concentration in the culture medium should be 0.5 μM per antagomir.
   - If using multiple antagomirs, combine each at 0.5 μM for a multiplexed treatment.
   - Prepare a scrambled or nontargeting control antagomir at the same concentration for comparison.

3. **Antagomir Treatment**
   - Remove existing media from cell cultures.
   - Add fresh complete medium containing the diluted antagomir(s).
   - Incubate cells at 37°C in 5% CO_2 for 24–72 h, depending on downstream assay timing.
   - For long-term cultures (>72 h), repeat the antagomir treatment every 48–72 h with fresh media.

4. **Posttreatment Applications**
   - Proceed with your experimental treatment (e.g., radiation, hypoxia, drug exposure, differentiation induction).
   - Harvest samples at appropriate time points for:
     - RNA or protein extraction
     - Functional assays (e.g., proliferation, migration, apoptosis)
     - Microscopy or immunostaining
   - For DNA damage assays (e.g., 53BP1 staining), fix cells at 90 min postexposure.

3.4.2 In Vivo Antagomir Treatment in Mice

1. **Antagomir Preparation**
   - Select miRNA targets for in vivo inhibition (*See* **Note 1**).
   - Prepare antagomirs at a dose of 10 mg/kg per antagomir.
   - Combine multiple antagomirs into a single dose if needed (e.g., 3 miRNAs = 30 mg/kg total dose).
   - Resuspend in sterile 1× PBS immediately before injection.

2. **Injection Schedule**
   - Administer antagomirs via intraperitoneal (IP) injection every 3 days (*See* **Note 2**).
   - Begin treatment 1 day prior to experimental setup (e.g., hindlimb unloading, irradiation, drug treatment).
   - Continue dosing on a 3-day interval until 1 day before the final experimental endpoint or as per study design.
3. **Monitoring and Controls**
   - Include a scrambled antagomir control group to account for nonspecific effects (*See* **Note 3**).
   - Monitor mice daily for weight, grooming, activity, and signs of toxicity.
   - No toxicity has been observed at the 10 mg/kg dosing schedule in previous studies.
4. **Experimental Readouts**
   - Collect tissues or fluids for analysis at endpoint (e.g., blood, muscle, brain, liver).
   - Perform qPCR, ddPCR, histology, or functional assays to assess antagomir impact.

4 Notes

1. Antagomir stability and efficacy may vary between miRNA targets; confirm knockdown via qPCR or ddPCR.
2. Optimal timing and frequency of antagomir administration should be tailored to each experimental model.
3. Use appropriate controls (vehicle, scrambled) to ensure data reliability.

Acknowledgments

A.B. was supported by research funding from the NASA E.9 Space Biology Research Studies 2023 grant.

References

1. Afshinnekoo E, Scott RT, MacKay MJ, Pariset E, Cekanaviciute E, Barker R, Gilroy S, Hassane D, Smith SM, Zwart SR, Nelman-Gonzalez M, Crucian BE, Ponomarev SA, Orlov OI, Shiba D, Muratani M, Yamamoto M, Richards SE, Vaishampayan PA, Meydan C, Foox J, Myrrhe

J, Istasse E, Singh N, Venkateswaran K, Keune JA, Ray HE, Basner M, Miller J, Vitaterna MH, Taylor DM, Wallace D, Rubins K, Bailey SM, Grabham P, Costes SV, Mason CE, Beheshti A (2020) Fundamental biological features of spaceflight: advancing the field to enable deep-space exploration. Cell 183:1162–1184. https://doi.org/10.1016/j.cell.2020.10.050
2. Mason CE, Green J, Adamopoulos KI, Afshin EE, Baechle JJ, Basner M, Bailey SM, Bielski L, Borg J, Borg J, Broddrick JT, Burke M, Caicedo A, Castañeda V, Chatterjee S, Chin C, Church G, Costes SV, De Vlaminck I, Desai RI, Dhir R, Diaz JE, Etlin SM, Feinstein Z, Furman D, Garcia-Medina JS, Garrett-Bakelman F, Giacomello S, Gupta A, Hassanin A, Houerbi N, Irby I, Javorsky E, Jirak P, Jones CW, Kamal KY, Kangas BD, Karouia F, Kim J, Kim JH, Kleinman A, Lam T, Lawler JM, Lee JA, Limoli CL, Lucaci A, MacKay M, McDonald JT, Melnick AM, Meydan C, Mieczkowski J, Muratani M, Najjar D, Othman MA, Overbey EG, Paar V, Park J, Paul AM, Perdyan A, Proszynski J, Reynolds RJ, Ronca AE, Rubins K, Ryon KA, Sanders LM, Glowe PS, Shevde Y, Schmidt MA, Scott RT, Shirah B, Sienkiewicz K, Sierra M, Siew K, Theriot CA, Tierney BT, Venkateswaran K, Hirschberg JW, Walsh SB, Walter C, Winer DA, Yu M, Zea L, Mateus J, Beheshti A (2024) A second space age spanning omics, platforms, and medicine across orbits. Nature 632:995–1008. https://doi.org/10.1038/s41586-024-07586-8
3. da Silveira WA, Fazelinia H, Rosenthal SB, Laiakis EC, Kim MS, Meydan C, Kidane Y, Rathi KS, Smith SM, Stear B, Ying Y, Zhang Y, Foox J, Zanello S, Crucian B, Wang D, Nugent A, Costa HA, Zwart SR, Schrepfer S, Elworth RAL, Sapoval N, Treangen T, MacKay M, Gokhale NS, Horner SM, Singh LN, Wallace DC, Willey JS, Schisler JC, Meller R, McDonald JT, Fisch KM, Hardiman G, Taylor D, Mason CE, Costes SV, Beheshti A (2020) Comprehensive multi-omics analysis reveals mitochondrial stress as a central biological hub for spaceflight impact. Cell 183:1185–1201 e20. https://doi.org/10.1016/j.cell.2020.11.002
4. McDonald JT, Kim J, Farmerie L, Johnson ML, Trovao NS, Arif S, Siew K, Tsoy S, Bram Y, Park J, Overbey E, Ryon K, Haltom J, Singh U, Enguita FJ, Zaksas V, Guarnieri JW, Topper M, Wallace DC, Meydan C, Baylin S, Meller R, Muratani M, Porterfield DM, Kaufman B, Mori MA, Walsh SB, Sigaudo-Roussel D, Mebarek S, Bottini M, Marquette CA, Wurtele ES, Schwartz RE, Galeano D, Mason CE, Grabham P, Beheshti A (2024) Space radiation damage rescued by inhibition of key spaceflight associated miRNAs. Nat Commun 15:4825. https://doi.org/10.1038/s41467-024-48920-y
5. Malkani S, Chin CR, Cekanaviciute E, Mortreux M, Okinula H, Tarbier M, Schreurs AS, Shirazi-Fard Y, Tahimic CGT, Rodriguez DN, Sexton BS, Butler D, Verma A, Bezdan D, Durmaz C, MacKay M, Melnick A, Meydan C, Li S, Garrett-Bakelman F, Fromm B, Afshinnekoo E, Langhorst BW, Dimalanta ET, Cheng-Campbell M, Blaber E, Schisler JC, Vanderburg C, Friedlander MR, McDonald JT, Costes SV, Rutkove S, Grabham P, Mason CE, Beheshti A (2020) Circulating miRNA spaceflight signature reveals targets for countermeasure development. Cell Rep 33(10):108448. https://doi.org/10.1016/j.celrep.2020.108448
6. Paul AM, Cheng-Campbell M, Blaber EA, Anand S, Bhattacharya S, Zwart SR, Crucian BE, Smith SM, Meller R, Grabham P, Beheshti A (2020) Beyond low-Earth orbit: characterizing immune and microRNA differentials following simulated deep spaceflight conditions in mice. iScience 23:101747. https://doi.org/10.1016/j.isci.2020.101747
7. Wuu Y-R, Hu B, Okunola H, Paul AM, Blaber EA, Cheng-Campbell M, Beheshti A, Grabham P (2020) LET-dependent low dose and synergistic inhibition of human angiogenesis by charged particles: validation of miRNAs that drive inhibition. iScience 23:101771. https://doi.org/10.1016/j.isci.2020.101771
8. Beheshti A, Ray S, Fogle H, Berrios D, Costes SV (2018) A microRNA signature and TGF-beta1 response were identified as the key master regulators for spaceflight response. PLoS One 13:e0199621. https://doi.org/10.1371/journal.pone.0199621
9. Jaquenod De Giusti C, Roman B, Das S (2018) The influence of microRNAs on mitochondrial calcium. Front Physiol 9:1291. https://doi.org/10.3389/fphys.2018.01291
10. Duarte FV, Palmeira CM, Rolo AP (2014) The role of microRNAs in mitochondria: small players acting wide. Genes (Basel) 5:865–886. https://doi.org/10.3390/genes5040865
11. Camera A, Tabetah M, Castañeda V, Kim J, Galsinh AS, Haro-Vinueza A, Salinas I, Seylani A, Arif S, Das S, Mori MA, Carano A, de Oliveira LC, Muratani M, Barker R, Zaksas V, Goel C, Dimokidis E, Taylor DM, Jeong J, Overbey E, Meydan C, Porterfield DM, Díaz JE, Caicedo A, Schisler JC, Laiakis EC, Mason CE, Kim MS, Karouia F, Szewczyk NJ, Beheshti A (2024) Aging and putative frailty biomarkers are altered by spaceflight. Sci Rep 14:13098. https://doi.org/10.1038/s41598-024-57948-5
12. Capri M, Conte M, Ciurca E, Pirazzini C, Garagnani P, Santoro A, Longo F, Salvioli S, Lau P, Moeller R, Jordan J, Illig T, Villanueva M-M, Gruber M, Bürkle A, Franceschi C, Rittweger J (2023) Long-term human

spaceflight and inflammaging: does it promote aging? Ageing Res Rev 87:101909. https://doi.org/10.1016/j.arr.2023.101909
13. Husna N, Aiba T, Fujita S-I, Saito Y, Shiba D, Kudo T, Takahashi S, Furukawa S, Muratani M (2024) Release of CD36-associated cell-free mitochondrial DNA and RNA as a hallmark of space environment response. Nat Commun 15:4814. https://doi.org/10.1038/s41467-023-41995-z
14. Wallace DC (2018) Mitochondrial genetic medicine. Nat Genet 50:1642–1649. https://doi.org/10.1038/s41588-018-0264-z
15. Wallace DC (2013) A mitochondrial bioenergetic etiology of disease. J Clin Invest 123:1405–1412. https://doi.org/10.1172/JCI61398
16. López-Otín C, Blasco MA, Partridge L, Serrano M, Kroemer G (2023) Hallmarks of aging: an expanding universe. Cell 186:243–278. https://doi.org/10.1016/j.cell.2022.11.001
17. Lima T, Li TY, Mottis A, Auwerx J (2022) Pleiotropic effects of mitochondria in aging. Nat Aging 2:199–213. https://doi.org/10.1038/s43587-022-00191-2
18. Vanderburg C, Beheshti A (2020) MicroRNAs (miRNAs), the final frontier: the hidden master regulators impacting biological response in all organisms due to spaceflight. Environ Sci Biol Med 1–39.
19. Corti G, Kim J, Enguita FJ, Guarnieri JW, Grossman LI, Costes SV, Fuentealba M, Scott RT, Magrini A, Sanders LM, Singh K, Sen CK, Juran CM, Paul AM, Furman D, Calleja-Agius J, Mason CE, Galeano D, Bottini M, Beheshti A (2024) To boldly go where no microRNAs have gone before: spaceflight impact on risk for small-for-gestational-age infants. Commun Biol 7:1–22. https://doi.org/10.1038/s42003-024-06944-6
20. McDonald JT, Enguita FJ, Taylor D, Griffin RJ, Priebe W, Emmett MR, Sajadi MM, Harris AD, Clement J, Dybas JM, Aykin-Burns N, Guarnieri JW, Singh LN, Grabham P, Baylin SB, Yousey A, Pearson AN, Corry PM, Saravia-Butler A, Aunins TR, Sharma S, Nagpal P, Meydan C, Foox J, Mozsary C, Cerqueira B, Zaksas V, Singh U, Wurtele ES, Costes SV, Davanzo GG, Galeano D, Paccanaro A, Meinig SL, Hagan RS, Bowman NM, Consortium UC-P, Wolfgang MC, Altinok S, Sapoval N, Treangen TJ, Moraes-Vieira PM, Vanderburg C, Wallace DC, Schisler JC, Mason CE, Chatterjee A, Meller R, Beheshti A (2021) Role of miR-2392 in driving SARS-CoV-2 infection. Cell Rep 37:109839. https://doi.org/10.1016/j.celrep.2021.109839
21. Kang W, Eldfjell Y, Fromm B, Estivill X, Biryukova I, Friedländer MR (2018) miRTrace reveals the organismal origins of microRNA sequencing data. Genome Biol 19:213. https://doi.org/10.1186/s13059-018-1588-9
22. Friedländer MR, Mackowiak SD, Li N, Chen W, Rajewsky N (2012) miRDeep2 accurately identifies known and hundreds of novel microRNA genes in seven animal clades. Nucleic Acids Res 40:37–52. https://doi.org/10.1093/nar/gkr688
23. Langmead B, Trapnell C, Pop M, Salzberg SL (2009) Ultrafast and memory-efficient alignment of short DNA sequences to the human genome. Genome Biol 10:R25. https://doi.org/10.1186/gb-2009-10-3-r25
24. Lorenz R, Bernhart SH, Höner zu Siederdissen C, Tafer H, Flamm C, Stadler PF, Hofacker IL (2011) ViennaRNA package 2.0. Algorithms Mol Biol 6:26. https://doi.org/10.1186/1748-7188-6-26
25. Zhang J, Storey KB (2018) RBiomirGS: an all-in-one miRNA gene set analysis solution featuring target mRNA mapping and expression profile integration. PeerJ 6:e4262. https://doi.org/10.7717/peerj.4262
26. Aparicio-Puerta E, Gómez-Martín C, Giannoukakos S, Medina JM, Scheepbouwer C, García-Moreno A, Carmona-Saez P, Fromm B, Pegtel M, Keller A, Marchal JA, Hackenberg M (2022) sRNAbench and sRNAtoolbox 2022 update: accurate miRNA and sncRNA profiling for model and non-model organisms. Nucleic Acids Res 50:W710–W717. https://doi.org/10.1093/nar/gkac363
27. Fehlmann T, Kern F, Laham O, Backes C, Solomon J, Hirsch P, Volz C, Müller R, Keller A (2021) miRMaster 2.0: multi-species non-coding RNA sequencing analyses at scale. Nucleic Acids Res 49:W397–W408. https://doi.org/10.1093/nar/gkab268
28. Shannon P, Markiel A, Ozier O, Baliga NS, Wang JT, Ramage D, Amin N, Schwikowski B, Ideker T (2003) Cytoscape: a software environment for integrated models of biomolecular interaction networks. Genome Res 13:2498–2504. https://doi.org/10.1101/gr.1239303
29. Brown KR, Otasek D, Ali M, McGuffin MJ, Xie W, Devani B, Toch ILvan, Jurisica I (2009) NAViGaTOR: network analysis, visualization and graphing Toronto. Bioinformatics 25:3327–3329. https://doi.org/10.1093/bioinformatics/btp595
30. Gebre SG, Scott RT, Saravia-Butler AM, Lopez DK, Sanders LM, Costes SV (2025) NASA open science data repository: open science for life in space. Nucleic Acids Res 53:D1697–D1710. https://doi.org/10.1093/nar/gkae1116
31. Bray NL, Pimentel H, Melsted P, Pachter L (2016) Near-optimal probabilistic RNA-seq quantification. Nat Biotechnol 34:525–527. https://doi.org/10.1038/nbt.3519

32. Dobin A, Davis CA, Schlesinger F, Drenkow J, Zaleski C, Jha S, Batut P, Chaisson M, Gingeras TR (2013) STAR: ultrafast universal RNA-seq aligner. Bioinformatics 29:15–21. https://doi.org/10.1093/bioinformatics/bts635
33. Korotkevich G, Sukhov V, Budin N, Shpak B, Artyomov MN, Sergushichev A (2021) Fast gene set enrichment analysis. bioRxiv 060012. https://doi.org/10.1101/060012
34. Kuleshov MV, Jones MR, Rouillard AD, Fernandez NF, Duan Q, Wang Z, Koplev S, Jenkins SL, Jagodnik KM, Lachmann A, McDermott MG, Monteiro CD, Gundersen GW, Ma'ayan A (2016) Enrichr: a comprehensive gene set enrichment analysis web server 2016 update. Nucleic Acids Res 44:W90–97. https://doi.org/10.1093/nar/gkw377
35. Chang L, Xia J (2023) MicroRNA regulatory network analysis using miRNet 2.0. Methods Mol Biol 2594:185–204. https://doi.org/10.1007/978-1-0716-2815-7_14
36. Chang L, Zhou G, Soufan O, Xia J (2020) miRNet 2.0: network-based visual analytics for miRNA functional analysis and systems biology. Nucleic Acids Res 48:W244–W251. https://doi.org/10.1093/nar/gkaa467
37. Ru Y, Kechris KJ, Tabakoff B, Hoffman P, Radcliffe RA, Bowler R, Mahaffey S, Rossi S, Calin GA, Bemis L, Theodorescu D (2014) The multiMiR R package and database: integration of microRNA–target interactions along with their disease and drug associations. Nucleic Acids Res 42:e133. https://doi.org/10.1093/nar/gku631
38. Ru Y, Kechris KJ, Tabakoff B, Hoffman P, Radcliffe RA, Bowler R, Mahaffey S, Rossi S, Calin GA, Bemis L, Theodorescu D (2014) The multiMiR R package and database: integration of microRNA-target interactions along with their disease and drug associations. Nucleic Acids Res 42:e133. https://doi.org/10.1093/nar/gku631
39. Hauschild A-C, Pastrello C, Ekaputeri GKA, Bethune-Waddell D, Abovsky M, Ahmed Z, Kotlyar M, Lu R, Jurisica I (2023) MirDIP 5.2: tissue context annotation and novel microRNA curation. Nucleic Acids Res 51:D217–D225. https://doi.org/10.1093/nar/gkac1070
40. Tastsoglou S, Skoufos G, Miliotis M, Karagkouni D, Koutsoukos I, Karavangeli A, Kardaras FS, Hatzigeorgiou AG (2023) DIANA-miRPath v4.0: expanding target-based miRNA functional analysis in cell-type and tissue contexts. Nucleic Acids Res 51:W154–W159. https://doi.org/10.1093/nar/gkad431
41. Kozomara A, Birgaoanu M, Griffiths-Jones S (2019) miRBase: from microRNA sequences to function. Nucleic Acids Res 47:D155–D162. https://doi.org/10.1093/nar/gky1141
42. Lu T-P, Lee C-Y, Tsai M-H, Chiu Y-C, Hsiao CK, Lai L-C, Chuang EY (2012) miRSystem: an integrated system for characterizing enriched functions and pathways of microRNA targets. PLoS One 7:e42390. https://doi.org/10.1371/journal.pone.0042390
43. McGeary SE, Lin KS, Shi CY, Pham TM, Bisaria N, Kelley GM, Bartel DP (2019) The biochemical basis of microRNA targeting efficacy. Science 366:eaav1741. https://doi.org/10.1126/science.aav1741

Chapter 7

The Cytoskeleton in Space

Angela Catizone, Marco Tafani, Marika Berardini and Giulia Ricci

Abstract

This chapter aims to synthesize the various methods and technologies employed in space-based research missions to study the cellular cytoskeleton alteration in the spaceflight environmental conditions. We describe interdisciplinary approaches combining cell imaging and live-cell imaging, mechanical assays, and molecular analyses.

Keywords Cytoskeleton, Altered gravity, Cell imaging, Live-cell imaging, Mechanical assays, Molecular analyses

1 Introduction

Scientific community has described the five hazards for human health associated with spaceflight that are: distance from Earth, confinement and isolation, hostile and closed environment, altered gravity (microgravity, ipogravity hypergravity), and space radiation [1]. The last two hazards directly impact on cell physiology altering cell cytoskeletal features.

1.1 The Cell Gravity Sensor: The Cytoskeleton

The cytoskeleton is a dynamic network of actin filaments, microtubules, and intermediate filaments that functionally interact with each other and with many other proteins. It plays vital roles in cell physiology, being responsible for chromosomal segregation during mitosis, providing mechanical support to dividing cells, maintaining cell shape, and spatially organizing proteins and organelles within the cell cytoplasm. Additionally, the cytoskeleton is involved in cell motility, membrane trafficking, signal transduction, and cell adhesion. Furthermore, cytoskeletal proteins can transduce and amplify signals captured by membrane receptors, transmitting this information to the nucleus, and ultimately regulating gene expression [2, 3] (Fig. 1).

Mariano Bizzarri (ed.), *Biomedicine and Biotechnology in Space*, Methods in Molecular Biology, vol. 3000,
https://doi.org/10.1007/978-1-0716-5174-2_7,

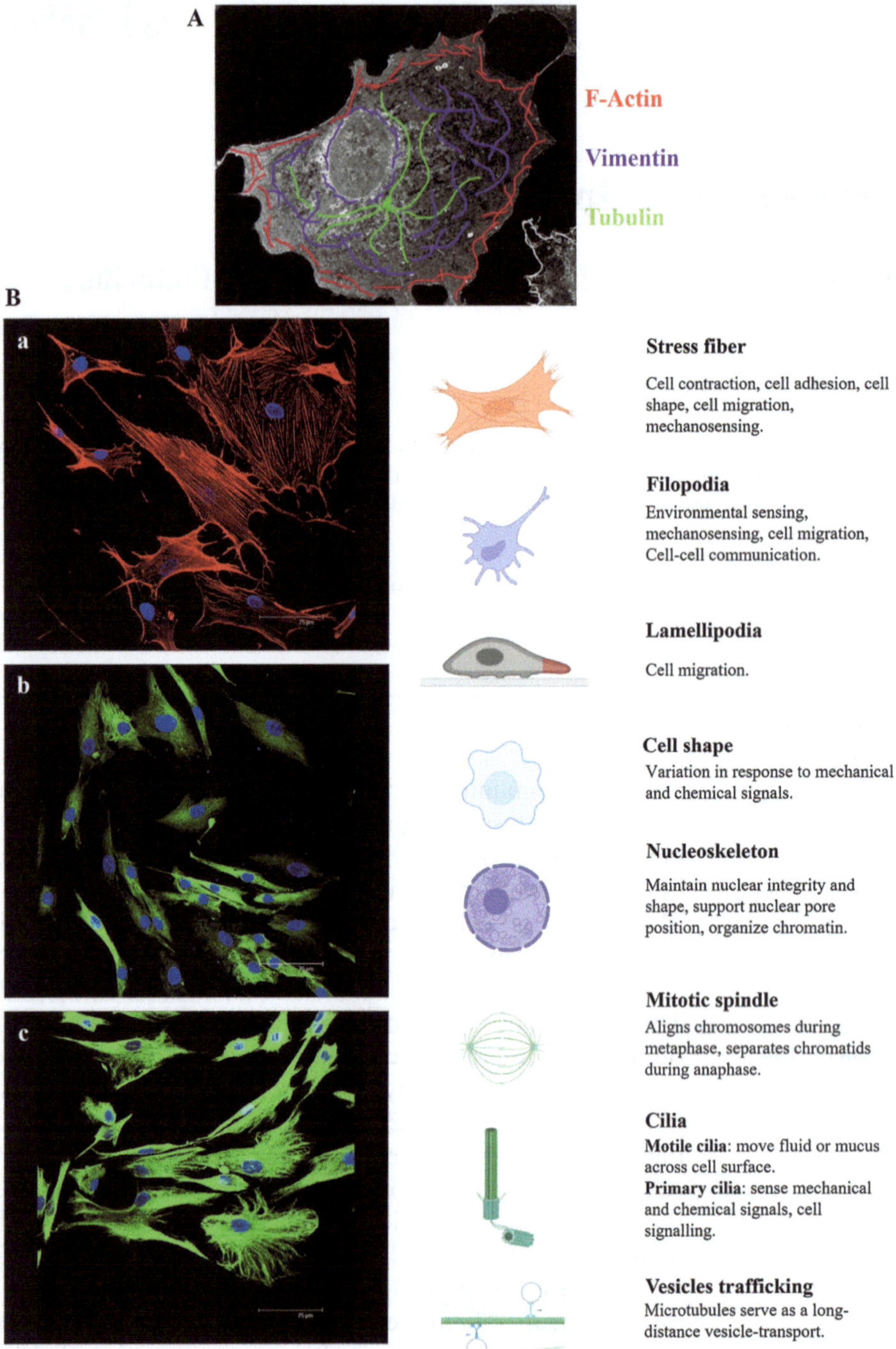

Fig. 1 (**A**) TEM image with schematic illustration of cytoskeletal filaments. (**B**) Representative confocal microscopy images of microfilaments (**a**: F-actin), microtubules (**b**: tubulin), and intermediate filaments (**c**: vimentin) in fibroblasts cultured under microgravity microgravity conditions (RPM) for 48 h (scale bars 75 μ m) nuclear staining was performed using TO-PRO3. The graphic representation on the right of the image was created in BioRender (Negri R (2025) https://BioRender.com/x6dvnih)

It is well known that mutations or abnormalities in cytoskeletal proteins and their related proteins are closely associated with several congenital diseases [4]. Given these observations, it is clear why cytoskeletal disorganization can compromise numerous cellular functions, sometimes leading to cell death.

Human cells can perceive mechanical forces through the cytoskeleton [2, 5, 6] and this discovery has given rise to the branch of cell biology called "mechanobiology." Research has highlighted that these mechanosensitive cytoskeletal structures play a key role in cell gravity sensing, allowing them to adapt to changes in gravitational environments [7, 8]. Altered gravity—whether microgravity or real microgravity (μg), (r-μg), hypergravity, profoundly impacts cytoskeletal organization, triggering cascades of cellular changes with implications for human health, space exploration, and disease mechanisms. This concept has been supported by studies on cells exposed to real or simulated altered gravity.

Early observations of cytoskeletal alterations under microgravity emerged from rocket experiments by Rijken et al., who documented increased filamentous actin density and disrupted stress fibers organization in A431 cells [9]. Subsequent studies revealed cell-type-specific responses: osteoblasts exposed to real microgravity on the STS-56 shuttle flight showed reduced stress fibers formation after 4 days [10], while monocytes exhibited rapid actin cytoskeleton density reduction within 24 h [11]. Endothelial cells exposed to simulated microgravity, demonstrated significant cytoskeletal alterations, including thinning and redistribution of microfilaments, as well as enhanced cell migration and repair mechanisms under hypogravity conditions [12]. Similarly, research on SH-SY5Y neuroblastoma cells revealed that microgravity influences microtubule dynamics, leading to bending and loop formation, while hypergravity promotes actin-driven lamellipodia protrusions [13]. In TCam-2, cell microgravity induces cytoskeleton modification and autophagy [14]. Also, transient EMT in human keratinocytes by early downregulation of e-cadherin was observed [15]. In a 3D coculture model, microgravity modifies the phenotype of fibroblast and promotes remodeling of the fibroblast–keratinocyte interaction [16]. In breast cancer cells, such as MCF-7, exposure to real microgravity induces rapid cytoskeletal changes, including disturbances in F-actin bundles and the formation of filopodia- and lamellipodia-like structures. These alterations are accompanied by changes in focal adhesion molecules and extracellular matrix components, suggesting a common mechanism of gravity perception across different cell types [17]. Furthermore, studies on Jurkat T lymphocytes have shown that gravitational forces rapidly influence gene expression profiles within seconds of microgravity exposure, underscoring the cytoskeleton's role in transducing mechanical signals into cellular responses [18]. Collectively, these findings underscore the cytoskeleton's role as a dynamic gravity sensor, with responses varying across cell types and exposure.

1.2 The Cytoskeleton and Space Radiation

The most studies performed on real spaceflights, revealed the effects of the combination of microgravity and space radiations on cell physiology, showing in the most cases alterations of cell migration, cell adhesion, cytoskeletal stiffness (thanks to variations in F-actin), broader distribution of microtubules and intermediate filaments within the cytoplasm, activation and or increase of cell primary cilia [19, 20]. In these studies, it is impossible to determine what are the specific effects of space radiations on cell physiology with respect to other hazards. For this reason, a systematic and comprehensive investigation on the effects of space radiation is still lacking. This happens even because the simulation of Cosmic radiation is very complex (see below).

The most well-documented radiation-induced cell alteration in the literature is DNA damage [21, 22]. However, an increasing body of evidence has shown that ionizing radiations, which partially simulate cosmic radiation on Earth, impact directly and/or indirectly on cytoskeleton integrity and functions [23]. The emerging picture tells us that ionizing radiations alter cytoskeletal protein expression, their polymerization, and these effects in turn cause alteration of cell stiffness [24–26], as well as cell capability to recruit cytoskeletal adaptors such as RhoA, ROCK1, Rac1 [27], and cofilin [28]. Notably, ionizing radiation triggers very different effects depending on the cell types exposed to them. For instance, it has been reported that actin networks of endothelial cells were impaired after the radiation treatment, whereas melanoma-derived cells exhibited a significant thickening in the marginal actin fibers [25, 29].

It is well known that ionizing radiation can generate reactive oxygen species (ROS) [21, 30] and even ionizing radiation-triggered DNA damage is partially due to the generation of free radicals. The increase of ROS levels in the cells also causes lipid peroxidation, aberrant post-translational modification of proteins, and therefore, it can in turn damage cytoskeletal components leading to cellular dysfunction. Microtubules and actin microfilaments contain certain amino acid residues that are susceptible to oxidation: this condition reduces the ability of cytoskeletal proteins to polymerize [31].

Furthermore, the cytoskeleton is involved in intracellular signaling by recruiting different adaptors. Radiation can interfere with these processes, influencing the expression and distribution pattern of the cytoskeleton linked proteins [32–35].

2 Materials

In this paragraph are summarized, the main platforms that can be used to culture cells and tissue to study the cell cytoskeleton modification in the Space environment.

2.1 Cell Culture in Simulated Space Environment

2.1.1 Ground Altered Gravity Simulation

There are three main microgravity simulators: (1) 1D/2D Clinostat/RWV (Rotating Wall Vessel), (2) 3D Clinostat/RPM (random positioning machine), and (3) Diamagnetic Levitation (for details see [36]). These devices make it possible to simulate altered gravity on 2D and 3D culture cells for variable times. After culture, the cells can be analyzed for cytoskeletal alterations (see methods paragraph). As controls, the same cells must be cultured at the same time in unitary gravity conditions.

2.1.2 Ground Space Radiation Simulation

Cosmic radiation is composed of Galactic Cosmic Rays (GCR) and Solar Particle Events (SPE), and therefore, their simulation on Earth is very hard. Notably, the demand of ground analog for space radiation simulation is high, and therefore, this makes scientists reduce their requirements for high fidelity at least for now [37]. Indeed, the ground cosmic radiation simulation experiments are performed on Earth using particle accelerators (cyclotrons, synchrotrons, and linear accelerators), with several limitations:

(i) the radiation field is different from the real space environment in direction, composition, and rate; (ii) the typical irradiation time in space is extremely long, and this raises experimental problems in the simulation.

It is worth mentioning that ground particle accelerators have the advantage of being of easy and controlled use, and this makes them helpful to study the ionizing radiation biological effects. These accelerators can be used to treat some cancer lesions, and the same machines can be used on 2D or 3D cell cultures [38]. As well as altered gravity simulators, after exposure the cells can be analyzed for cytoskeletal alterations (see methods paragraph), choosing as controls cell cultures, cultured at the same time, in a neutral radiation environment.

2.2 Cell Culture in Real Space Environment

The International Space Station (ISS) is a unique laboratory for scientific research being the only platform where real space environment conditions can be experimented [36]. On board ISS cell cultures are carried out, however, it is worth mentioning that there are several limitations to cell biology experiments in this environment:

(1) the access to the ISS is very limited due to the limited number of space missions and to the costs of launching and returning of the experimental material;

(2) there is prelaunch time (from 4 days to longer than a week), in which the cells prepared for the experiments on the ISS are subjected to a noncontrolled temperature, therefore, the

experiment payload must be designed to overcome this non-physiological condition, but not all the cell lines are able to bear this condition;

(3) even if cell cultures can be performed on board the ISS, most of the methods needed to study cytoskeleton must be carried out on Earth, since the ISS laboratory is not equipped with all the devices needed, and the crew is not expert in cell biology;

(4) the design of control cell culture is very difficult to make;

(5) some aldehyde-based fixatives cannot be used in the ISS laboratory, and this gives some limitations in the morphology and antigen maintenance of cells cultured on board the ISS.

3 Methods to Study the Cytoskeleton Under Altered Gravity Conditions

To study the cytoskeleton under altered gravity conditions (e.g., microgravity or hypergravity), researchers employ interdisciplinary approaches combining cell imaging, and live-cell imaging, mechanical assays, and molecular analyses.

3.1 Cell Imaging: Immunofluorescence and Confocal Analysis

This method can be applied both to samples cultured in vitro under simulated gravity conditions and to samples cultured in the real space environment. The samples are then processed on Earth. Before proceeding with direct or indirect immunofluorescence, the samples must be fixed and permeabilized.

3.1.1 Fixation Methods

While formalin remains the gold standard for tissue fixation due to its excellent preservation of cytoplasm, nuclear contours, and overall tissue architecture, alternatives fixative was designed to preserve tissue samples without the use of formaldehyde, which is toxic and carcinogenic. NoToX offers similar morphological preservation with improved safety profiles. NoToX is a noncarcinogenic, non-noxious, noncorrosive fixative free of heavy metals and chloride ions, suitable for fixing all sizes of specimens from small sections to whole organs. Formalin fixation is rapid and effective but involves health hazards, whereas NoToX provides a nontoxic, noncarcinogenic option that maintains tissue integrity suitable for routine analysis NoToX offers a safer alternative to traditional formalin fixation, while maintaining tissue morphology and antigenicity for downstream analyses, such as histology and immunohistochemistry. It is used for routine tissue fixation, either manually or automatically, and can also serve as a transport medium for samples prior to histological [39, 40]. NoTox could potentially be suitable for space-related biological sample fixation, where safety and ease of handling are critical. However, there is no direct evidence or documented application of NoTox fixation specifically in space or microgravity conditions based on the current search results.

3.1.2 Permeabilization Methods

To highlight cytoskeletal structures, it is necessary to permeabilize the cells. There are several methods of permeabilization:

1. Alcohol-based method
   A widely used single-step method involving cold methanol (−20°C) for 6–10 min effectively fixes and permeabilizes mammalian cells, while preserving structural integrity. This approach works well for cytoplasmic and nuclear targets in HeLa cells and other adherent lines.
2. Detergent-based methods
   Triton X-100 (0.07%–0.5% in PBS) provides gentle permeabilization for membrane proteins, though prolonged exposure (>2 h) can damage membranes.
   Saponin (0.2% in PBS) preserves membrane structures, while allowing antibody access to intracellular epitopes.
   Studies demonstrate that methanol-based protocols yield superior image clarity compared to paraformaldehyde/Triton combinations in some systems [41].

3.1.3 Immunofluorescence Staining

This protocol provides a reliable framework for immunofluorescence staining of cells to visualize cytoskeletal protein localization and expression patterns.

1. Seed cells on coverslips or chamber slides and culture under unitary gravity and altered gravity conditions.
2. Fix with 4% PFA for 15–30 min at 4°C or 100% ice-cold methanol for 5 min at −20°C. Avoid overnight fixation to reduce autofluorescence and sample damage.
3. Permeabilize in PBS supplemented with 1% BSA and 0.1% Triton X-100 for 2 h at room temperature (RT).
4. Incubate with PBS 1% BSA, 0.1% Triton X-100, and 5% Serum from the animal, in which the secondary antibody has been developed for at least 2 h at RT.
5. Wash cells three times with PBS with 1% BSA 0.1% Triton X-100.
6. Dilute primary antibody in antibody dilution buffer (e.g., PBS with 1% BSA 0.1% Triton X-100).
7. Incubate cells with primary antibody for 1–3 h at room temperature or overnight at 4°C.
8. Wash three times in PBS with 1% BSA and 0.1% Triton X-100 for 30 min at RT.
9. Incubate with the fluorochrome-conjugated secondary antibody for 2 h. Include controls such as secondary antibody only (no primary) to check for nonspecific staining.
10. Wash cells three times with PBS with 1% BSA and 0.1% Triton X-100 to remove unbound antibodies.

11. Use FITC/TRITC Phalloidin for actin visualization and TO-PRO-3 iodide fluorescent dye or Hoechst for nuclei staining for 15–20 min at RT in the dark.
12. Wash cells twice with PBS.
13. Remove excess liquid carefully and mount coverslips or chamber slides with antifade mounting medium.
14. Store samples at 4°C protected from light until imaging.
15. Analyze the samples using a confocal microscope. For a careful observation of any changes in the cytoskeletal arrangement, it is important to perform optical spatial series with a step size of 1 μm and perform quantitative analysis using specialized software provided by the microscope manufacturer.

3.2 Live-Cell Imaging

This method applied to Automated Culture Systems, maintains cell viability while monitoring cytoskeletal dynamics in real-time. It revealed rapid cytoskeletal changes (e.g., actin depolymerization, loss of filopodia) during parabolic flights or aboard the International Space Station (ISS) [42]. Lifeact-GFP or mCherry-α-tubulin tags allow visualization using a Fluorescence Microscopy Analysis System (FLUMIAS FEI Munich GmbH) a confocal laser spinning disc microscope [5] optimized for live-cell imaging under microgravity or ground-based simulations. Spinning Disc Technology enhances temporal resolution by rapidly scanning laser light through a rotating disc, minimizing photobleaching and phototoxicity [42, 43]. FLUMIAS presents compatibility with microgravity in fact its compact design is suitable for integration into spaceflight platforms (e.g., ISS) or parabolic flights [42]. Confocal Sectioning reduces background fluorescence, enabling high-contrast imaging of F-actin (via Lifeact-GFP) and microtubules (via mCherry-Tubulin). This method eliminates fixation artifacts and allows real-time observation of dynamic processes (e.g., F-actin, microtubules polymerization, microtubule reorientation, or cytoskeletal remodelling during mechanical stress) (e.g., microgravity) [42, 44, 45].

In this method, cells were stably transfected using a pcDNA3.1(+)-based vector encoding:

- Lifeact-GFP: A 17-amino-acid peptide fused to GFP, enabling live-cell visualization of F-actin dynamics without disrupting actin polymerization or cellular processes.
- mCherry-Tubulin: A fluorescent tag for microtubules, allowing parallel tracking of both actin and tubulin cytoskeletal components [42].

Lifeact-GFP avoids steric clashes with actin-binding proteins (e.g., cofilin, myosin) compared to bulkier actin-GFP chimeras. mCherry-Tubulin provides dual-color imaging to study cytoskeletal crosstalk (e.g., actin-microtubule interactions during cell migration or division) [42]. While Lifeact-GFP is minimally disruptive, studies highlight dose-dependent artefacts: overexpression risks: high Lifeact-GFP levels can stabilize F-actin, alter stress fibers thickness, and induce morphological changes (e.g., increased cell spread area) [46]. Furthermore, binding competition can occur: lifeact occupies a hydrophobic site on F-actin also used by endogenous proteins (e.g., cofilin), potentially skewing cytoskeletal dynamics; it is necessary. Use low-to-moderate Lifeact-GFP intensity to avoid artefacts, while maintaining signal clarity and compare results with actin-GFP or phalloidin staining to confirm specificity [43, 46, 47].

3.3 Mechanical Assay

Mechanical assays are essential tools to investigate how the cytoskeleton responds to changes in mechanical forces, such as those experienced under altered gravity conditions like microgravity or hypergravity. These assays provide quantitative and qualitative data on the mechanical properties and behavior of cells, particularly focusing on the cytoskeleton's structure and function. Mechanical assays measure the stiffness, elasticity, viscoelasticity, and contractile forces of cells. This helps researchers understand how the cytoskeleton adapts or reorganizes in response to altered gravitational forces.

Under microgravity or hypergravity, mechanical assays reveal changes in cytoskeletal stiffness, contractility, and force transmission. For example, microgravity often leads to cytoskeletal softening and reduced cellular tension, while hypergravity can increase stiffness and force generation. These mechanical changes correlate with alterations in cell shape, motility, and signaling pathways.

Mechanical sensors are used to measure forces on the cytoskeleton during space experiments. These sensors are often integrated into complex measurement systems capable of operating in space or simulated environments, where they must ensure precision, reliability, and resistance to extreme conditions such as vacuum, radiation, and temperature variations. Other common techniques include:

1. *Atomic Force Microscopy (AFM):* AFM systems have been developed and operated to study cellular mechanoresponse and cell shape changes under variable gravity conditions, including microgravity and hypergravity. For example, a novel AFM system operating in liquid has been used to measure cell shape changes in vitro under hypergravity by mounting the AFM on a centrifuge, and similar approaches are applicable for microgravity studies to understand cytoskeletal and cellular mechanical responses. Additionally, AFM instruments have been included in space missions, such as the MIDAS (Micro-Imaging Dust Analysis System) AFM onboard the Rosetta orbiter, demonstrating the feasibility of AFM in space environments. AFM uses a cantilever with a fine tip to physically probe the cell surface. By indenting the cell and measuring the force

response, AFM quantifies cell stiffness and elasticity, reflecting cytoskeletal integrity and remodelling [48, 49].

2. *Micropipette Aspiration*: This technique involves applying suction to a small portion of the cell membrane using a micropipette. The degree of membrane deformation under controlled pressure reveals mechanical properties related to cytoskeletal tension and membrane-cytoskeleton interactions [50].
3. *Magnetic or Optical Tweezers:* These tools apply localized forces to magnetic or optically trapped beads attached to the cell surface or cytoskeleton. The resulting bead displacement and force measurements provide insights into cytoskeletal connectivity and mechanical resistance [51].
4. *Traction Force Microscopy (TFM):* Cells are cultured on flexible substrates embedded with fluorescent beads. By tracking bead displacement caused by cellular traction forces, TFM quantifies the forces generated by the cytoskeleton during adhesion and motility [52].

3.4 Molecular Analyses

3.4.1 Western Blot Assay

Western Blot technique can be used to determine cytoskeleton proteins content and alterations in samples subjected to spaceflight environmental conditions. In this context, it is important to compare them to the same samples kept on ground.

For this technique, solutions are prepared in bidistilled water and stored at room temperature or at 4°C.

Lysis buffer composition (store at 4°C): 50 mM Tris-Cl, 250 mM sodium chloride, 5 mM ethylenediaminetetraacetic acid (EDTA), 0.1% Triton® X-100, and 0.1 mM dithiothreitol (DTT) plus 1 mM phenylmethylsulphonyl fluoride (PMSF), protease inhibitor cocktail, 1 mM sodium orthovanadate (NA_3VO_4), and 10 mM sodium fluoride (NaF).

SDS-PAGE: Running gel (Total volume 10 mL/gel): Acrylamide 30%, 4× Tris-HCl 1.5 M pH 8.8, SDS 10%, APS10%, TEMED, distilled water. Dilute reagents accordingly to reach the desired acrylamide percentage.

Stacking gel (Total volume 5 mL/gel): Acrylamide 30%, 4× Tris-HCl 0.5 M pH 6.8, SDS 10%, APS 10%, TEMED, distilled water. Dilute reagents accordingly to reach a 4% acrylamide.

Precast SDS-PAGE: Use 4%–15% Mini-PROTEAN® TGX™ Precast Gel.

Tris-Glycine Ethanol: Dilute 100 mL of 10× Tris-Glycine in 900 mL bidistilled water.

Tris-Glycine Ethanol: Dilute 100 mL of 10× Tris-Glycine in 700 mL bidistilled water. Add 200 mL Ethanol.

PBST (0.1%): Dissolve 5 PBS tablets in 1 L of distilled water. Add 1 mL of TWEEN 20.

Experimental procedure

(1) Collect cells and lyse them by adding an adequate amount of lysis buffer (50 μL/1 × 10^6 cells).

(2) Keep lysates on ice for 30 min, mixing them every 5/10 min.

(3) Centrifuge lysates at 13,000 rpm for 10 min at 4°C. Collect supernatants and transfer them to a new microtube.

(4) Determine protein concentration by Bradford or Bicinchoninic acid assay.

(5) Boil an equivalent amount of proteins for 5 min in 1.5 mL microtubes and then transfer them to room temperature.

(6) Add a loading buffer and load the proteins from each sample onto a denaturing SDS-PAGE gel (precast or homemade). Remember to also load a prestained molecular weight marker.

(7) Run the samples in the gel at 80 V for about 2 h using the Tris-Glycine buffer. It is a good practice to keep the running system in the fridge to avoid excessive temperature increase.

(8) Remove the gel from the casting structure and move it to a transfer system. Apply a 0.45 or 0.22 μm nitrocellulose or PVDF membrane on top of the gel paying attention to remove any air bubbles present.

(9) Transfer the proteins to the membrane at 80 V for about 90 min at room temperature using the Tris-Glycine Ethanol buffer.

(10) Recover the membrane with the transferred proteins and marker on it and incubate for 1 h in 5% milk (fat-free dry milk) to reduce nonspecific binding of the antibody to the membrane.

(11) Incubate the membrane in a solution of 5% milk and diluted primary antibody for 1 h r.t. to o/n at 4°C.

(12) Wash the membrane three times in PBST for 10 min each.

(13) Incubate the membrane in a solution of 5% milk and diluted secondary antibody for 1 h r.t.

(14) Wash the membrane three times in PBST for 10 min each.

(15) Remove excessive PBST from the membrane and add an ECL solution. Adjust the dilution of ECL and the incubation time accordingly.

(16) Image acquisition with a Chemidoc instrument or developing blue films.

3.4.2 Proximity Ligation Assay (PLA)

This technique is applied to fixed (NoToX) cellular samples and combines the signals from immunofluorescence assay with the specificity of oligonucleotides amplification and ligation.

PLA can be used to detect protein–protein interaction or protein posttranslational modifications. This can be achieved by using two different primary antibodies from two different species (e.g., mouse and rabbit). The rest of the procedure requires the Duolink in Situ reagents based on in situ PLA® containing two oligonucleotide labeled secondary antibodies (PLA probes) that will generate a fluorescence signal only if in proximity (nanometers). The fluorescent signals will appear as a series of single fluorescence spots that can be acquired through a confocal microscope and then counted [53].

Experimental procedure:

(1) Cells are fixed and permeabilized as indicated in the immunofluorescence procedure.
(2) Incubate fixed and permeabilized cells with the primary antibody against the first protein for 2 h at room temperature.
(3) Wash in PBS three times.
(4) Incubate with the primary antibody against the second protein for 2 h at room temperature.
(5) Wash in PBS three times.
(6) Add the two secondary antibodies labeled with matching fluorescent oligos (plus and minus) dilute 1:5. Incubate for 1 h at 37°C.
(7) Wash twice for a total of 10 min.
(8) Proceed with the ligation step by diluting the Ligase enzyme 1:40 in the ligation buffer. The ligation buffer is obtained by diluting ligation solution 1:5 in ultrapure water.
(9) Incubate 30 min at 37°C.
(10) Wash in PBS.
(11) Proceed with the amplification step by diluting the Polymerase 1:80 in the amplification buffer. The amplification buffer is obtained by diluting the amplification solution 1:5 in ultrapure water.
(12) Incubate 100 min at 37°C in the dark.
(13) Wash in PBS.
(14) Labeled cells can be counter-stained with DAPI and coverslips mounted on a cover glass.
(15) Observe the cells using a fluorescence or confocal microscope.

If studying posttranslational modifications (phosphorylation, acetylation, etc.) of cytoskeletal proteins, it is necessary to employ one primary antibody binding to a nonmodified epitope and a primary antibody binding to the phosphorylated, acetylated, etc. epitope. Again, the two antibodies must be from different species.

At this point, only if the two chosen proteins are at a distance of a few nanometers, the fluorescent oligos on the secondary antibodies will match and will be amplified revealing this interaction as a bright dot inside the cell. Dots can be counted to distinguish the effect of a treatment on the cells.

3.4.3 Real Time PCR (RT-PCR)

Analysis of the RNA transcripts can be achieved through RNA isolation from samples followed by reverse transcription to obtain the more stable cDNA that can then be amplified through a series of polymerase reactions to obtain a detectable quantity of the RNA of interest.

This technique can be used to detect a single RNA through single reactions or a large number of mRNA using commercially available RNA cards or customized RNA cards designed by the operator to include the genes of interest.

An important point is to maintain sterility by using a sterile hood and avoid touching samples without gloves, since mRNA is rapidly degraded by RNase.

Finally, such technique is usually associated with the Western blot assay to compare mRNA and protein expression of the same gene so to understand if there is a cellular control of a gene at the transcription or translation level.

Also, in this case, the procedure can start with a fresh or frozen cellular pellet.

Experimental procedure

(1) Collect fresh or frozen cells and add an adequate amount of TRIZOL (e.g., 250 μL for 10^6 cells in each microtube). Pay attention to dissolve the cells completely.

(2) Prepare one tube with Ethanol 75% and one tube with Isopropanol. Keep them in ice under the hood.

(3) Add 50 μL Chloroform to the microtubes with TRIZOL, mix and leave for 5 min at RT.

(4) Centrifuge samples for 15 min at 12,000 × *g* at 4°C. At the end of the centrifugation, there will be three separate phases: (i) upper transparent phase containing the RNA, (ii) a central white ring with proteins, (iii) a lower pink phase with phenols.

(5) Carefully collect the upper phase with RNA and transfer to a new microtube.

(6) Add 2–3 μL of glycogen to each microtube to facilitate RNA precipitation.

(7) Add 125 μL of Isopropanol at RT to precipitate RNA.

(8) Keep 10 min at RT.

(9) Transfer the microtubes to ice and then centrifuge at 12,000×*g* for 10 min at 4°C.

(10) Remove and discard the supernatant and wash twice with 500 μL EtOH 75% at RT.

(11) Centrifuge, after each wash, at 7500 rpm for 8 min at 4°C.

(12) Transfer the microtubes under the hood and open them to dry the pellet (RNA) so that it becomes transparent and gelatinous. Do not over-dry the RNA otherwise, it will be very difficult to dissolve it in water. The little amount of EtOH that is still present in the pellet and then in the water will be removed in Step 14.

(13) Resuspend the RNA in 11 μL of water so as to use 1 μL for quantization and 10 μL for amplification.

(14) Turn on the heating block and set the temperature to 55°C. Insert the microtubes in the heating block for 10 min. In this way, the EtOH still present in the sample from Step 12 will evaporate.

(15) Determine the quantity and purity of the RNA (1 μL) using a Nanodrop instrument. The purity is measured as RNA/proteins ratio, which at 260/280 must be as close as possible to 2, and at 260/230 must be as close as possible to 2,2. The Nanodrop instrument will also determine the RNA concentration as ng/μL.

RNA to cDNA retro transcription:

(1) Use 1–1,5 μg of RNA obtained as described in the previous steps.

(2) Employ a retro transcription kit (we use one from INVITROGEN) to obtain cDNA from RNA as described below.

(3) Prepare a 20 μL reaction composed of: 10 μL of 2× first Strand Reaction Mix, 2 μL of Enzyme Mix, 1 μL of Oligo dT, 1 μL of annealing buffer, and 6 μL of RNA and water.

(4) Transfer the tubes with the 20 μL to a thermocycler set with the following program: 50 min to 50°C followed by 5 min to 85°C and then to 4°C.

(5) Determine the cDNA concentration with a Nanodrop or a Spectrophotometer and dilute it in DNase/RNase-free water to a final concentration of 7.5 ng/μL.

Real Time PCR:

Perform the reaction in duplicate or triplicate.

(1) For each tube, if using a single assay, or well, if using a 96-well plate for multiple assays, prepare:

- 8 μL of sample reaction composed of 1 μL annealing buffer, 1 μL Oligo, and 6 μL of cDNA (7,5 ng/μL) in water.
- 12 μL of a mix composed of 1 μL of Forward Primer (4 μM), 1 μL of Reverse Primer (4 μM), 10 μL of SYBR GREEN 2×.

Keep the tube separate and mix them in the well of the plate by adding the 8 μL first followed by the 12 μL of the SYBR mix.

Cover the plate with a protective film and then centrifuge using a rotor that can adapt 96-well plates.

Insert the plate in an RT-PCR machine (such as qPCR Quant Studio 3) and use the preset amplification program.

It is also important to include more than one housekeeping gene, such as 18 S, Actin, and GAPDH, as internal control. These genes must be amplified within the first ten PCR cycles. The rest of the genes must be amplified within 30–35 cycles.

References

1. Afshinnekoo E et al (2020) Fundamental biological features of spaceflight: advancing the field to enable deep-space exploration. Cell 183(5):1162–1184. https://doi.org/10.1016/J.CELL.2020.10.050
2. Moujaber O, Stochaj U (2020) The cytoskeleton as regulator of cell signaling pathways. Trends Biochem Sci 45(2):96–107. https://doi.org/10.1016/j.tibs.2019.11.003
3. Etienne-Manneville S (2025) Cytoplasmic intermediate filaments in cell biology. Annu Rev Cell Dev Biol 34:1–28. https://doi.org/10.1146/annurev-cellbio-100617-062534
4. Li M et al (2023) Roles of the cytoskeleton in human diseases. Mol Biol Rep 50(3):2847–2856. https://doi.org/10.1007/s11033-022-08025-5
5. Corydon TJ et al (2016) Alterations of the cytoskeleton in human cells in space proved by life-cell imaging. Sci Rep 6:20043. https://doi.org/10.1038/srep20043
6. Crawford-Young SJ (2006) Effects of microgravity on cell cytoskeleton and embryogenesis. Int J Dev Biol 50(2–3):183–191. https://doi.org/10.1387/ijdb.052077sc
7. Takahashi K et al (2021) Gravity sensing in plant and animal cells. NPJ Microgravity 7(1):2. https://doi.org/10.1038/s41526-020-00130-8
8. Vorselen D, Roos WH, MacKintosh FC, Wuite GJL, Van Loon JJWA (2014) The role of the cytoskeleton in sensing changes in gravity by nonspecialized cells. FASEB J 28:536–547. https://doi.org/10.1096/fj.13-236356
9. Rijken PJ, de Groot RP, Kruijer W, de Laat SW, Verkleij AJ, Boonstra J (1992) Identification of specific gravity sensitive signal transduction pathways in human A431 carcinoma cells. Adv Space Res 12(1):145–152. https://doi.org/10.1016/0273-1177(92)90277-5
10. Hughes-Fulford M, Lewis ML (1996) Effects of microgravity on osteoblast growth activation. Exp Cell Res 224(1):103–109. https://doi.org/10.1006/EXCR.1996.0116
11. Meloni MA, Galleri G, Pani G, Saba A, Pippia P, Cogoli-Greuter M (2011) Space flight affects motility and cytoskeletal structures in human monocyte cell line J-111. Cytoskeleton 68(2):125–137. https://doi.org/10.1002/cm.20499
12. Vorselen D, Roos WH, MacKintosh FC, Wuite GJL, Van Loon JJWA (2014) The role of the cytoskeleton in sensing changes in gravity by nonspecialized cells. FASEB J 28:536–547. https://doi.org/10.1096/fj.13-236356
13. Rösner H, Wassermann T, Möller W, Hanke W (2006) Effects of altered gravity on the actin and microtubule cytoskeleton of human SH-SY5Y neuroblastoma cells. Protoplasma 229(2–4): 225–234. https://doi.org/10.1007/s00709-006-0202-2
14. Ferranti F et al (2014) Cytoskeleton modifications and autophagy induction in TCam-2 Seminoma cells exposed to simulated microgravity. BioMed Res Int 2014:904396. https://doi.org/10.1155/2014/904396
15. Ricci G et al (2021) Article microgravity induces transient EMT in human keratinocytes by early down-regulation of e-cadherin and cell-adhesion remodeling. Appl Sci (Switz) 11(1):1–19. https://doi.org/10.3390/app11010110

16. Fedeli V et al (2022) Microgravity modifies the phenotype of fibroblast and promotes remodeling of the fibroblast–keratinocyte interaction in a 3D co-culture model. Int J Mol Sci 23(4): 2163. https://doi.org/10.3390/ijms23042163
17. Nassef MZ et al (2019) Real microgravity influences the cytoskeleton and focal adhesions in human breast cancer cells. Int J Mol Sci 20(13):3156. https://doi.org/10.3390/ijms20133156
18. Thiel CS et al (2020) Rapid cellular perception of gravitational forces in human Jurkat T cells and transduction into gene expression regulation. Int J Mol Sci 21(2):514. https://doi.org/10.3390/ijms21020514
19. Lin X et al (2020) The impact of spaceflight and simulated microgravity on cell adhesion. Int J Mol Sci 21(9):3031. https://doi.org/10.3390/IJMS21093031
20. Barravecchia I et al (2022) Microgravity and space radiation inhibit autophagy in human capillary endothelial cells, through either opposite or synergistic effects on specific molecular pathways. Cell Mol Life Sci 79(1):28. https://doi.org/10.1007/S00018-021-04025-Z
21. Ward JF (1988) DNA damage produced by ionizing radiation in mammalian cells: identities, mechanisms of formation, and reparability. Prog Nucleic Acid Res Mol Biol 35(C):95–125. https://doi.org/10.1016/S0079-6603(08)60611-X
22. Mavragani IV, Nikitaki Z, Kalospyros SA, Georgakilas AG (2019) Ionizing radiation and complex DNA damage: from prediction to detection challenges and biological significance. Cancers 11(11):1789. https://doi.org/10.3390/CANCERS11111789
23. La Verde G, Artiola V, Panzetta V, Pugliese M, Netti PA, Fusco S (2021) Cytoskeleton response to ionizing radiation: a brief review on adhesion and migration effects. Biomedicines 9(9):1102. https://doi.org/10.3390/BIOMEDICINES9091102
24. Woloschak GE, Shearin-Jones P, Chang-Liu C-M (1990) Effects of ionizing radiation on expression of genes encoding cytoskeletal elements: kinetics and dose effects. Mol Carcinog 3(6):374–378. https://doi.org/10.1002/MC.2940030609
25. Jasińska-Konior K et al (2019) Increased elasticity of melanoma cells after low-LET proton beam due to actin cytoskeleton rearrangements. Sci Rep 9(1):7008. https://doi.org/10.1038/S41598-019-43453-7.
26. Mohammadkarim A, Tabatabaei M, Parandakh A, Mokhtari-Dizaji M, Tafazzoli-Shadpour M, Khani MM (2018) Radiation therapy affects the mechanical behavior of human umbilical vein endothelial cells. J Mech Behav Biomed Mater 85:188–193. https://doi.org/10.1016/j.jmbbm.2018.06.009
27. Amano M, Nakayama M, Kaibuchi K (2010) Rho-kinase/ROCK: a key regulator of the cytoskeleton and cell polarity. Cytoskeleton 67(9):545–554. https://doi.org/10.1002/CM.20472
28. Huang Q et al (2020) Low-dose X-ray irradiation induces morphological changes and cytoskeleton reorganization in osteoblasts. Exp Ther Med 20(6):1–1. https://doi.org/10.3892/ETM.2020.9413
29. Onoda JM, Kantak SS, Diglio CA (1999) Radiation induced endothelial cell retraction in vitro: correlation with acute pulmonary edema. Pathology and Oncology Research 5(1): 49–55. https://doi.org/10.1053/PAOR.1999.0049
30. Sinclair WK, Fry RJM (1987) Mechanisms of radiation interaction with DNA: potential implications for radiation protection. Radiat Res 112(3):407–417. https://doi.org/10.2307/3577094
31. Wilson C, González-Billault C (2015) Regulation of cytoskeletal dynamics by redox signaling and oxidative stress: implications for neuronal development and trafficking. Front Cell Neurosci 9:381. https://doi.org/10.3389/FNCEL.2015.00381
32. Meineke V et al (2002) Ionizing radiation modulates cell surface integrin expression and adhesion of COLO-320 cells to collagen and fibronectin in vitro. Strahlenther Onkol 178(12):709–714. https://doi.org/10.1007/S00066-002-0993-9
33. Schaller MD (2001) Biochemical signals and biological responses elicited by the focal adhesion kinase. Biochim Biophys Acta, Mol Cell Res 1540(1):1–21. https://doi.org/10.1016/S0167-4889(01)00123-9
34. Zheng Q et al (2015) X-ray radiation promotes the metastatic potential of tongue squamous cell carcinoma cells via modulation of biomechanical and cytoskeletal properties. Hum Exp Toxicol 34(9):894–903. https://doi.org/10.1177/0960327114561664
35. Lee SH, Cheng H, Yuan Y, Wu S (2014) Regulation of ionizing radiation-induced adhesion of breast cancer cells to fibronectin by alpha5beta1 integrin. Radiat Res 181(6):650–658. https://doi.org/10.1667/RR13543.1
36. Ferranti F, Del Bianco M, Pacelli C (2021) Advantages and limitations of current microgravity platforms for space biology research. Appl Sci 11(1):68. https://doi.org/10.3390/APP11010068
37. Walsh L et al (2019) Research plans in Europe for radiation health hazard assessment in exploratory space missions. Life Sci Space Res (Amst) 21:73–82. https://doi.org/10.1016/j.lssr.2019.04.002

38. Miller J, Guetersloh SB, Heilbronn LH, Zeitlin CJ (2005) Simulating elements of the space radiation environment on earth. IEEE Aerosp Conf Proc 2005:698–703. https://doi.org/10.1109/AERO.2005.1559360
39. Meyer R, Niedobitek F, Wenzelides WK (1996) Experiences with the NoToX formalin substitute solution. Pathologe 17:130–132
40. Prento P, Lion H (1997) Commercial formalin substitutes for histopathology. Biotech Histochem 72(5):273–282
41. Piña R et al (2022) Ten approaches that improve immunostaining: a review of the latest advances for the optimization of immunofluorescence. Int J Mol Sci 23(3):1426. https://doi.org/10.3390/IJMS23031426/S1
42. Riedl J et al (2008) Lifeact: a versatile marker to visualize F-actin. Nat Methods 5(7):605–607. https://doi.org/10.1038/nmeth.1220
43. Sliogeryte K, Thorpe SD, Wang Z, Thompson CL, Gavara N, Knight MM (2016) Differential effects of LifeAct-GFP and actin-GFP on cell mechanics assessed using micropipette aspiration. J Biomech 49(2):310–317. https://doi.org/10.1016/j.jbiomech.2015.12.034
44. Thiel CS et al (2020) Rapid cellular perception of gravitational forces in human Jurkat T cells and transduction into gene expression regulation. Int J Mol Sci 21(2):514. https://doi.org/10.3390/ijms21020514
45. Nassef MZ et al (2019) Real microgravity influences the cytoskeleton and focal adhesions in human breast cancer cells. Int J Mol Sci 20(13):3156. https://doi.org/10.3390/ijms20133156
46. Flores LR, Keeling MC, Zhang X, Sliogeryte K, Gavara N (2019) Lifeact-GFP alters F-actin organization, cellular morphology and biophysical behaviour. Sci Rep 9(1):9507. https://doi.org/10.1038/s41598-019-40092-w
47. Belyy A, Merino F, Sitsel O, Raunser S (2020) Structure of the Lifeact–F-actin complex. PLOS Biol 18(11) https://doi.org/10.1371/journal.pbio.3000925
48. Haase K, Pelling AE (2015) Investigating cell mechanics with atomic force microscopy. J R Soc Interface 12(104):20140970. https://doi.org/10.1098/rsif.2014.0970
49. Riedler W et al (2007) MIDAS—the micro-imaging dust analysis system for the Rosetta mission. Space Sci Rev 128(1–4):869–904. https://doi.org/10.1007/S11214-006-9040-Y/METRICS
50. Guevorkian K, Maître JL (2017) Micropipette aspiration: a unique tool for exploring cell and tissue mechanics in vivo. Methods Cell Biol 139:187–201. https://doi.org/10.1016/bs.mcb.2016.11.012
51. Sumbul F, Rico F (2019) Single-molecule force spectroscopy: experiments, analysis, and simulations. Methods Mol Biol 1886:163–189. https://doi.org/10.1007/978-1-4939-8894-5_9
52. Polacheck WJ, Chen CS (2016) Measuring cell-generated forces: a guide to the available tools. Nat Methods 13(5):415–423. https://doi.org/10.1038/NMETH.3834
53. Alam MS (2022) Proximity Ligation Assay (PLA). Methods Mol Biol 2422:191–201. https://doi.org/10.1007/978-1-0716-1948-3_13

Check for updates

Chapter 8

Epigenetics in Space: The Dynamic Cellular Response

Andrea Fuso

Abstract

Space exploration is moving toward the next milestone: space colonization. The risks for human health, associated to the permanence in the space, are already object of studies by several experiments both in orbit and, under simulated space conditions, on hearth. Now, these assessments will need to be extended over the increased permanence, in view of the stable life on other planets. It is well known that the environmental conditions present in the space, primarily related to the altered gravity, introduce a physical stress and the lack of a fundamental bond that contributes to the spatial organization of the cells. This status further induces possible alteration in cell metabolism and in gene expression. Unexpectedly, despite their fundamental involvement in the regulation of gene expression and their responsiveness to stresses, epigenetics modifications have been poorly studied in space and in altered gravity environment. Epigenetic modifications, such as DNA methylation, histones modifications, chromatin remodeling, and non-coding RNAs, can exert the role of mediator of the specific environmental stimuli encountered in space, resulting in modulations of the epigenome, possibly leading to altered cell response. Increasing evidence point out that this three-way interaction—environment–epigenome–cell—is associated to cell and tissue dyshomeostasis and possibly to disease onset. For this reason, increasing attention to the epigenetic modifications induced by the spatial environment and more comprehensive epigenetic studies are warranted. Unfortunately, the study of the epigenetic modifications in space presents some technical difficulty mainly due to the lability of some of these marks, particularly the most studied, i.e., the DNA methylation. Profiling the epigenetic modification often requires long and complex protocols that implies the use of specific instruments. These studies can be, on the other hand, performed on biological specimens collected during the space missions and returned to hearth. For the most labile epigenetic marks, using simulated microgravity on ground may represent a valid alternative approach. This chapter is therefore aimed at introducing the basics of the epigenetic modifications, the already known epigenetic changes associated to the space environment and the techniques used to study these marks.

Keywords Epigenetics, DNA methylation, Histone modifications, Chromatin remodeling, MicroRNAs

1 Introduction

Epigenetics, as a field of biological research, at the crossroad between biology, biochemistry, and molecular biology, was born in 1942 after the definition of Conrad Waddington, an embryologist studying the mechanisms of cell differentiation. He defined "epigenetics" as

Mariano Bizzarri (ed.), *Biomedicine and Biotechnology in Space*, Methods in Molecular Biology, vol. 3000, https://doi.org/10.1007/978-1-0716-5174-2_8,

"the branch of biology that studies the causal interactions between the genes and their products, which give rise to the phenotype." At that time, without knowing either the biochemical nature of genes nor their role in the flux of genetic information, Waddington described a conceptual model to figure out that the different interactions between the genes and their environment could give rise to different phenotypes, although on the basis of the identical genetic material. This model was further depicted in the metaphor of the "epigenetic landscape" used to explain the processes leading to cell differentiation. In this metaphoric view, the fate of the different cells of an organism across the developmental process is determined in a similar way to a marble rolling down a slope of a mountain: the increasing irreversibility of cellular phenotypes acquired with differentiation, due to the effects exerted by the surrounding cellular environment, is compared to the effects of ridges and valleys the marble encounters along the slope, which can direct different stones in different directions as they roll [1].

In more recent times, epigenetics has been formally defined as "the study of the temporal and spatial control of genetic activity during the development of complex organisms" [2]. According to this definition, however, the term "epigenetic" could be virtually referred to any cellular process or activity retaining the capability to drive differentiation, physiology, and fate of any cell or of the whole organism.

The current use of the term "epigenetics" is actually more restricted and refers to "epigenetic modifications" of a given domain of the DNA sequence, capable of influencing the reading of genes, characterized by heritability (over cycles of cell divisions and, possibly, transgenerationally) and that do not involve changes in the underlying DNA sequence. On this basis, epigenetics has recently been defined as "the study of any potentially stable and, ideally, heritable change in gene expression or cellular phenotype that occurs without changes in DNA base pairing" [3].

Epigenetic modifications, as we refers to them today, include methylation of cytosine residues in DNA and posttranslational modifications of histone proteins. Specific combinations of epigenetic modifications also determine the accessibility of chromatin, thus having the potential to regulate the transcriptional potential of associated genes. Chromatin remodeling is, therefore, usually indicated as a further epigenetic modification per se, since involving many other nuclear proteins involved in the modulation of chromatin accessibility [4].

Finally, non-coding RNAs (ncRNAs) are usually included among the epigenetic factors although some lack of consensus exists on that assignment, mainly due to the lack, so far, of evidence about their heritability [5]. This class of molecules includes long (lnc-RNAs [6]) circular (cRNAs [7]) and particularly micro (miRNAs, or miR [8])

RNAs that can interact with the messenger RNAs (mRNAs) reducing their translation rate or inducing their degradation, thus resulting in decreased gene expression [9].

In simple terms, the genome of our cells is identical, whether we consider a muscle cell or a neuron. What makes them distinguish and develop different functions is, therefore, not the sequence of genes, but "which" of these genes are "expressed" (and, therefore, translated into proteins) and "how much" they are expressed. Epigenetic mechanisms do just this: they make genes silent or active and regulate their activity levels.

This effect, as we have said, occurs normally and physiologically during the development of an organism, a period during which different tissues originate from identical cells. But the same can occur after development, when epigenetic mechanisms continue to regulate different cellular functions. This also occurs in response to environmental factors (nutrients, pollutants, exercise, physical and psychological stress, etc.), with epigenetic factors acting as intracellular "mediators" of environmental stimuli [10, 11].

What happens if epigenetic factors become deregulated or abnormal? In this case, epigenetics moves from having a physiological role to a potentially pathological role. As a matter of fact, there are many known pathologies for which the epigenetic basis, or at least an involvement of epigenetic factors, in their etiogenesis has been ascertained [12]. The reasons for the pathological outcomes may be due to mutations that alter the genes responsible for the production of enzymes and proteins involved in epigenetic mechanisms (e.g., in Fragile X Syndrome or Rett Disease [13–15]) or to epigenetic-dependent regulation of genes. In other words, the regulatory regions of some genes could undergo changes in their epigenetic profile and thus be expressed more or less, respect to the normal level, triggering potentially pathological processes. Many of the pathologies that show an epigenetic contribution of this type are chronic-degenerative pathologies, particularly those related to aging, including many pathologies of the nervous system [16–19]. In general, the decrease of the DNA methylation potential is observed with aging which, aggravated by environmental factors, could exacerbate a condition prodrome of neurodegeneration.

One interesting thing about epigenetic modifications is that, being reversible and inducible, we can imagine to define interventions through drugs or supplements exerting epigenetic action, aimed at restoring a "healthy" epigenetic profile, which can assist in the prevention, if not even the therapy of pathologies with an epigenetic basis.

There are many molecular actors listed among epigenetic regulators. Enzymes of this class are classified as "writers," which produce modifications to histones and DNA, "erasers," which remove modifications and "readers," which recognize and interpret specific epigenetic modifications [20]. The main epigenetic modifications are shortly described in the next sections.

2 Epigenetic Mechanisms

2.1 DNA Methylation

DNA methylation is an extremely conserved process, in fact, it is shared by both eukaryotic and prokaryotic organisms.

This modification is carried out by DNA methyltransferase enzymes (DNMTs) and consists in the substitution of hydrogen with a methyl group ($–CH_3$) linked to carbon in position 5 of cytosine; the functions associated with methylation consist in the modulation of gene transcription during embryonic development [21], genomic imprinting, cellular reprogramming, stability and differentiation, X inactivation in mammals, RNA splicing, and DNA repair [22].

The methylation state of DNA varies according to the development processes of the organism, aging, and according to the interaction with the environment. The cellular methylation pattern, at a given time, is the result of three processes: maintenance methylation, de novo methylation, and demethylation.

Maintenance methylation occurs in conjunction with the DNA replication process and allows the heritability of the modification on the CpG dinucleotide. In fact, following replication, DNA is in a state defined as "hemimethylated" since only the template strand presents methylation, which is instead lacking in the newly synthesized strand. To ensure the methylation of both strands, the enzyme DNA methyltransferase 1 (DNMT1) is recruited. DNMT1, by means of the adaptor protein Ubiquitin-like containing PHD and RING finger domains 1 (UHRF1), specifically binds hemimethylated CpG dinucleotides, located at the replication fork, and H3 histones whose tails have di- or tri-methylation on lysine 9 (H3K9me2 and H3K9me3, respectively). DNMT1, recruited to the DNA by UHRF1, at this point, exerts its methylation activity by copying the methylation pattern of the parental strand also on the newly synthesized one [23].

De novo methylation, as the name suggests, creates new methylation patterns, thanks to the action of the enzymes DNMT3A and DNMT3B. These operate in proximity to sites where lysine 4 of histone H3 (H3K4) is not methylated. In fact, DNMT3 has a specific affinity with this conformation of the histone tail, so much so that its methylation inhibits its binding. The main "de novo" methylation events occur during early embryonic development and during the formation of germinal cells, characterized by a specific methylation pattern [24].

DNA methylation is a reversible process: DNA can be demethylated, therefore, deprived of methyl groups. Demethylation can be divided into passive and active processes.

Passive demethylation derives from the lacking of DNMT1 methylating activity or of methyl-donor molecules. In these conditions, the newly synthesized DNA strand is not added with methyl groups and consequently, after repeated replications, DNA becomes demethylated on both strands. Active demethylation consists in the removal and replacement of the methylated cytosine by DNA repair systems. Different putative demethylating processes have been investigated, all involving, as a final step, the BER (Base Excision-Repair) system: a DNA repair system that leads to the removal of the residue by the TDG (Thymidine DNA Glycosidase) enzyme so as to replace it with a cytosine [25]. Today, it is believed that the main demethylation pathway is the one that begins with hydroxymethylation of methylated cytosines by TETs enzymes. Although the idea that hydroxymethylation may represent an epigenetic modification in itself, capable of modulating gene expression per se, is still under discussion, the speed of the reactions downstream of hydroxymethylation resulting in demethylation, suggest that hydroxymethylation is the first step of active demethylation [26].

The change in DNA methylation status occurs during both physiological events, such as embryonic development, and pathological events. Since methylated cytosine represents a binding site for many proteins associated with chromatin modifications, DNA methylation is involved in the process of inhibition of gene expression, which can act by inhibiting transcription factors that bind DNA or by influencing the structure of chromatin itself. The way in which DNA methylation influences the change in gene expression depends on the position occupied by the methylated residue. In general, methylation in the promoter region leads to a silencing of the relative gene present downstream. Methylation of cytosines in the body of the gene leads to an increase in its expression. The repression of intragenic cryptic promoters, i.e., of sequences that, by simulating the structure of the promoter, causes the recruitment of the transcriptional machinery and in this way interfere with the transcription of the full-length transcript. Finally, DNA methylation modulates gene expression also through the facilitation of the transcriptional elongation process and of cotranscriptional splicing [27].

Some transcriptional factors show different affinity for DNA based on its methylation state, which implies both that transcriptional factors capable of repressing transcription can associate with methylated DNA, and that activating transcriptional factors associate with DNA when it is not methylated.

2.2 Histone Modifications

Histones are small basic nuclear proteins capable of binding negatively charged DNA with high affinity and condensing nucleic acid. They can be classified into two classes: linker histones, which include H1, and core histones, which include H2A, H2B, H3, and H4. Core histones form the protein octamer around which DNA wraps to form the nucleosome, the fundamental unit of chromatin.

Chromatin is represented by the complex of DNA and proteins in the chromosome. The basic structural unit of chromatin is called the nucleosome. Each nucleosome contains 147 base pairs of DNA that wrap 1.7 times around a histone octamer, called the histone core, composed of two heterodimers of histones H2A and H2B and a tetramer of histones H3 and H4 [28, 29]. Each core histone contains two different functional domains: a highly conserved, C-terminal, globular motif, called the histone fold, involved in the formation of the histone core, and an N-terminal end (tail), unique to each histone species, which protrudes from the nucleosome in a loose form.

Histone posttranslational modifications (PTMs) mainly involve the different amino acids of the N-terminal tails and include: acetylation, methylation, phosphorylation, poly(ADP-ribosylation), ubiquitination, and sumolylation [30–32]. Each of them and their combinations are induced by particular stimuli and have a different effect on gene expression, either directly, by altering chromatin compaction, or indirectly, by recruiting effector proteins [33]. Several enzymes participate in epigenetic regulation: acetyltransferases (HATs), deacetylases (HDACs), methyltransferases (HMTs), demethylases (HDMs), protein kinases (PKs), protein phosphatases (PPs), and enzymes that provide ubiquitination and sumolylation. The enzymes responsible for epigenetic regulation can act independently or in synergy, defining the so-called "histone code" [34–36].

In general, acetylation and phosphorylation have been associated with transcriptional activation because they destabilize the electrostatic attraction between DNA and histones in the nucleosome. Methylation has been found in both transcriptionally active and repressed regions, depending on the amino acid involved and the number of methyl groups, which can vary from one to three [35, 36]. Acetylation and deacetylation on ε-NH2 of lysine residues are mediated by histone acetyltransferases (HATs) and histone deacetylases (HDACs), respectively, which can modify not only histones but also non-histone proteins [37]. Acetylation on histone H3 usually occurs on lysines K9, K14, K18, and K23, while for histone H4 it occurs on K5, K8, K12, and K16.

Methylation can involve arginine (R) and lysine (K) residues. Arginine can undergo symmetric or asymmetric monomethylation. Lysine residues can not only be mono-, di-, trimethylated but can also be subject to acetylation. These last two types of post-translational modifications (–CH3 and –Ac) in cis are mutually exclusive and, for example, considering lysine 9 of histone H3 (H3K9), produce opposite effects. Indeed, H3K9 methylation has been associated with transcriptional repression, whereas its acetylation has been associated with transcriptional activation [36].

2.3 Chromatin Remodeling

DNA methylation is, therefore, able to drive histone deacethylation, thus resulting in a stronger connection between DNA and nucleosomes and in more dense nucleosomes packing. In other terms, DNA methylation and histone modifications interact to induce chromatin condensation, which is associated to a more stable repression of gene expression. This process involves the ability of methylation to induce chromatin condensation in two ways. The first way involves the DNMTs themselves that deliver chromatin remodeling factors to the target site with which they are complexed. For example, the chromatin remodeling proteins Lymphocyte-Specific Helicase (LSH), H3K9 methyltransferase, and histone deacetylase. The second mechanism involves the binding of methyl-CpG-binding domain (MDB) proteins to methylated DNA that leads to the recruitment of histone deacetylation and nucleosome remodeling complexes [38]. To enable dynamic access to packaged DNA and to tailor nucleosome composition in chromosomal regions, cells have evolved a set of specialized chromatin remodeling complexes (remodelers). Remodelers use the energy of ATP hydrolysis to move, destabilize, eject, or restructure nucleosomes [39]. Among these, the most know is the SWI/SNF complex, a 2 MDa multi-subunit DNA-dependent ATPase that contributes to the regulation of gene transcription by altering chromatin structure. The SWI/SNF complex is targeted to promoters via direct interactions with transcription activators, alters nucleosome structure and contributes to the remodeling of chromatin [40].

2.4 microRNAs

The ncRNAs modulate gene expression by altering the stability of target mRNAs or their accessibility by the ribosomal proteins, i.e., exerting their repression at transcriptional or at translational level.

miRNAs are the most abundant and the most studied ncRNAs and have been recently associated to a pletora of cellular processes and to several pathological pathways.

miRNAs are small non-coding RNA molecules that direct post-transcriptional repression of mRNA targets through the RNA-induced silencing complex (RISC). Characterized by a length of approximately 22–25 nucleotides, miRNAs count at least one conserved binding site hosted by over 60% of human protein-coding genes [41]. miRNAs can target specific mRNA sequences, thus fine-tuning the expression of genes involved in various cellular processes.

Despite the debate on their heritability [5], these molecules can be associated to the epigenetic mechanisms at least for their cooperation with other epigenetic modifications like DNA methylation. miRNAs can target components of the DNA methylation or histone modification machinery, creating a complex network of interactions that collectively control gene expression in cells. The intricate crosstalk between DNA methylation, histone modifications, and microRNAs is crucial to clarify how alterations in these mechanisms contribute to various diseases, including cancers and neurodegenerative disorders. Their coordinated actions orchestrate the dynamic regulation of gene expression essential for proper cellular function and development.

Genes encoding for microRNAs are typically transcribed by RNA polymerase II (Pol II), generating an initial longer primary transcript characterized by a local hairpin structure within which miRNA sequences are embedded [42]. While specific classification rules are not universally standardized, it is commonly anticipated that the precise sequence encompassing nucleotides 2–8, known as the "seed sequence," of the mature miRNA belongs to the same "miRNA family." Each genetic locus produces two mature miRNAs: one derived from the 5' strand (plus) and the other from the 3' strand (minus) of the precursor molecule [43].

In humans, most of canonical miRNAs are encoded within introns of non-coding or coding transcripts. However, some miRNAs originate from exonic regions [44]. miRNAs within the same cluster are generally cotranscribed, even though individual miRNAs can function independently at a post-transcriptional level. Due to variations in their transcription start sites, promoters of intronic miRNAs often rely on the promoters of their host genes [45].

The biogenesis of miRNA can proceed via either a canonical or a non-canonical pathway [46]. In the canonical pathway, the primary miRNA (pri-miRNA) undergoes several maturation steps after its transcription. Pri-miRNA, typically spanning over 1 kb, is characterized by a local stem–loop structure that harbors mature miRNA sequences. A standard pri-miRNA includes a stem of 33–35 base pairs, a terminal loop, and single-stranded RNA segments on both the 5' and 3' ends. Remaining within the cell nucleus, pri-miRNAs are processed into pre-miRNAs by the microprocessor complex, comprising a ribonuclease III enzyme known as Drosha [47].

After export to the cytoplasm, pre-miRNA undergoes processing by Dicer in the vicinity of the terminal loop, resulting in the release of a small RNA duplex. Dicer comprises C-terminal tandem RNA III domains, functioning akin to those found in Drosha, and an N-terminal helicase domain responsible for facilitating the interaction of pre-miRNA with the terminal loop, aiding in its recognition [48]. The resultant miRNA duplex is loaded into the Argonaute (AGO) protein with the assistance of ATP-dependent chaperone proteins [49].

In addition to the canonical miRNA biogenesis pathway, alternative processes exist for producing miRNAs or analogous small molecules. Generally, non-canonical biogenesis pathways can be categorized into Drosha/DGCR8-independent and Dicer-independent pathways.

The prevailing findings from numerous studies regarding the mechanism of indicate that miRNAs predominantly bind to specific sequences located in the 3' untranslated region (UTR) of target mRNAs. This interaction typically leads to translational repression and degradation of the mRNA through deadenylation and decapping processes [50]. The fundamental unit responsible for miRNA-mediated silencing, known as the miRNA-induced silencing complex (miRISC), comprises the guide strand and AGO protein. Target specificity of miRISC arises from its recognition and binding to complementary sequences on the target mRNA, referred to as miRNA response elements (MREs).

3 Epigenetic Studies in Microgravity and Simulated Microgravity

Physical and behavioral stress are included among the environmental factors that are able to modulate the epigenetic markers and therefore potentially altering the physiology of the organisms and eventually leading to the onset of epigenetic-associated pathologies. Despite the cosmic radiation, microgravity and space magnetic fields and, in general, the conditions associated to the space travel and the permanence in the space environment represent significant and challenging sources of stress, the studies aimed at ascertaining the epigenetic variations in these conditions have been very limited so far. A simple PubMed search reveals, in fact, that for about 14,000 publications dealing with microgravity, only 70 investigate epigenetic modifications.

3.1 Space Epigenetics on Plants

Plants represented one of the easiest organism models to be studied in the space. Plants also retain many of the epigenetic signatures present in mammals although not completely identical, particularly for the DNA methylation that also occurs on other bases [51].

One of the first studies evidenced extensive alteration in both DNA methylation and gene expression in rice plants subjected to a spaceflight, at transposable elements and cellular genes. In this specific case, all detected alterations were hypermethylation at CpG and non-CpG (specifically, CNG) residues. Furthermore, a set of genes encoding for the various putative DNA methyltransferases, 5-methylcytosine DNA glycosylases, the SWI/SNF chromatin remodeller (DDM1), and siRNA-related proteins resulted extremely sensitive to perturbation by spaceflight, suggesting the cause for the altered methylation patterns observed [52].

Arabidopsis thaliana is one of the mainly used plant models for epigenetic studies and it has been, in fact, used in conditions of real and simulated microgravity. A first study on *Arabidopsis* cells cultured in simulated microgravity reported alterations in the epigenetic modifications of chromatin, potentially associated to disturbance of cell proliferation and growth [53]. Epigenetic changes, in terms of reduced methylation at CpG and non-CpG residues, were also observed at genomic level in *Arabidopsis* seedlings in microgravity conditions onboard a satellite. These changes were associated to DNA methylation-associated genes, hormone signaling related genes, cell-wall modification genes and transposable elements, evidencing mechanisms of plant adaptation to microgravity [54]. In a second experiment, the same group evidenced that some phenotypic differences still existed in the F1 and F2 generations, indicating that these epigenetic DNA methylation modifications were partially retained, resulting in phenotypic differences in the offspring. Furthermore, some of the spaceflight-induced heritable differentially methylated regions (DMRs) were retained. Changes in epigenetic modifications caused by spaceflight affected the growth of two future seed generations [55].

When plants are grown in the novel environment of spaceflight aboard the International Space Station (ISS), an environment completely outside their evolutionary history, they respond with unique alterations to their gene expression profile. Whole genome bisulfite sequencing of DNA of *Arabidopsis* grown on the ISS from seed, revealed organ-specific patterns of differential methylation compared to ground controls. Interestingly, the overall CpG and non-CpG methylation profiles were similar between flight and ground DNA, however, thousands of specifically differentially methylated cytosines were discovered, and there were clear organ-specific differences in methylation patterns. Based on these results, it was possible to argue that gene expression associated with physiological adaptation to spaceflight is regulated in part by methylation strategies similar to those engaged with terrestrial stress responses [56].

More recently, the "Advanced Plant Experiment-04—Epigenetic Expression experiment" examined the role of cytosine methylation in spaceflight adaptation onboard the International Space Station, and evaluated the spaceflight-altered, genome-wide methylation profiles of two methylation-regulating gene mutants [methyltransferase 1 (met1-7) and elongator complex subunit 2 (elp2-5)] along with a wild-type Col-0 control. It was demonstrated that plants suffered in their physiological adaptation to spaceflight in that their roots failed to extend away from the seed and the overall development of the plants was greatly impaired in space. *Arabidopsis* physiological adaptation to spaceflight was associated with differential DNA methylation in an organ-specific manner. Disruption of Met1 methyltransferase function does not dramatically affect spaceflight growth or morphology, yet met1-7 reprograms the spaceflight transcriptomic response in a unique manner [57].

3.2 Space Epigenetics on Cell Cultures

The majority of the biomolecular experiments in conditions of simulated microgravity or on-flight on the ISS or satellites have been performed on cultured mammal or human cells, either as stabilized cell lines or animal and human primary cells. Few of these experiments, as stated above, have been directed at investigating how the spaceflight conditions can impact on the epigenetic traits, either at local (sequence specific) or global (genome) level.

Circulating blood cells (immune cells) may represent a relevant target for the assessment of the epigenetic changes since these cells are actively produced and renewed in the adult organisms and, therefore, the bone marrow can be more sensitive. Moreover, altered immune function has been demonstrated in astronauts during spaceflights related to the Apollo and Skylab programs [58]. One of the first studies were denoted to investigate whether simulated microgravity alters DNA methylation and histone acetylation, as well as the expression of the genes regulating these processes, namely DNMT1, DNMT3a, DNMT3b, and HDAC1. Using a rotary cell culture system (RCCS) simulating microgravity, microgravity-induced DNA methylation changes were detected in human T-cells were analyzed by methylation sensitive-random amplified polymorphic DNA (MS-RAPD) analysis of genomic DNA. The analysis indicated that simulated microgravity exposure results in DNA hypomethylation [59]. A similar conclusion was reached with a further analysis, also conducted on ground in conditions of simulated microgravity. In this case, the whole genome DNA methylation and hydroxymethylation were studied by next generation sequencing (NGS) in human lymphoblastoid cells [60]. Also, in this case, in fact, the authors observed hypomethylation and hyperhydroxymethylation at DMRs. These two results are concordant since, as above discussed, hydroxymethylation is considered mainly as the first step of active DNA demethylation rather than a distinct epigenetic factor per se [26]. Histone modifications and miRNAs expression have been also studied in relation to altered gravity in immune cells. A recent study taking advantage of parabolic flights to reproduce microgravity, evidenced the downregulation of H3K4me3 binding to immunoregulatory genes primary human M1 macrophages. The genomic distribution of the downregulated H3K4me3 loci correlates very well to immunoregulatory genes [60]. Finally, human leukocytes stimulated with mitogens onboard the International Space Station showed altered expression of different miRNAs, suggesting that T-cell activation is regulated by transcription promotion and by modulating translation via ncRNA mechanisms [58].

Other areas of investigation in the field of epigenetic in space raised from the observation of the major physical alterations the astronauts show after the permanence in space, i.e., muscle and bone tissues impairment [61]. It is well known that muscle differentiation is driven by hypomethylation of a set of myogenic genes. One experiment performed with rat L6 myoblasts in simulated microgravity suggested that the expression of Myod1, a master gene in muscle differentiation, may be regulated in microgravity [62]. Noteworthy, this remains the only study exploring the epigenetic changes on myogenic genes in microgravity and suggests that this condition induces DNA hypomethylation. This result is in contrast with the other here described (in different cellular models) indicating increased methylation associated to microgravity and may depend on the timing of exposure. The expression of osteoblasts differentiation markers, for instance, resulted inhibited by simulated microgravity in association with mitochondrial alterations and increased histone methylation [63]. These results are coupled with previous data showing the inhibition of embryonic osteoblast markers as Sox2 and Oct3/4 in blood-derived stem cells, an osteogenic differentiation model, in the ISS respect to the cell cultured on hearth [64].

More studies have been performed on tumor cell lines, with the double attempt to disclose potential tumorigenic effects of the microgravity and to better understand the molecular basis of the tumor development. One of the first studies in this sense evidenced that we found that exposing human keratinocytes to simulated microgravity induced the epithelial–mesenchymal transition (EMT) through the expression of the typical EMT transcription factors and markers, such as Snail1, Snail2 and ZEB2, metalloproteases, mesenchymal adhesion molecules, and cytoskeletal components [65]. In this study, it was formulated the idea that, modifying a physical constraint, microgravity may exert epigenetic effects and reset the cell phenotype at cell division [66]. One of the epigenetic alterations able to modify the cancer cells response to microgravity has been individuated in the miRNAs. In fact, the exposure of A549 cells to microgravity, causing the generation of polynucleated cells, cell cycle imbalance, growth inhibition, and the presence of highly damaged mitochondria., is associated with the expression of a pool of miRNAs mainly involved in cell cycle regulation, apoptosis, and stress response [65]. More recently, it was evidence through "omic" approaches that microgravity exposure is associated to alteration of DNA methylation, histone modifications, and miRNAs expression, stressing the relevance of this kind of studies to depict a comprehensive epigenetic response of cancer cells to space environment [67–70].

3.3 Space Epigenetics on Animal Models

As for all the fields of the biomedical studies in space-related conditions, the studies on the epigenetic signature in animal models are more limited respect to the studies on cellular models, due to the difficulty to set, for example, a simulated microgravity environment for animals or to launch live animals on the ISS.

Therefore, it is not coincidence that some epigenetic study has been performed on relatively non-complex organism including *C. elegans*. In this worm, it was demonstrated that after simulated microgravity treatment, 19 miRNAs were dysregulated. Among these, miRNAs, let-7, mir-54, mir-67, mir-85, mir-252, mir-354, mir-789, mir-2208, and mir-5592 were required for the toxicity induction of simulated microgravity in suppressing locomotion behavior [71].

Furthermore, also histone modification were dysregulated in microgravity, leading to altered expression of group of 39 genes [72]. The major limitation of these animals in pedogenetic studies is, however, represented by the different DNA methylation processes, that makes the results on *C. elegans* not easily comparable to what is observed in other mammals.

Mice are the most used animal model in science, nevertheless growing them in real or simulated microgravity is not an easy task. Few experiments have been performed. In one of them, the mice were shipped onboard the ISS and kept in microgravity or artificial 1 g gravity [73]. The transcriptomic analysis revealed the differential expression of genes related to the atrophy of skeletal muscle and evidenced that artificial gravity was able to prevent muscle atrophy induced by microgravity. The data suggested that chromatin remodeling possibly associated to the altered gene expression. In a second analysis performed on the mice returned to heart, the same research group observed that male mice that had experienced spaceflight showed alterations in binding of transcription factor ATF7, a regulator of heterochromatin formation, on promoter regions in testis, and altered ncRNA expression in spermatozoa. These further results suggest that spaceflight has intergenerational effect possibly mediated by, or at least associated to, epigenetic factors [74].

More recently, other studies evidenced the alteration of miRNAs in the brain of rats kept in conditions of simulated microgravity [75] and in the bone of mice after spaceflight [76], confirming that miRNAs represent a dynamic epigenetic regulator that can be influenced by the spaceflight and therefore worthing further analysis to understand, and possibly contrast, the induced negative effects.

Finally, a very recent work attempted to better understand the role of cosmic beans on DNA methylation in mice, evidencing differential hypermethylation at the long interspersed nucleotide element 1 (LINE-1) in the lung of male mice and a trend to hypomethylation in the lung of female mice [77]. The authors suggest that LINE-1 methylation may be used as a marker of cosmic radiations exposure, but this result also reveal a further layer of complexity, suggesting that the epigenetic response to spaceflight conditions may be sex-specific.

3.4 Space Epigenetics on Human Samples

So far, only three studies explored the effects of the space environment on epigenetic markers in human samples.

The first analyzed the epigenetic aging in leukocytes from participants to the on ground, simulated space travel, MARS-500 mission [78]. In this study, five blood DNA-methylation–based metrics were measured longitudinally: DNAmGrimAge, DNAmPhenoAge, DNA-methylation–based estimator of telomere length (DNAmTL), mitotic divisions (epigenetic mitotic clock [epiTOC2]), and pace of aging (PoA). The results evidenced that, relative to baseline, mission duration was associated with significant decreases in epigenetic aging, although these markers returned to the baseline in few days after the end of the mission, indicating the reversibility of the epigenetic changes. A second study in simulated spaceflight conditions, the Deep Immersion (DI) model, evidenced a set of miRNAs altered in platelets after the 3-day DI protocol, suggesting that platelets may be used as a biomarker for epigenetic variations [79].

Finally, a study analyzed the biomedical profiles from 59 astronauts and data from NASA's GeneLab to determine transcriptomic, proteomic, metabolomic, and epigenetic responses to spaceflight [80]. Overall pathway analyses on the multi-omics datasets showed significant enrichment for mitochondrial processes, specifically suggesting that mitochondrial damage is associated to epigenetic changes during spaceflight and could have lasting consequences after the return to hearth.

Taken together, these data identify epigenetic, gene expression, inflammatory, and metabolic responses to spaceflight, facilitating a better biomedical roadmap of features that should be monitored and safeguarded in upcoming missions [81].

4 Methods to Study Epigenetic Markers in Space: Considerations and Perspectives

Studying the epigenetic modifications in samples sent to the space, either cells or animal or even body fluids and biopsies from astronauts, presents a major limitation due to the reversible nature of the epigenetic marks [82]. This is particularly true for the DNA methylation and miRNAs modulation, while histone modification and chromatin remodeling represent more stable modifications. This means that when cells/animals/astronauts return on ground, their miRNAs expression profile [79] and their methylome [83] can rapidly return to a control-like status. Setting the right conditions for the collection, conservation, and treatment of the biological samples, is a mandatory step [84].

A second limitation is represented by the complexity of the analysis necessary to study the epigenetic marks, not allowing these assessment are performed in automated manner or by the limited means of, for example, the ISS. The establishment of a permanent extraterrestrial base, with a fully equipped biochemical and biomolecule laboratory will allow, in the future, the execution of this kind of analysis.

Histone modifications and chromatin remodeling represent a more stable epigenetic modification that can be assessed on hearth, after the experimental models return from spaceflight. Cell cultures can be fixed by formaline analogues (e.g., NoTox [66]), whereas cells/tissues from animal models or human samples can be collected after the return to the hearth. The assessment of these epigenetic modification mainly requires immune-based assays as western blotting (WB), Chromatin ImmunoPrecipitation (ChIP), and derived targeted or "omic" analyses.

miRNAs expression evaluation can be reasonably performed on fixed cells after the return to hearth, or on frozen or fixed animal and human samples, collected during the spaceflight and stabilized. Several commercial kits, for instance, may be easily used to stabilize blood samples. This analysis is normally performed by qPCR (quantitative real-Time PCR) or ddPPCR (digital droplet PCR) for the analysis of specific targets [85], or by next generation sequencing (NGS) or microarrays for the "omic" analyses aimed at evaluating the miRNome or the transcriptome [86]. On the other hand, the idea of studying the miRNAs in animals or human subjects after the return to hearth should take into account the possible bias represented by the reversion of the modulation previously induced in space.

The assessment of DNA methylation is largely the most complicated determination to be performed among the above-mentioned epigenetic marks. The vaste majority of the analysis are based on the reaction of genomic DNA with sodium bisulfite [87], which modifies the unmethylated cytosines and transform them into thymines. Methylated cytosines are protected from being modified by bisulfite. After this reaction, the unmodified cytosines can be then detected by a variety of methods, depending on the desired level of the information. PCR amplification and further sequencing of bisulfite-modified DNA returns the methylation level of each cytosine in the investigated target sequence [88]. This method allows the precise mapping of a target sequence, typically a gene promoter. Since this is a time-consuming and complex protocol, other methods have been developed to assess the methylation status of selected cytosines or subgroups in a sequence. High resolution melting (HRM), an example of this kind of analysis, can detect the methylation status by assessing on bisulfite-modified PCR-amplified DNA the differential melting temperature depending on the transformation of C–G couples (with three hydrogen bonds) in T–A couples (with two hydrogen bonds) [89]. The faster assessment, due to the lack of the sequencing, allowed by this and other methods, is counterbalanced by the lower resolution (few cytosines assessed and often grouped). Finally, bisulfite modification is also the basis for NGS or microarray analyses, aimed at mapping the whole methylome or large set of genes [90, 91].

The complexity of the above-described methods makes it really hard to imagine their application directly in space. On the other hand, the idea to assess methylation after cells, tissues, or animals return to hearth present other critical points. As already stated, DNA methylation/demethylation process are exerted in part by active and rapid molecular mechanisms and are both reversible processes [82]. Therefore, the risk exists that hyper- or hypo-methylation induced by the spaceflight can rapidly return to control-like conditions as soon as the normal gravity is restored. Fixing the cells by using analogs of formaline (e.g., the NoTox fixative) at the end of the experimental protocol, yet during the spaceflight, is a common protocol for many cell cultures launched on orbit, but could not be effective when applied to DNA methylation analysis. DNA methylation, in fact, may be unstable after prolonged cell or tissue fixing and storage [92]. Specifically, formaldehyde and derivatives can induce DNA demethylation, altering the analysis [93]. More generally, the interval between the end of the experiment or the isolation of the tissue and the storage is fundamental to prevent post-experimental, unwanted alterations of the epigenetic marks [94].

Paradoxically, one of the most recent and advanced sequencing technology is a good candidate for assessing DNA methylation while onboard the ISS or future extraterrestrial bases. Third generation NGS (next generation sequencing) platform allow the sequencing at genome-wide level by direct reading of the DNA sequence, without the need of amplification or genomic library construction [95]. Among these, the Nanopore Technologies workflow can detect, besides the genetic sequence, also the presence of methylated cytosines. The Oxford Nanopore Technologies MinION has been already tested during a parabolic flight to understand the effects of variable gravity on the instrument and data. During this proof, the instrument-generated DNA reads over the course of the flight, including the first ever sequenced in microgravity. A protocol to facilitate Nanopore sequencing aboard the ISS and in other microgravity environments that has been optimized [96].

5 Conclusions

Studying the epigenetic modifications occurring when an organism is subjected to an extraterrestrial environment, like a spaceflight or permanence on an orbiting base or, like planned in the future, in a lunar or martian base, appears fundamental. The exposure to lower or null gravity and to cosmic radiations, may induce epigenetic modifications that could affect the physiologic cellular process and eventually cause the onset of pathologies. The epigenetic modification, in fact, act as mediator of the environmental stimuli and regulates gene expression. Evidence already exist that microgravity alters some specific epigenetic trait, leading to altered gene expression. However, systematic studies aimed at deciphering the epigenome and their variation in the space are still lacking. Gene-specific studies, evidencing epigenetic changes at target sequences level are also very limited. On the contrary, it appears of great relevance to profile the epigenetic changes induced by the spatial environments, in order to understand the molecular causes of the physical alteration observed in the astronauts. Moreover, since these changes may be reverted by administration of drugs and supplements with epigenetic action, such studies would also suggest possible intervention aimed at contrasting the detrimental effect of altered epigenetics.

Unfortunately, these considerations are contrasted by the often complex and time-consuming procedures necessary to perform epigenetic assessment, which could prevent the execution of these assay directly during the permanence in the space. On the other hand, waiting to perform the epigenetic assessment after the return to hearth, may induce the reversal of the epigenetic changes, nullifying the scope of the analysis.

References

1. Waddington CH (1942) The epigenotype. Endeavour 1:18–20
2. Holliday R (1990) Mechanisms for the control of gene activity during development. Biol Rev Camb Phil Soc 65:431–471
3. Goldberg AD, Allis CD, Bernstein E (2007) Epigenetics: a landscape takes shape. Cell 128:635–638
4. Geiman TM, Robertson KD (2002) Chromatin remodeling, histone modifications, and DNA methylation-how does it all fit together? J Cell Biochem 87:117–125
5. Fuso A, Troen AM (2021) Environmental epigenetics: myth and reality. Organisms. J Biol Sci 4:31–34
6. Gardini A, Shiekhattar R (2015) The many faces of long noncoding RNAs. FEBS J 282:1647–1657
7. Chen LL, Yang L (2015) Regulation of circRNA biogenesis. RNA Biol 12:381–388
8. Lu TX, Rothenberg ME (2018) MicroRNA. J Allergy Clin Immunol 141:1202–1207
9. Krol J, Loedige I, Filipowicz W (2010) The widespread regulation of microRNA biogenesis, function and decay. Nat Rev Genet 11:597–610
10. Fuso A, Lucarelli M (2019) CpG and non-CpG methylation in the diet-epigenetics-neurodegeneration connection. Curr Nutr Rep 8:74–82
11. Stoccoro A, Coppedè F (2024) Exposure to metals, pesticides, and air pollutants: focus on resulting DNA methylation changes in neurodegenerative diseases. Biomolecules 14:1366
12. Tollefbol T (eds) (2023) Epigenetics of human disease, 3rd edn. Elsevier
13. Hendrich B (2000) Methylation moves into medicine. Curr Biol 10:R60–R63.
14. Khalifa MM, Reiss AL, Migeon BR (1990) Methylation status of genes flanking the fragile site in males with the fragile-X syndrome: a test of the imprinting hypothesis. Am J Hum Genet 46:744–753
15. Van den Veyver IB, Zoghbi HY (2000) Methyl-CpG-binding protein 2 mutations in Rett syndrome. Curr Opin Genet Dev 10:275–279
16. Men J, Wang X, Zhou Y et al (2025) Neurodegenerative diseases: epigenetic regulatory mechanisms and therapeutic potential. Cell Signal 131:111715
17. van Zundert B, Montecino M (2025) Epigenetics in neurodegenerative diseases. SubCell Biochem 108:73–109
18. Prasanth MI, Sivamaruthi BS, Cheong CSY et al (2024) Role of epigenetic modulation in neurodegenerative diseases: implications of phytochemical interventions. Antioxidants (Basel) 13:606
19. Bou Ghanem A, Hussayni Y, Kadbey R et al (2024) Exploring the complexities of 1C metabolism: implications in aging and neurodegenerative diseases. Front Aging Neurosci 15:1322419
20. Jakovcevski M, Akbarian S (2012) Epigenetic mechanisms in neurological disease. Nat Med 18:1194–1204
21. Okano M, Bell DW, Haber DA et al (1999) DNA methyltransferases Dnmt3a and Dnmt3b are essential for de novo methylation and mammalian development. Cell 99:247–257
22. Ndlovu MN, Denis H, Fuks F (2011) Exposing the DNA methylome iceberg. Trends Biochem Sci 36:381–387
23. Ehrlich M, Lacey M (2013) DNA methylation and differentiation: silencing, upregulation and modulation of gene expression. Epigenomics 5:553–568
24. Yagi M, Kabata M, Ukai T (2019) De novo DNA methylation at imprinted loci during reprogramming into naive and primed pluripotency. Stem Cell Reports 12:1113–1128
25. Li Y, Kumar S, Qian W (2018) Active DNA demethylation: mechanism and role in plant development. Plant Cell Rep 37:77–85
26. Poole CJ, Lodh A, Choi JH et al (2019) MYC deregulates TET1 and TET2 expression to control global DNA (hydroxy)methylation and gene expression to maintain a neoplastic phenotype in T-ALL. Epigenetics Chromatin 12:41

27. Greenberg M, Teissandier A, Walter M et al (2019) Dynamic enhancer partitioning instructs activation of a growth-related gene during exit from naïve pluripotency. eLife 8:e44057.
28. Luger K, Rechsteiner TJ, Flaus AJ et al (1997) Characterization of nucleosome core particles containing histone proteins made in bacteria. J Mol Biol 272:301–311
29. Davey CA, Sargent DF, Luger K et al (2002) Solvent mediated interactions in the structure of the nucleosome core particle at 1.9A resolution. J Mol Biol 319:1097–1113
30. Rice JC, Briggs SD, Ueberheide B et al (2003) Histone methyltransferases direct different degrees of methylation to define distinct chromatin domains. Mol Cell 12:1591–1598
31. Martin C, Zhang Y (2005) The diverse functions of histone lysine methylation. Nat Rev Mol Cell Biol 6:838–849
32. Kouzarides T (2007) Chromatin modifications and their function. Cell 128:693–705
33. Sampath SC, Marazzi I, Yap KL et al (2007) Methylation of a histone mimic within the histone methyltransferase G9a regulates protein complex assembly. Mol Cell 27:596–608
34. Strahl BD, Allis CD (2000) The language of covalent histone modifications. Nature 403:41–45
35. Shilatifard A (2006) Chromatin modifications by methylation and ubiquitination: implications in the regulation of gene expression. Annu Rev Biochem 75:243–269
36. Latham JA, Dent SY (2007) Cross-regulation of histone modifications. Nat Struct Mol Biol 14:1017–1024
37. Yang XJ, Seto E (2008) Lysine acetylation: codified crosstalk with other posttranslational modifications. Mol Cell 31:449–461
38. Greenberg MVC, Bourc'his D (2019) The diverse roles of DNA methylation in mammalian development and disease. Nat Rev Mol Cell Biol 20:590–607
39. Peterson CL, Workman JL (2000) Promoter targeting and chromatin remodeling by the SWI/SNF complex. Curr Opin Genet Dev 10:187–192
40. Clapier CR, Cairns BR (2009) The biology of chromatin remodeling complexes. Annu Rev Biochem 78:273–304
41. Friedman RC, Farh KK, Burge CB et al (2009) Most mammalian mRNAs are conserved targets of microRNAs. Genome Res 19:92–105
42. Lee Y, Kim M, Han J et al (2013) MicroRNA genes are transcribed by RNA polymerase II. EMBO J 23:4051–4060
43. García-López J, Brieño-Enríquez MA, Del Mazo J (2013) MicroRNA biogenesis and variability. Biomol Concepts 4:367–380
44. Lee Y, Jeon K, Lee JT et al (2002) MicroRNA maturation: stepwise processing and subcellular localization. EMBO J 21:4663–4670
45. Monteys AM, Spengler RM, Wan J et al (2010) Structure and activity of putative intronic miRNA promoters. RNA 16:495–505
46. Han J, Lee Y, Yeom KH et al (2004) The Drosha-DGCR8 complex in primary microRNA processing. Genes Dev 18:3016–3027
47. Denli AM, Tops BB, Plasterk RH et al (2004) Processing of primary microRNAs by the Microprocessor complex. Nature 432:231–235
48. Tsutsumi A, Kawamata T, Izumi N et al (2011) Recognition of the pre-miRNA structure by *Drosophila* Dicer-1. Nat Struct Mol Biol 18:1153–1158
49. Yoda M, Kawamata T, Paroo Z et al (2010) ATP-dependent human RISC assembly pathways. Nat Struct Mol Biol 17:17–23
50. Jo MH, Shin S, Jung SR et al (2015) Human Argonaute 2 has diverse reaction pathways on target RNAs. Mol Cell 59:117–124
51. Richards EJ (1997) DNA methylation and plant development. Trends Genet 13:319–323
52. Ou X, Long L, Zhang Y et al (2009) Spaceflight induces both transient and heritable alterations in DNA methylation and gene expression in rice (*Oryza sativa* L.). Mutat Res 662:44–53
53. Kamal KY, Herranz R, van Loon JJWA et al (2018) Simulated microgravity, Mars gravity, and 2g hypergravity affect cell cycle regulation, ribosome biogenesis, and epigenetics in *Arabidopsis* cell cultures. Sci Rep 8:6424
54. Xu P, Chen H, Jin J et al (2018) Single-base resolution methylome analysis shows epigenetic changes in *Arabidopsis* seedlings exposed to microgravity spaceflight conditions on board the SJ-10 recoverable satellite. NPJ Microgravity 4:12
55. Xu P, Chen H, Hu J et al (2021) Potential evidence for transgenerational epigenetic memory in *Arabidopsis thaliana* following spaceflight. Commun Biol 4:835
56. Zhou M, Sng NJ, LeFrois CE et al (2019) Epigenomics in an extraterrestrial environment: organ-specific alteration of DNA methylation and gene expression elicited by spaceflight in *Arabidopsis thaliana*. BMC Genom 20:205
57. Paul AL, Haveman N, Califar B et al (2021) Epigenomic regulators *Elongator Complex Subunit 2* and *Methyltransferase 1* differentially condition the spaceflight response in *Arabidopsis*. Front Plant Sci 12:691790
58. Singh KP, Kumari R, Dumond JW (2010) Simulated microgravity-induced epigenetic changes in human lymphocytes. J Cell Biochem 111:123–129
59. Chowdhury B, Seetharam A, Wang Z et al (2016) A study of alterations in DNA epigenetic modifications (5mC and 5hmC) and gene expression influenced by simulated microgravity in human lymphoblastoid cells. PLoS One 11:e0147514

60. Vahlensieck C, Thiel CS, Christoffel S et al (2022) Rapid downregulation of H3K4me3 binding to immunoregulatory genes in altered gravity in primary human M1 macrophages. Int J Mol Sci 24:603
61. Williams DR (2020) A historical overview of Space Medicine. McGill J Med 6:62–65
62. Furukawa T, Tanimoto K, Fukazawa T et al (2018) Simulated microgravity attenuates myogenic differentiation via epigenetic regulations. NPJ Microgravity 4:11
63. Shi Q, Song Y, Cao J et al (2025) Inhibition of mitochondrial fission reverses simulated microgravity-induced osteoblast dysfunction by enhancing mechanotransduction and epigenetic modification. Research (Wash D C) 8:0602
64. Gambacurta A, Merlini G, Ruggiero C et al (2019) Human osteogenic differentiation in Space: proteomic and epigenetic clues to better understand osteoporosis. Sci Rep 9:8343
65. Ranieri D, Proietti S, Dinicola S et al (2017) Simulated microgravity triggers epithelial mesenchymal transition in human keratinocytes. Sci Rep 7:538
66. Masiello MG, Verna R, Cucina A et al (2018) Physical constraints in cell fate specification. A case in point: microgravity and phenotypes differentiation. Prog Biophys Mol Biol 134:55–67
67. Degan P, Cortese K, Pulliero A et al (2021) Simulated microgravity effects on human adenocarcinoma alveolar epithelial cells: characterization of morphological, functional, and epigenetic parameters. Int J Mol Sci 22:6951
68. Graf J, Schulz H, Wehland M et al (2024) Omics studies of tumor cells under microgravity conditions. Int J Mol Sci 25:926
69. Beheshti A, McDonald JT, Hada M et al (2021) Genomic changes driven by radiation-induced DNA damage and microgravity in human cells. Int J Mol Sci 22:10507
70. Schulz H, Abdelfattah F, Heinrich A et al (2025) Omics investigations of prostate cancer cells exposed to simulated microgravity conditions. Biomolecules 15:303
71. Sun L, Li W, Li D et al (2020) microRNAs involved in the control of toxicity on locomotion behavior induced by simulated microgravity stress in *Caenorhabditis elegans*. Sci Rep 10:17510
72. Higashitani A, Hashizume T, Takiura M et al (2021) Histone deacetylase HDA-4-mediated epigenetic regulation in space-flown *C. elegans*. NPJ Microgravity 7:33
73. Yoshida K, Fujita SI, Isotani A et al (2021) Intergenerational effect of short-term spaceflight in mice. iScience 24:102773
74. Okada R, Fujita SI, Suzuki R et al (2021) Transcriptome analysis of gravitational effects on mouse skeletal muscles under microgravity and artificial 1 g onboard environment. Sci Rep 11:9168
75. Rasheed M, Wang H, Wang C et al (2023) Opposing manner of miR-455-3p against NR2B-PSD-95-nNOS complex in the cortex and hippocampus of depressive rats under simulated complex space environment. J Neurochem 165:391–412
76. Gautam A, Chakraborty N, Dimitrov G et al (2024) Microgravity's effects on miRNA-mRNA regulatory networks in a mouse model of segmental bone defects. PLoS One 19:e0313768
77. Clement K, Nemec-Bakk AS, Jun SR et al (2025) Long-term effects of combined exposures to simulated microgravity and galactic cosmic radiation on the mouse lung: sex-specific epigenetic reprogramming. Radiat Environ Biophys 64:17–27
78. Nwanaji-Enwerem JC, Nwanaji-Enwerem U, Van Der Laan L et al (2020) A longitudinal epigenetic aging and leukocyte analysis of simulated space travel: the Mars-500 mission. Cell Rep 33:108406
79. Twomey L, Navasiolava N, Robin A et al (2021) A dry immersion model of microgravity modulates platelet phenotype, miRNA signature, and circulating plasma protein biomarker profile. Sci Rep 11:21906
80. da Silveira WA, Fazelinia H, Rosenthal SB et al (2020) Comprehensive multi-omics analysis reveals mitochondrial stress as a central biological hub for spaceflight impact. Cell 183:1185–1201.e20
81. Iosim S, MacKay M, Westover C et al (2019) Translating current biomedical therapies for long duration, deep space missions. Precis Clin Med 2:259–269
82. Szyf M (2007) The dynamic epigenome and its implications in toxicology. Toxicol Sci 100:7–23
83. Lucarelli M, Fuso A, Strom R et al (2001) The dynamics of myogenin site-specific demethylation is strongly correlated with its expression and with muscle differentiation. J Biol Chem 276:7500–7506
84. Rizzardi LF, Kunz H, Rubins K et al (2016) Evaluation of techniques for performing cellular isolation and preservation during microgravity conditions. NPJ Microgravity 2:16025
85. Hindson CM, Chevillet JR, Briggs HA et al (2013) Absolute quantification by droplet digital PCR versus analog real-time PCR. Nat Methods 10:1003–1005
86. Git A, Dvinge H, Salmon-Divon M et al (2010) Systematic comparison of microarray profiling, real-time PCR, and next-generation sequencing technologies for measuring differential microRNA expression. RNA 16:991–1006
87. Clark SJ, Harrison J, Paul CL et al (1994) High sensitivity mapping of methylated cytosines. Nucleic Acids Res 22:2990–2997
88. Fuso A, Ferraguti G, Scarpa S et al (2015) Disclosing bias in bisulfite assay: MethPrimers underestimate high DNA methylation. PLoS One 10:e0118318.

89. Migheli F, Stoccoro A, Coppedè F et al (2013) Comparison study of MS-HRM and pyrosequencing techniques for quantification of *APC* and *CDKN2A* gene methylation. PLoS One 8:e52501
90. Lee EJ, Luo J, Wilson JM et al (2013) Analyzing the cancer methylome through targeted bisulfite sequencing. Cancer Lett 340:171–178
91. Moran S (2017) Technologies for deciphering epigenomic DNA patterns. Adv Exp Med Biol 978:477–488
92. García-Giménez JL, Mena-Mollá S, Beltrán-García J et al (2017) Challenges in the analysis of epigenetic biomarkers in clinical samples. Clin Chem Lab Med 55:1474–1477
93. Li T, Wei Y, Qu M et al (2021) Formaldehyde and de/methylation in age-related cognitive impairment. Genes (Basel) 12:913
94. Ferrer I, Martinez A, Boluda S et al (2008) Brain banks: benefits, limitations and cautions concerning the use of post-mortem brain tissue for molecular studies. Cell Tissue Bank 9:181–194
95. Fuso A (2013) The 'golden age' of DNA methylation in neurodegenerative diseases. Clin Chem Lab Med 51:523–534
96. McIntyre ABR, Rizzardi L, Yu AM et al (2016) Nanopore sequencing in microgravity. NPJ Microgravity 2:16035

Part IV

Tissue and Systems

Chapter 9

Modulation of Cell-to-Cell Adhesion and Interaction with the Extracellular Matrix in Microgravity Conditions

Elena Di Nisio, Emanuele Cacci, Giuseppe Lupo and Rodolfo Negri

Abstract

We present an experimental strategy to dissect the effects of simulated microgravity on the interaction between cells and extracellular matrix. The protocol encompasses morphological and transcriptomic analyses on human neural progenitor cells (hNSPCs) cultured in flasks coated with poly-L-ornithine and laminin matrix at low density and subjected to microgravity simulation. High-density cultures subjected to microgravity or grown under normal gravity without a matrix are also analyzed as controls. The experimental approach could be adapted to different cell types by changing the extracellular matrix composition and the culture medium accordingly.

Keywords Microgravity, Space conditions, Human neural stem progenitor cells, Focal adhesion, Extracellular matrix, Laminin

1 Introduction

The relevance of the extracellular matrix composition and properties in the physiological and pathological effects of microgravity on human organs is well established [1, 2].

Many studies showed that different mammalian cell types cultured in simulated microgravity, both in random positioning machine apparatus (RPM) and in rotating vessels, undergo dramatic morphological changes, leading to the simultaneous presence of two alternative phenotypes, adherent cells and floating cell clumps, in the same culture [3–9]. This is a reversible process; when a floating aggregate cell population is seeded in normal gravity, it goes back to a fully adherent phenotype; when reseeded in microgravity condition, a subset of the cells detach from the substrate generating nonadherent aggregates [7]. In our model, human neural progenitor cells (hNSPCs) are cultured in flasks coated with poly-L-ornithine and laminin matrix at low density. The interaction with the extracellular matrix was previously shown to be crucial for the morphological development of these cells [10].

Mariano Bizzarri (ed.), *Biomedicine and Biotechnology in Space*, Methods in Molecular Biology, vol. 3000,
https://doi.org/10.1007/978-1-0716-5174-2_9,

We have developed a protocol to analyze the switch from a laminin-adherent phenotype to the formation of floating cell aggregates, when hNSPCs are subjected to microgravity simulation in an RPM apparatus. The results show that the switch involves at least 60% of the whole population after 24 h [11]. We, then, investigated the transcriptomic changes associated with the morphological change and found that several genes belonging to the focal adhesion pathway are downregulated (Fig. 1). This is particularly relevant since focal adhesion components were previously considered as biomechanical sensors of the microenvironment [12]. Some focal adhesion genes were previously found to be modulated in human bone marrow mesenchymal stromal cells after different periods of simulated microgravity [13], but such systematic modulation has not been previously described. Our interpretation is that the transcriptomic changes are due to a feedback transcriptional regulation of the focal adhesion pathway in response to the decreased cell interactions with the extracellular matrix underlying, which, in turn, was causing cell detachment and floating aggregate formation. Figure 2 summarizes this working hypothesis. To support this hypothesis, it is crucial to test the cell morphology and the expression of the genes modulated by RPM treatment in cultures grown without the extracellular matrix, but under normal gravity. Indeed, in the control experiments, the cells efficiently form floating cell aggregates and show downregulation of several focal adhesion genes (TUBB, ACTN, MYL9, VIM, ZYX). It would be important to understand if these gene modulations are observed on most of the cells subjected to simulated microgravity or only on those that form the aggregates. For this purpose, RNA-Seq analysis should be performed on isolated floating aggregates. An important variable to be considered in these experiments is cell density. When hNSPCs were grown closer to confluence and subsequently subjected to simulated microgravity conditions, they did not show significant floating aggregate formation and significant modulation of focal adhesion genes [11], showing that in dense cultures, cell–cell interactions are stronger and can successfully counteract the weakening of cell–matrix interactions due to simulated microgravity. For this reason, cultures with different cell densities should be included in the experimental plan. Here, we describe a general protocol that can be adapted to different cell types by changing the extracellular matrix composition and the culture medium according to the investigated cell type.

2 Materials

2.1 Cell Cultures and Exposure to Simulated Microgravity Conditions

1. DMEM/F12 supplemented with: 50× B27 minus vitamin A, 100× P/S, 100× Glutamax, 10 ng/mL bFGF, 10 ng/mL EGF, and 1 mM Hepes
2. Cell Culture Flask, with a Canted Neck Plug Seal Cap.

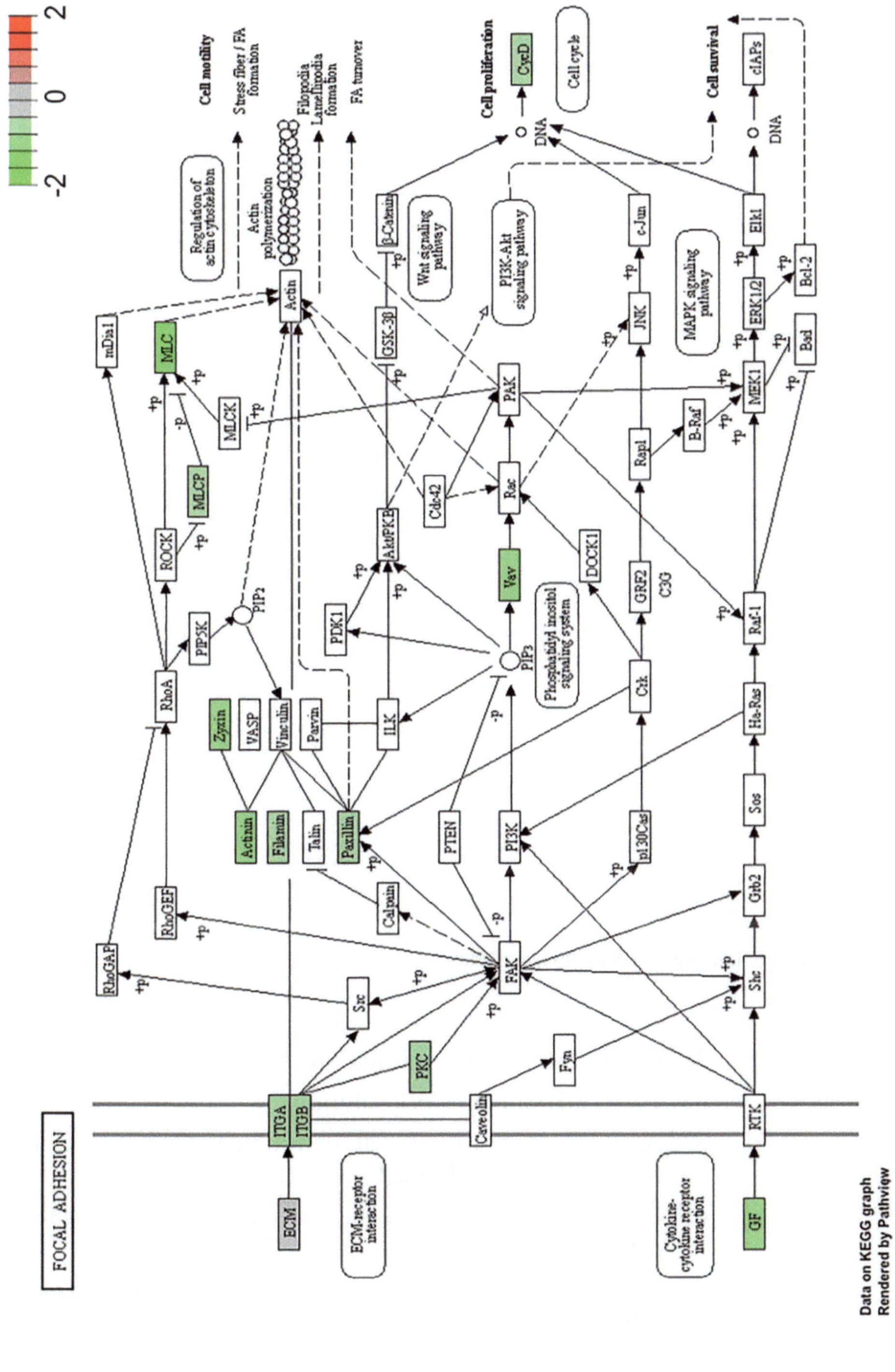

Fig. 1 DEGs at 24 h comparison that belong to focal adhesion KEGG pathway. Downregulated genes are colored in green; no genes of this pathway were significantly upregulated (readapted from [11])

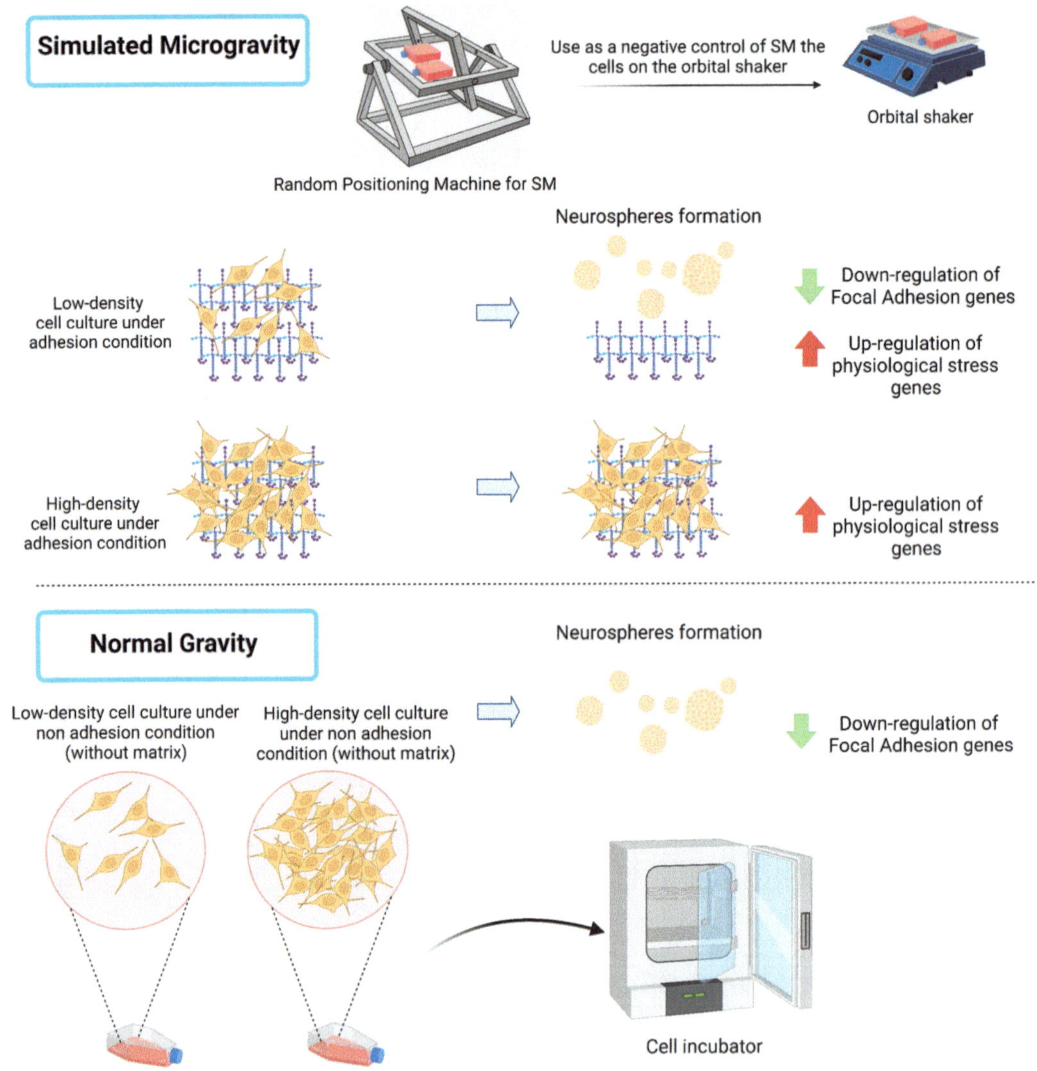

Fig. 2 Schematic representation of the working hypothesis

3. 0.1 μg/mL poly-L-ornithine stock solution.
4. 10μg/mL laminin stock solution.
5. Accutase ready-to-use cell detachment solution (Corning Life Sciences) (*see* **Note 1**).
6. PBS w/o calcium and magnesium (sterile and suitable for cell cultures).
7. T75 flasks, 50 mL tubes, 15 mL tubes, 1.5 mL RNase-free tubes.
8. Cell cultures incubator.
9. Cell counter and Trypan blue dye.

10. Refrigerated centrifuges.
11. Phase-contrast Microscope.
12. Random Positioning Machine located in a dedicated 37°C controlled room.
13. Mechanical agitator with nonorbital motion, preferably located in the same 37°C controlled room.

2.2 RNA Purification and RT-qPCR

1. RNA extraction kit.
2. RNase free water.
3. cDNA Synthesis Kit.
4. SYBR Hi-ROX mix.
5. Primers.
6. qPCR instrument.

3 Methods

3.1 Flasks Coating

1. Prepare the poly-ornithine working solution by diluting the poly-ornithine stock solution (1 mg/mL) 1:10,000 in PBS to have a final concentration of 0.1 μg/mL.
2. Add a suitable volume of poly-ornithine working solution to the flasks (*see* **Note 2**) and incubate them for 4 h at 37°C in the cell incubator.
3. Before removing the poly-ornithine solution, prepare the laminin working solution by diluting the stock solution (1.5 mg/mL) 1:150 in PBS to obtain a final concentration of 10 μg/mL.
4. Remove the poly-ornithine working solution and wash the flasks twice with room temperature PBS.
5. Add a suitable volume of laminin working solution to the flasks (*see* **Note 2**) and incubate them overnight in the cell incubator at 37°C.
6. Remove the laminin solution and add PBS preventing the drying out of the flasks (*see* **Note 3**).
7. Remove the PBS and proceed with plaiting the cells.

3.2 Exposure to Simulated Microgravity

1. Plate the hNSPCs in T75 flasks (*see* **Notes 4 and 5**) coated with a matrix consisting of poly-ornithine and laminin (see Sect. 3.1), at low or high density (ranging from 10,000 to 40,000 cells/cm^2) (*see* **Note 6**) and culture the cells for 5 days at 37°C and 5% CO_2 in proliferation-supporting medium, as described in Sect. 2.

2. Before starting the experiment, check the cells with a microscope and then fully fill the flasks with complete cell culture medium (*see* **Note** 7).
3. Place the fully filled flasks in the cell incubator at 37°C and 5% CO_2, in a vertical position and with the screw cap slightly unscrewed for 1 h before the experiment to equilibrate the cell culture medium with the CO_2.
4. Remove the flasks from the incubator, tightly screw the caps, and transfer them to the RPM in a temperature-controlled room (37°C) to maintain the cells under a simulated microgravity condition (*see* **Note 8**) for 6 and/or 24 h; include sham controls by placing the control flasks on a mechanical agitator with nonorbital motion at 100 rpm, at the same temperature and same time intervals.
5. Set up the spinning velocity of the frames at $60° s^{-1}$. For routine conditions, set a value of 0.003 g (*see* **Note 9**).

3.3 Morphological Analysis

1. For each treatment condition and time point, morphologically analyze the cultures by phase-contrast microscopy (*see* **Note 10**) to evaluate any possible morphological change (*see* **Note 11**).
2. Collect images of the cells at different magnifications (*see* **Note 12**), then immediately proceed with the downstream analyses (see Sect. 3.4).

3.4 Cell Counting and RNA Purification

1. Collect all the cell medium to harvest any floating clumped cells in 50 mL tubes and keep the tubes on ice until centrifugation (*see* **Note 13**). Follow the same protocol also for the controls, even in case floating cells are absent.
2. Centrifuge the medium at 800 g, 4°C for 5 min.
3. Carefully aspirate the supernatant (*see* **Note 14**).
4. Resuspend the first pellet using 2.5 mL of warm Accutase solution, then use this volume to sequentially collect all the other pellets of the same experimental condition (*see* **Note15**).
5. Aspirate the PBS from the flask and add the 2.5 mL of warm Accutase solution including the floating cells to the flask to detach all the remaining adherent cells; keep the flask at 37°C for 2–3 min.
6. Check the microscope for cell detaching, then recover in a 15 mL tube all the detached cells using 11.5 mL of cold PBS to dilute the Accutase enzyme mixture reducing its proteolytic activity.
7. Centrifuge at 800 × *g*, 4°C for 5 min.
8. Carefully aspirate the supernatant.
9. Resuspend the cell pellet in 1 mL of cold PBS and transfer it to a 1.5 mL tube (RNase-free certified).

10. Transfer 20 μL of the resuspended cells in a new tube for the cell count; after a proper dilution, take 20 μL and add 20 μL of Trypan Blue dye (*see* **Note 16**).
11. Centrifuge the 1.5 mL tubes to pellet the cells at 500 g for 10 min at 4°C, then proceed with viable cell counting (*see* **Note 17**).
12. Keep the tubes on ice, aspirate the supernatant, and immediately proceed to RNA purification according to the RNA extraction kit's manufacturer guidelines (*see* **Note 18**).

3.5 RNA-Seq Analysis

1. After total RNA purification and quantification, prepare the RNA samples for RNA-Seq libraries preparation by keeping the same amount of RNA for each sample according to the manufacturer's instructions (*see* **Note 19**).
2. Proceed with the sequencing of the amplified fragmented cDNAs of 300 bp in size with a read length of 100 bp.
3. Evaluate the sequence read quality using FastQC v.0.11.8 (Babraham Institute, Cambridge, UK), followed by trimming using BBDuk software to remove adapter sequences, poly-A tails, and low-quality end bases (Q <20).
4. Map and align to the human genome reference build GRCh38 (obtained from Ensembl database) using STAR v.2.5.0a [14].
5. Analyze the differentially expressed genes (DEGs) and cluster them by functional annotation in gene ontology and pathway enrichment analysis using Bioconductor R package clusterProfiler v.4.2.0 [15] with annotation of Gene Ontology Database and with annotation of Kyoto Encyclopedia of Genes and Genomes (KEGG) [16] for pathways (*see* **Note 20**).

3.6 Validation of RNA-Seq Results by RT-qPCR

1. Perform the reverse transcription of RNA samples by using a cDNA synthesis kit.
2. Proceed with the qPCR in a 20 μL reaction system containing forward and reverse primers (*see* **Note 21**), cDNA, and SensiFAST SYBR Hi-ROX using a 96-well plate and a qPCR instrument.
3. Calculate the target mRNA levels via the $2^{-\Delta\Delta Ct}$ method, using an appropriate endogenous calibrator gene.

4 Notes

1. The Accutase is a ready-to-use mixture with proteolytic enzyme activity used as a direct replacement of trypsin; the formulation of the Accutase solution is usually secretary; therefore, we suggest following the guidelines in the relevant datasheet of the product.

2. For a T75 flask use 4 mL of 0.1 μg/mL poly-ornithine working solution and 6 mL of 10 μg/mL laminin working solution.
3. It is possible to prepare the coated flasks in advance and store them at −20°C up to a few weeks before use, although freshly prepared coated flasks are recommended.
4. The stationary motion of the fluid-filled culture flask during random rotation could cause shear stress and increased cell deformations, which could affect cell-matrix interactions, especially in low-density cultures [17]. T75 flasks have a larger surface area (around 75 cm^2) compared to T25 flasks (around 25 cm^2); smaller flasks usually experience higher shear stress because they contain a smaller volume of medium, which leads to increased fluid movement and potentially faster fluid flow near the cell surface. In contrast, larger flasks such as T75, with a higher volume of cell culture medium, help distribute the shear forces more evenly, resulting in lower shear stress. However, often using small flasks is more convenient and cheaper, but always consider including at least two different systems to produce simulated microgravity conditions, such as the random positioning machine and the rotating vessel, to rule out confounding shear stress effects.
5. Always use flasks with screw caps without filters to avoid medium leakage and seal with parafilm.
6. It would be wise to perform experiments by seeding different cell densities to consider the possible contribution of shear stress to the observed phenotypes [17].
7. Measure the volume of cell culture medium required to fill the in-use flasks during the experiment settings fully. Prepare the complete cell culture medium according to the needed volume in advance. Filling the flasks as much as possible with the cell culture medium is crucial to avoid air bubbles and foam formation during RPM exposure.
8. Wrap the flasks with absorbent paper and secure them with adhesive tape to the RPM. Do the same with the sham flasks placed on the mechanical agitator with a 2D motion.
9. Set random separate rotations of each frame driven by dedicated software to obtain a constant simulation at the chosen value of the g vector, according to the manufacturer's instructions.
10. For a direct comparison of sham control cells and simulated microgravity-incubated cells, remember to collect microscopy images not only at each experiment time point (up to 24 h of microgravity treatment), but also at time zero, before starting the experiment. If the same flasks will be used also for other functional applications (such as cell viability evaluation,

FACS analysis, and/or RNA purification) keep the flasks on ice during the image acquisition and try to be as quick as possible; indeed, any delay could impact on the downstream analyses.

11. Incubation in simulated microgravity conditions often induces dramatic morphological changes, leading to the formation of "floating cell clumps"; however, consider that the entity of these morphological changes will depend on the experimental conditions; in dense cultures, the cell–cell interactions are stronger and can successfully counteract the weakening of cell–matrix interactions due to simulated microgravity.
12. Dedicating a flask for each condition to a deeper morphological analysis would be recommended. In that case, it is possible to culture the cells directly in flasks on slide, commercially available. Otherwise, quickly taking a representative picture of each condition using at least two magnifications would be enough, then immediately proceed with the other functional applications.
13. Once collected all the medium do not let the cells dry, but add 5 mL of cold PBS to each flask.
14. Aspirate on the opposite side of the cell pellets that are not always visible because of the variable number of floating clumped cells.
15. If the focus is on the features of floating clumped cells versus the attached cells, it is possible to isolate and analyze the floating clumps by collecting and spinning down the cell media; if that is the case, perform all the following protocol steps without mixing the collected floating cells and the ones accutase-detached.
16. Trypan blue selectively colors dead cells blue. This staining method cannot distinguish between necrotic and apoptotic cells. Add the Trypan blue dye to the cell suspension in a volume ratio of 1:1 immediately before each cell counting, otherwise, the viability count will not be reliable. Perform each cell count in a double technical replicate.
17. Knowing the number of viable cells is beneficial to evaluate the reliability of the downstream results; moreover, according to the starting number of cells, different commercial RNA purification kits can be used to maximize the yield. If other applications are needed besides RNA purification, take the appropriate volume of cells for each application.
18. To reduce the variability of RNA purification across different biological replicates, snap-freeze the harvest cell pellets and store them at −80°C until processing. Be sure to dry-out the cell pellets before the snap-freezing to avoid cell lyse.

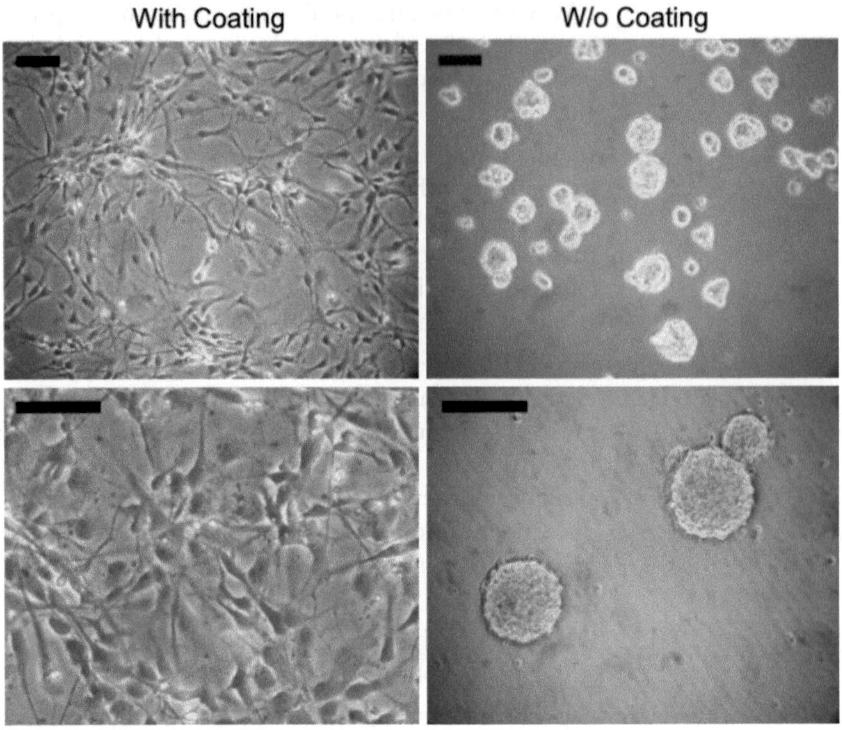

Fig. 3 Human NSPCs under adherent and normal gravity conditions (on the left); human NSPCs under nonadherent and normal gravity conditions (on the right). Scale bars 100 μm; (readapted from [11])

19. Follow the guidelines of the RNA-Seq service company for RNA-Seq sample preparation.
20. Human NSPC culture under nonadherent, normal gravity conditions may induce transcriptional changes similar to those observed upon RPM treatment. To assess the relevance of hNSPC adhesion to a laminin-coated substrate with the observed transcriptional modulations, consider plating the cells on uncoated flasks and culturing in proliferation-supporting media under normal gravity for 5 days, allowing the formation of floating cell aggregates (as shown in Fig. 3) and compare their gene expression levels with those from adherent hNSPC cultures plated at the same density on flasks coated with poly-L-ornithine and laminin in normal gravity.
21. Any residual genomic DNA could interfere with the qPCR results; therefore, it is recommended to perform DNase treatment before proceeding to cDNA synthesis reactions. However, to rule out the contribution of genomic DNA to the amplification plots, use primer pairs spanning on exon–exon junctions, whenever possible. Also, consider adding negative reverse transcription controls in the qPCR to evaluate the genomic DNA contamination of each sample.

Acknowledgments

This work has been supported by the grant: PNRR-PE15-Spoke 9 Attività spaziali_2022-04-25_MP_EZ-1.

References

1. Pacelli C, Ferranti F, Del Bianco M (2024) Special issue: "Advances in space biology". Life (Basel) 14:931. https://doi.org/10.3390/life14080931
2. Andreeva E, Matveeva D, Zhidkova O, Zhivodernikov I, Kotov O, Buravkova L (2022) Real and simulated microgravity: focus on mammalian extracellular matrix. Life (Basel) 12:1343. https://doi.org/10.3390/life12091343
3. Masiello MG, Cucina A, Proietti S, Palombo A, Coluccia P, D'Anselmi F, Dinicola S, Pasqualato A, Morini V, Bizzarri M (2014) Phenotypic switch induced by simulated microgravity on MDA-MB-231 breast cancer cells. BioMed Res Int 2014:652434. https://doi.org/10.1155/2014/652434
4. Kopp S, Slumstrup L, Corydon TJ, Sahana J, Aleshcheva G, Islam T, Magnusson NE, Wehland M, Bauer J, Infanger M et al (2016) Identifications of novel mechanisms in breast cancer cells involving duct-like multicellular spheroid formation after exposure to the random positioning machine. Sci Rep 6:26887. https://doi.org/10.1038/srep26887
5. Kopp S, Warnke E, Wehland M, Aleshcheva G, Magnusson NE, Hemmersbach R, Corydon TJ, Bauer J, Infanger M, Grimm D (2015) Mechanisms of three-dimensional growth of thyroid cells during long-term simulated microgravity. Sci Rep 5:16691. https://doi.org/10.1038/srep16691
6. Grimm D, Wehland M, Pietsch J, Aleshcheva G, Wise P, van Loon J, Ulbrich C, Magnusson NE, Infanger M, Bauer J (2014) Growing tissues in real and simulated microgravity: new methods for tissue engineering. Tissue Eng Part B Rev 20:555–566. https://doi.org/10.1089/ten.TEB.2013.0704
7. Po A, Giuliani A, Masiello MG, Cucina A, Catizone A, Ricci G, Chiacchiarini M, Tafani M, Ferretti E, Bizzarri M (2019) Phenotypic transitions enacted by simulated microgravity do not alter coherence in gene transcription profile. NPJ Microgravity 5:27. https://doi.org/10.1038/s41526-019-0088-x
8. Bizzarri M, Monici M, van Loon JJWA (2015) How microgravity affects the biology of living systems. BioMed Res Int 2015:863075. https://doi.org/10.1155/2015/863075
9. Redden RA, Doolin EJ (2011) microgravity assay of neuroblastoma: in vitro aggregation kinetics and organoid morphology correlate with MYCN expression. In Vitro Cell Dev Biol Anim 47:312–317. https://doi.org/10.1007/s11626-011-9393-8
10. Harper MM, Ye E-A, Blong CC, Jacobson ML, Sakaguchi DS (2010) Integrins contribute to initial morphological development and process outgrowth in rat adult hippocampal progenitor cells. J Mol Neurosci 40:269–283. https://doi.org/10.1007/s12031-009-9211-x
11. Wang W, Di Nisio E, Licursi V, Cacci E, Lupo G, Kokaia Z, Galanti S, Degan P, D'Angelo S, Castagnola P et al (1827) Simulated microgravity modulates focal adhesion gene expression in human neural stem progenitor cells. Life (Basel) 2022:12. https://doi.org/10.3390/life12111827
12. Kechagia JZ, Ivaska J, Roca-Cusachs P (2019) Integrins as biomechanical sensors of the microenvironment. Nat Rev Mol Cell Biol 20:457–473. https://doi.org/10.1038/s41580-019-0134-2
13. Gershovich PM, Gershovich JG, Zhambalova AP, Romanov YuA, Buravkova LB (2012) Cytoskeletal proteins and stem cell markers gene expression in human bone marrow mesenchymal stromal cells after different periods of simulated microgravity. Acta Astronaut 70:36–42. https://doi.org/10.1016/j.actaastro.2011.07.028
14. Dobin A, Davis CA, Schlesinger F, Drenkow J, Zaleski C, Jha S, Batut P, Chaisson M, Gingeras TR (2013) STAR: ultrafast universal RNA-seq aligner. Bioinformatics 29:15–21. https://doi.org/10.1093/bioinformatics/bts635
15. Wu T, Hu E, Xu S, Chen M, Guo P, Dai Z, Feng T, Zhou L, Tang W, Zhan L et al (2021) clusterProfiler 4.0: a universal enrichment tool for interpreting omics data. Innovation (Camb) 2:100141. https://doi.org/10.1016/j.xinn.2021.100141
16. Ogata H, Goto S, Sato K, Fujibuchi W, Bono H, Kanehisa M (1999) KEGG: kyoto encyclopedia of genes and genomes. Nucleic Acids Res 27:29–34. https://doi.org/10.1093/nar/27.1.29
17. Mansour J, Berwanger C, Jung M, Eichinger L, Fabry B, Clemen CS (2023) Clinorotation inhibits myotube formation by fluid motion, not by simulated microgravity. Eur J Cell Biol 102:151330. https://doi.org/10.1016/j.ejcb.2023.151330

Chapter 10

Histochemistry and Immunohistochemistry in Space Biology: Analysis of the Functional Morphology of the Tissue Microenvironment

Dmitrii Atiakshin, Viktoriya Shishkina, Igor B. Buchwalow and Markus Tiemann

Abstract

Approaches based on analytical technologies of molecular histology are of basic importance, among the analytical algorithms for studying the functional morphology of tissue microenvironment in space biology. This is due to both the unique ability to visualize the phenomena under study in situ in the spatial aspect, and the great potential to develop promising innovative protocols for the detection of cells and extracellular structures. This chapter is dedicated to the possible applications of histochemical and immunohistochemical analysis of the tissue microenvironment in space biology.

Keywords Space biology, Microgravity, Tissue microenvironment, Histochemistry and immunohistochemistry, Intercellular matrix, Mast cells

1 Introduction

Rapid development of cosmonautics poses new challenges for the scientific community due to the increased duration of orbital flights and the approaching perspectives of interplanetary flights. The most crucial issue of space biology is adaptation of living organisms to the factors of orbital flight, weightlessness being the most critical challenge among them for the majority of the body functional systems. In particular, a deeper understanding of the mechanisms of extracellular matrix remodeling in zero gravity conditions is essential to improve the effectiveness of preventive measures on board orbital stations. It should be noted that, despite the widespread introduction of omics technologies into research practice, including genomics, transcriptomics, lipidomics, metabolomics, interactomics, etc., histochemical and immunohistochemical approaches in molecular histology have a number of significant advantages over other

Mariano Bizzarri (ed.), *Biomedicine and Biotechnology in Space*, Methods in Molecular Biology, vol. 3000,
https://doi.org/10.1007/978-1-0716-5174-2_10,

technologies for analyzing the functional morphology of the tissue microenvironment, primarily the state of collagen and elastic fibers, mechanisms of collagen fibrillogenesis, interaction of cells and the extracellular matrix. This is due to both the unique ability to visualize the studied phenomena in situ and the great potential for expanding the staining design due to the development of new innovative protocols for detecting cells and extracellular structures. In this case, the morphological analysis of the cellular (immune, nonimmune, stromal) landscape and extracellular matrix of the tissue microenvironment acquires a multidimensional nature, allowing, with the involvement of artificial intelligence technologies, performing spatial phenotyping, profiling, and mapping of the tissue microenvironment to solve key problems of space biology.

The use of multiplex histochemical and immunohistochemical technologies for detecting tissue structures on a histological sample, and artificial intelligence technologies in bioinformatics analysis of big data have significant advantages in interpretation for multidimensional assessment of the tissue microenvironment compared to classical monoplex approaches. Technologies for multiplex staining of histological sections, spatial phenotyping, and digital profiling provide unique information on the state of the immune and stromal landscapes in situ, the extracellular matrix remodeling, as well as the role of intercellular interactions in the implementation of key mechanisms of morphogenesis in space flight conditions.

More detailed information on the methodological approaches to immunohistochemical analysis can be found in the relevant publications [1, 2].

2 Histochemistry

Histochemistry is a science about the structure and chemical composition of cells and tissues, which allows linking together the structure and function of biological objects. The rise of histochemistry as a leading analytical tool for space biologists, histologists, pathologists, and forensic scientists fell in the 70–90 s of the last century. However, the accumulated potential of histochemistry to study chemical components in situ at the tissue, cellular, and subcellular levels is still far from depleted. The introduction of innovative approaches combining well-established techniques of detecting the extracellular matrix, on the one hand, and metachromatic components, on the other hand, provides a multidimensional nature of histochemical analysis—the possibility of simultaneous detection of two or even three structural components. This fact significantly expands the horizons of the histochemical interpretation, since it becomes possible to simultaneously study collagen and elastic fibers with the cellular component of the tissue microenvironment, including mast cells.

Further photo documentation of the histological section using scanning systems allows for spatial phenotyping, profiling, and mapping of the content of search targets while preserving histotopography, which, with further bioinformatics analysis, allows for the identification of new functional relationships existing at the level of the tissue microenvironment. Thus, histochemistry continues to retain its important status in the analytical potential of molecular histology, and in numerous cases prove to be indispensable for solving fundamental and applied problems in space biomedicine.

3 Immunohistochemistry

By linking two scientific disciplines—immunology and morphology—immunohistochemistry has developed into an important tool in research and clinical pathology. In the last decade of the 20th century, the efforts of histochemists led to a significant improvement in reagents and staining protocols. Milestones in this development were the heat-induced antigen demasking technique, signal amplification, and multiple immunolabeling. In this chapter, the problems that are of particular importance in current immunohistochemical practice are discussed in the following order:

- Chemistry of fixation and its influence on the result of the immunohistochemical reaction.
- Procedures that can be used for antigen demasking in formaldehyde-fixed tissue.
- Signal amplification through chain polymer-conjugated technology and tyramide signal enhancement.
- Modern development of multiple immunolabeling.

4 Sample Preparation in Histochemistry and Immunohistochemistry

4.1 Fixation

Tissues subjected to immunohistochemical analysis must meet the following technical criteria:

- Cell and tissue preservation should be adequate to locate the localization of the antigens to be detected, as well as
- The antigenicity of the molecules to be detected should be preserved and accessible to the antibody.

To meet these demands, on the one hand, fixation should maintain the integrity of the tissue over the duration of the immunostaining method, and on the other hand, it should not destroy the antigen of interest. The choice of fixation is a matter of experience, as one must understand that the effect of Fixation cannot predict the preservation of antigenic properties. The epitopes (amino acid sequences) against which the antibody is directed may be hidden (masked), modified, or even completely lost due to fixation. There is no ideal fixative that can be universally recommended in immunohistochemistry. The choice of adequate fixation can only be evaluated through experiments.

Three important types of fixation are known in immunohistochemistry: air drying, shock freezing, and chemical fixation.

The air drying with or without subsequent chemical fixation is particularly applicable for cell smears, for cytospins, for cell monolayers, and for cryostat sections.

The shock freezing (usually in liquid nitrogen) can be routinely used for tissue and for samples suitable for cryosections. The tissue specimens are shock frozen with or without cryoprotectant embedding medium (Tissue Tek, Miles Laboratories). The sections are then stored at −80°C until analysis. Optionally, an aldehyde fixation can also be used for tissues and organs before the tissue pieces are shock frozen.

The chemical fixation is performed in aldehydes, acetone or alcohol (methanol, ethanol). Usually, cells or cryostat sections are fixed with acetone or alcohol for about 5–15 min to preserve the structure and enable the visualization of most antigens. A limiting factor of these fixatives is that soft tissue or fatty tissue cannot be adequately stabilized with acetone or alcohol, resulting in the loss of the morphological integrity of the tissue. This is often also associated with a loss of the antigen itself. Therefore, formaldehyde is currently the best fixative, provided, however, that the antigen to be investigated does not lose its antigenic properties due to structural changes induced by formaldehyde. Formaldehyde has various advantages over alcohol and acetone, especially, it is far superior in preserving the morphological details.

When the tissues are embedded in paraffin or synthetic plastic, a formalin fixation is the best choice. Formaldehyde, which was originally discovered as a tissue fixative by Ferdinand Blum in 1893, is now predominantly used in histological and pathological investigations. The formaldehyde fixative consists of a commercially available formalin (37%–40% solution of formaldehyde), which is diluted to a 10% solution (3.7%–4% formaldehyde), usually in phosphate-buffered saline (0.01 M phosphate buffer, 150 mM NaCl, pH 7.4; PBS). It can be recommended as a universal fixative for routine use. Tris–HCl buffer with or without isotonic saline can also be used.

Formalin solutions, which are produced by dissolving and depolymerizing paraformaldehyde (a homopolymer of formaldehyde with empirical formula $HO(CH_2O)_nH$, where $n \geq 6$), are free from impurities with methanol or formic acid. Depolymerized paraformaldehyde is useful in enzyme histochemistry, when the preservation of enzyme activity is of particular importance. It has no advantage over formalin solutions that are routinely used in immunohistochemistry.

In most cases, fixation is performed at room temperature. The duration of formalin fixation depends on the type and size of the tissue pieces and can vary between 15 min and 24 h. Longer fixation times can be associated with a partial loss of antigenicity. After formalin fixation, the tissues are washed three times in buffered saline solution (PBS) for 15 min to 2 h, preferably not longer, as the formaldehyde fixation is partially reversible. After washing in PBS, the tissues can be shock frozen in liquid nitrogen followed by the preparation of cryosections—or they are dehydrated and embedded in paraffin or synthetic plastic. Cell smears and cell monolayers are fixed in alcohol or acetone and after a short wash in buffered saline solution directly subjected to an immunohistochemical analysis.

4.2 Antigen Unmasking Techniques

The antigen–antibody reaction in immunohistochemistry usually takes place between two protein macromolecules: the antigen, which can be a glycoprotein, a lipoprotein or a free protein, and the antibody, which is a glycoprotein. An antigen unmasking is necessary on a formalin-fixed and paraffin-embedded tissue. While the structural integrity of the immunoglobulin protein is well preserved—in its native form—, the structure of the antigen protein in the tissue must be considered altered. The antigen has been exposed to the effect of the formaldehyde fixative and may be altered in its properties or in its structural conformation. Routine formaldehyde fixation and the other components of the procedure for paraffin embedding may, thus, have altered the protein macromolecule, resulting in a reduced intensity of the immunohistochemical reaction. To restore the antigenicity of the tissue sections, the tissue sections must, therefore, undergo antigen unmasking. In the current literature, antigen unmasking (Antigen Retrieval) is defined as a high-temperature heating method.

The development of antigen unmasking as a simple method of boiling archived, paraffin-embedded tissue sections in buffered saline solution was the result of pioneering work by Shi et al. [3], with the success of a signal amplification in immunohistochemistry. The goal here was a broad application of immunohistochemistry in routinely formalin-fixed and paraffin-embedded sections for research and clinical pathology. The wide application of this technique in pathology and other fields of morphology has shown that the antigen unmasking is applicable for a number of antigens, such as for Ki-67, MIB-1, estrogen, and androgen receptors, as well as for many cytokeratins and CD markers. These antigens can only be weakly or not at all represented in the immunohistochemical staining without antigen unmasking. Earlier unmasking techniques, such as enzymatic digestion, have not led to a satisfactory result for many antigens, not to mention that the enzymatic digestion of tissue can also lead to the destruction of the structural integrity of the tissues.

The hypothesis of heat-induced renaturation is based on a fundamental principle of immunology, which implies that antigen–antibody reactions depend on the protein structure. Antibodies recognize specific epitopes in a certain arrangement in the protein molecule. A structural change of the protein leads after formalin fixation to a masking of the epitopes and thus influences the antigenicity of the protein in this tissue. The antigen unmasking can lead to a renaturation or at least to a partial restoration of the protein structure.

For the antigen unmasking, the tissue is heated in a suitable vessel with antigen unmasking solution (e.g., 10 mM citrate, pH 6, or 1 mM EDTA buffer, pH 8 or 9) to 90–110°C (depending on the tissue). Microwave devices, steam cookers, normal pressure cookers, or autoclaves can be used for this purpose.

4.3 Signal Amplification

While the antigen unmasking technique is used to improve immunocytochemical signals in the predetection phase, the conventional method of signal amplification, such as the Avidin-Biotin Complex (ABC) or the Enzyme-Anti-Enzyme Immune Complex Techniques (PAP and APAAP), are used to amplify the signal in the phase of the detection process. The ABC method is essentially based on the high affinity that streptavidin (from Streptomyces avidinii) and avidin (chicken egg white) have for biotin. The PAP method uses a peroxidase-antiperoxidase complex. APAAP, finally, represents an alkaline phosphatase-anti-alkaline phosphatase complex. For many years, the PAP and APAAP methods were the most sensitive, and therefore, the most popular techniques in many pathology laboratories, not least because of their easy handling. Today, these techniques are used less frequently and are increasingly being replaced by more sensitive methods. A multiple higher antigen detection can be achieved, for example, by the chain polymer conjugated technology (EnVision system) developed by DakoCytomation. This technology uses an enzyme-labeled inert dextran molecule. The dextran can on average be loaded with 70 molecules of an enzyme (alkaline phosphatase or horseradish peroxidase), as well as ten molecules of antibodies. Since this system does not rely on streptavidin and biotin, unspecific staining by endogenous biotin can be avoided.

With the introduction of tyramine conjugates (tyramides) as a substrate for the horseradish peroxidase (HRP), the sensitivity of immunohistochemical methods was greatly improved and further applications of immunohistochemistry were opened up. In this procedure, known as Catalyzed Reporter Deposition (CARD) or tyramide signal amplification (TSA), HRP-labeled secondary antibodies or HRP-labeled avidin are used. HRP catalyzes the oxidation of the phenol unit of the labeled tyramine in the presence of H_2O_2. After oxidation by HRP, activated tyramine molecules quickly covalently bind to electron-rich amino acids of proteins that are in the immediate vicinity of the immune reaction. This method, therefore, allows an increase in the detectability of an antigen by a hundredfold, compared to the conventionally indirect method, without loss of resolution.

The tyramide signal amplification technology (TSA) was developed by the Litts group [4] at DuPont NEN (now part of the Perkin-Elmer Corporation) and licensed to Molecular Probes for in-cell and in-tissue applications.

While the fluorophore markers can be visualized in a fluorescence microscope, the biotin labeling can be further amplified via the ABC technique with HRP-conjugated avidin or streptavidin to then be visualized through chromogenic enzymatic reactions in a bright field microscope. Alternatively, the second layer of HRP labeling can continue to be amplified with the next layer of fluorophore tyramine. The combination of tyramide-based amplifications with the ABC method is referred to as a Blast amplification technique. It allows for further signal amplification up to a thousand times. This can even detect antigens with very low expression, where conventional indirect methods fail. In addition, the amplification allows the primary antibody to be highly diluted. The higher dilution results in less nonspecific background staining, while simultaneously intensifying specific signals.

These advanced immunohistochemical staining techniques in pathology and other areas of morphology have led to a significant improvement in immunohistochemical staining reactions on archived, formalin-fixed, and paraffin-embedded tissue sections and increased sensitivity.

4.4 Multiple Immunofluorescence Labeling

Multiple labels for the colocalization of different antigens were developed in two different lines—on the one hand the multicolor immunoenzyme methods and the immunofluorescence staining methods. While the multicolor immunoenzyme staining is only applicable for antigens with different locations, the fluorescence method allows antigens to be detected not only in the same cell, but even in the same cellular compartment.

Methods for the simultaneous immunofluorescence detection of different tissue antigens use in their simplest form primary antibodies that were developed in different species, and therefore, can be visualized by different species-specific labeled secondary antibodies using indirect immunocytochemical methods (two or more layers of antibodies). Unfortunately, appropriate combinations of primary antibodies from different species are not always available. A general problem results from the fact that the most available primary antibodies were developed in two animal species, namely the rabbit and the mouse. The use of primary antibodies of the same species, however, leads to cross-reactions of the secondary, species-specific antibody with the primary antibodies.

The cross-reaction of the secondary, species-specific antibodies with the primary antibodies of the same species can be avoided by a direct labeling method (one layer of antibodies). Recently, a direct method has been developed with primary antibodies that are covalently labeled by different fluorophores. They are, therefore, suitable for simultaneous detection of up to seven antigens.

The multicolor fluorescence immunolabeling not only allows the antigens to be recognized in the same cell, but also in the same cellular compartment. It should be noted here that the coincidence of two or more antigens in the same compartment cannot be convincingly represented by a composite image. For this purpose, the composite image is broken down into its individual color components.

Multiple immunofluorescence labeling methods represent a successful strategy to represent the spatial relationship between different antigens in histological section preparations.

Another approach to preventing cross-reactivity in multiple immunolabeling using antibodies of the same species is antibody elution. A number of buffers with different pH, osmolarity, detergent content, and denaturing properties were tested to remove the already bound antibody complex from previous IHC staining cycles. It has been reported that glycine–HCl/sodium dodecyl sulfate (SDS), pH 2, or 2-mercaptoethanol/SDS, pH 6, 75, are effective [5]. Recently, Zhang et al. [6] reported on multiple fluorescence immunolabeling using antibodies of the same species in combination with tyramide signal amplification. In this method, the first antigen on the tissue is detected by using a target-specific primary antibody and then a secondary HRP-labeled species-specific antibody, which in turn controls the in situ deposition of fluorophore-conjugated tyramide. The bound primary antibody and the secondary antibody-HRP complex are then eluted with a citrate/acetate-based buffer, pH 6, 0, containing 0.3% SDS (CC2 solution, available from Ventana cat# 950-223), at high temperature. The tyramide fluorophore molecule remains covalently bound to the tissue near the first target. The same procedure is repeated to detect the next targets. This technology allows the use of primary antibodies from the same species, e.g., mouse or rabbit.

4.4.1 Protocol for Multiple Immunostaining by Use of Monoclonal Primary Antibodies of Different Ig Isotypes

- Deparaffinize and rehydrate the tissue. Wash in distilled water for 5 min.
- Antigen unmasking: Heat section preparations in a suitable vessel with an antigen unmasking solution (e.g., 10 mM citric acid, pH 6) to 90–110°C (depending on the tissue) in a Microwave, a steam cooker, a pressure cooker, or autoclave.
- Washing and Diluting: Rinsing the sections in 10 mM PBS for 2–3 min. PBS is used for all rinses and dilutions. Other buffers such as Tris-buffered saline (TBS) can also be used.
- Primary Antibody: Incubation for 60 min at room temperature or overnight at 4°C with a mixture of the adequately diluted primary antibodies. Rinse the section preparations in PBS for 2–3 min.

- Secondary Antibody: Incubation of the sections for 60 min at room temperature with a mixture of the fluorophore-conjugated secondary antibodies, which are directed against the corresponding IgG isotypes (IgG1 or IgG3) of the primary antibody. Rinse the sections in PBS for 2–3 min.
- Counterstaining of Nuclei, if necessary with DAPI (5 μg mL—1 PBS). Rinse the sections in PBS for 2 × 3 min.
- Covering: Covering the sections in 80%–90% glycerin in PBS with an antifading substance (0.1% p-phenylenediamine or 2% n-propyl gallate). Molecular Probes (Leiden, The Netherlands) offers a wide selection of antifading mounting media. The authors have had good experiences with Vectashield mounting medium from Vector Laboratories (Burlingame, USA).
- All incubations are carried out at room temperature, unless otherwise stated.

Of course, this protocol can also be used for simpler immunostaining, if primary antibodies were produced in different species.

Micrographs with simultaneous visualization of tryptase and chymase from mast cells of Mongolian gerbils using diplex immunohistochemical staining, returning from a 12-day space flight, are presented in Atiakshin et al. [7, 8].

4.4.2 Multiple Immunoenzyme Labeling

Procedures that use simultaneous staining of two or more tissue antigens in immunoenzyme staining methods suffer from the fact that the chromogen deposits of the first antigen can mask the corresponding epitopes of the second antigen. This makes simultaneous detection of two antigens in the same cellular compartment (e.g., nucleus, cell membranes, or cytoplasm) unfeasible. Nevertheless, this approach is successfully used in cases where the antigens are located in different cellular compartments or—even simpler—in different cells.

When applying the method to frozen sections, the endogenous enzyme activity can occasionally produce nonspecific staining. To suppress, for example, the endogenous horseradish peroxidase (HRP) activity, the section preparations are previously treated with a 0.6% H_2O_2 solution in methanol. The endogenous alkaline phosphatase activity can be blocked by adding the inhibitor of alkaline phosphatase—levamisole—to the substrate solution. Additional information about enzyme activities can be found in corresponding textbooks of enzyme histochemistry [9–11] or in the manufacturer's specifications (Roche Molecular Biochemicals, Sigma, Vector Laboratories, DakoCytomation, Immunotech).

The enzymatic methods to detect two antigens in the same location can be divided into two principal categories:

- simultaneous immunoenzyme double staining and
- sequential immunoenzyme double staining.

Simultaneous immunoenzyme double staining is only applicable for primary antibodies that were produced in different species, as well as when using different enzyme systems to mark the antigens. The sequential immunoenzyme double staining is also applicable for primary antibodies that were produced in the same species (e.g., two mouse-monoclonal antibodies). As a rule, HRP represents the location of the corresponding antigen better than alkaline phosphatase. In double staining, the HRP should be developed before the alkaline phosphatase. The most strongly expressed antigen should be stained first, followed by the staining of the less expressed antigen.

4.4.3 Protocol for Simultaneous Immunoenzyme Double Staining Using Primary Antibodies Produced in Two Different Species

- Deparaffinize and rehydrate the tissue. Wash in distilled water for 5 min.
- Antigen unmasking: Heat section preparations in a suitable vessel with an antigen unmasking solution (e.g., 10 mM citrate, pH 6) |to 90–110°C (depending on the tissue) in a microwave, a steam cooker, a pressure cooker, or an autoclave.
- Washing and diluting: Rinse the sections in 10 mM PBS for 2–3 min. PBS is used for all rinses and dilutions unless otherwise stated. Other buffers, such as Tris-buffered saline (TBS), can also be used.
- Primary antibody: Incubate for 60 min at room temperature or overnight at 4°C with a mixture of the adequately diluted primary antibodies. Rinse the section preparations in PBS for 2 × 3 min.
- Secondary antibody: Incubate the sections for 60 min at room temperature with a mixture of the enzyme-conjugated secondary antibodies (e.g., HRP or AP) that are directed against the corresponding species-IgG of the primary antibody. Rinse the sections in the same buffer in which the first enzyme substrate is dissolved.
- Substrates: Incubate the sections with the first enzyme substrate. Rinse first with the same buffer and then with the buffer in which the second enzyme substrate is dissolved, for 2 × 3 min. Incubate the sections with the second enzyme substrate and rinse with the same buffer and then with distilled water.
- Counterstaining and mounting: Counterstain the nuclei (if necessary) and mount with a suitable aqueous or permanent mounting medium.

All incubations are performed at room temperature, unless otherwise stated.

4.4.4 Protocol for Sequential Immunoenzyme Double Staining

- Deparaffinize and rehydrate the tissue. Wash in distilled water for 5 min.
- Antigen unmasking: Heat section preparations in a suitable vessel with an antigen unmasking solution (e.g., 10 mM citrate, pH 6)

to 90–110°C (depending on the tissue) in a microwave, a steam cooker, a pressure cooker, or an autoclave.

- Washing and diluting: Rinse the sections in 10 mM PBS for 2–3 min. PBS is used for all rinses and dilutions unless otherwise stated. Other buffers, such as Tris-buffered saline (TBS), can also be used.
- The first immunolabeling sequence: Incubation for 60 min at room temperature or overnight at 4°C with the adequately diluted first primary antibody. Rinse the section preparations in PBS for 2 × 3 min.
- Secondary antibody: Incubate the sections for 60 min at room temperature with a corresponding enzyme-conjugated secondary antibody. Rinse the sections in the same buffer in which the first enzyme substrate is dissolved.
- Substrate: Incubate the sections with the first enzyme substrate. Rinse first with the same buffer and then with PBS for 2 × 3 min.
- The second immunolabeling sequence: To locate the second antigen, the sections are incubated with the second primary antibody and then with the second enzyme-conjugated secondary antibody. Rinse the sections in the same buffer, in which the second enzyme substrate is dissolved. Visualize the second enzyme marker and wash the sections with the same buffer, then wash with distilled water.
- Counterstaining and covering: Counterstain the Kernels (if necessary) and cover with suitable aqueous or permanent mounting medium.
- All incubations are performed at room temperature, unless otherwise stated.

A third antigen can be displayed (triple labeling) by simply repeating the protocol with an additional staining sequence that is suitable for the third antibody. Multiple labeling requires more time for antigen–antibody binding. If necessary, one should take a break between the staining of individual antibodies. Vector Laboratories recommend that if the staining protocol is interrupted, this should ideally be done after the substrate development of the first antigen. The section preparations should be stored in a buffer for up to 1 h at room temperature (20–25°C) during this time, without affecting the staining results. For a longer time, however, it is recommended to store the section preparations in PBS at 4°C. The success of the sequential immunoenzyme method for double staining with antibodies developed in the same species is based on the complete execution of the first antigen representation, so that the chromogen can completely cover these structures. However, there is a risk of possible cross-reactivity with the second antibody. The chain polymer-based technology of the EnVision system, developed by DakoCytomation, avoids this error. The application of DakoCytomation's Double-Stain-Block before the next step prevents possible cross-reactivity.

4.5 Background Staining and Blocking Steps

Unwanted background staining has added gray hair to the heads of many histochemists. In most cases, the background staining is not caused by a single factor. In addition to the Fc receptors, common causes of background staining are endogenous enzyme activity, when peroxidase or alkaline phosphatase is used as enzyme markers of the secondary antibodies. Also, endogenous biotin, when using a streptavidin or avidin reagent, leads to increased background.

4.5.1 Fc Receptors

It is assumed that a background staining as a result of a nonspecific antibody binding to endogenous Fc receptors (FcRs) occurs. In our study [12], cell and tissue samples were routinely processed with or without blocking step (serum or BSA). Surprisingly, no antibodies in samples that were processed without a blocking step showed a tendency to nonspecific binding, which led to background staining, implying that endogenous FcRs do not retain their ability to bind the Fc part of antibodies after standard fixation. Similarly, we found no nonspecific antibody binding that is either due to ionic or hydrophobic interactions. We have found that traditionally used protein blocking steps in the immunostaining of routinely fixed cell and tissue samples are unnecessary. Omitting the conventional protein blocking step can save both reagent costs and preparation time in research, as well as in immunohistopathology.

4.5.2 Endogenous Peroxidase

The use of HRP as an enzyme marker can lead to a strong nonspecific background staining of some cells and tissues that contain endogenous peroxidase. This nonspecific background can be reduced by pretreatment of cells and tissue sections with hydrogen peroxide (H_2O_2) can be eliminated.

Peroxidase Blocking Solution (0.6% H_2O_2 in Methanol or PBS).

Mix 1 mL of 30% H_2O_2 and 50 mL of Methanol or PBS well and store at 4°C. Perform blocking steps for 10–15 min before or after the primary antibody incubation.

Methanol is a better choice in most cases. Although methanol can reduce the staining of some cell surface antigens, such as CD4 and CD8, you are not doing anything wrong even for these sensitive antigens, if you use peroxidase-blocking solution with methanol after (not before) incubation with primary antibodies. The methanol treatment can also cause frozen sections to detach from slides. In such rare cases, the use of hydrogen peroxide in PBS is recommended.

4.5.3 Endogenous Alkaline Phosphatase

Alkaline phosphatase is an enzyme that occurs in various isoforms in many tissues, such as liver, bone, intestine, placenta, some tumors, and in leukocytes. The addition of 1 mM levamisole to the chromogen/substrate solution inhibits the endogenous activity of the alkaline phosphatase with the exception of the intestinal isoform. If required, this can be blocked by washing with a weak acid, such as 0.03–0.5 N HCl or 1 M citric acid.

1 M Citric Acid Solution

- Dissolve 192 g of citric acid (free acid) in 500 mL of distilled water.
- Dilute to 1 L with distilled water.

When using high-temperature antigen retrieval, you can omit the blocking of alkaline phosphatase, as some endogenous enzymes, such as alkaline and acid phosphatases, unlike peroxidase, are already destroyed by short-term boiling at 100°C.

4.5.4 Endogenous Biotin

Biotin is a B-vitamin that is present in some organs, such as kidney, liver, and spleen. It can be a cause of high nonspecific background staining when using the ABC method. To prevent this background staining, an avidin/biotin block can be applied before the incubation with biotinylated antibody on tissue sections that contain moderate to high amounts of this vitamin. Blocking of endogenous biotin can be done with commercially available kits, however, they are expensive.

Avidin/Biotin Blocking

- A: Avidin 0.01% (≈1.7 µM) in PBS
- B: Biotin 0.05% (2 mM) in PBS

Store blocking solutions at 4°C.

Incubate the sections with solutions A and B each for 10–15 min before the primary antibody application and rinse 2× with PBS between the steps.

Normally, avidin/biotin blocking is avoidable when working with paraffin tissue sections, as the tissue pretreatment in this case leads to a complete loss of endogenous biotin. However, you may also encounter significant endogenous biotin artifacts in paraffin sections, especially when signal amplification is used with biotinylated tyramide.

5 Remodeling of the Extracellular Matrix: Mast Cells and Collagen Fibrillogenesis

Aspects of regeneration of extracellular matrix components of the connective tissue have always been in the limelight of researchers. Advanced methodological techniques allowed us to gradually understand new patterns of the integrative buffer metabolic environment function in both physiological and pathological conditions [13–20]. The pioneering works in morphology aimed at the identification of the selective components of the extracellular matrix initiated a large number of studies for the fibers and the amorphous component detection [9, 10, 21–26]. This allowed designing numerous histochemical criteria of the extracellular matrix diagnostically significant for the analysis of adaptive processes and the pathological change development assessment. The use of electronic, scanning, and confocal microscopy has opened exceptional opportunities in the knowledge of the structural and physiological features of the connective tissue at the level of supramolecular aggregates [17–19].

Until now, ongoing work in this direction has provided new data expanding the range of regulatory mechanisms in updating the intercellular matrix components [18, 27–29]. Further clarification of tissue cellular composition functions in the amorphous matrix formation and fibrous element state of the specific tissue microenvironment, which is considered to be an important condition for solving these problems. In particular, the role of macrophages in the arrangement of newly formed collagen fibers was discussed. Mast cells (MC), which are distributed practically everywhere in the body and form organ-specific populations, should be considered as one of the promising objects of local homeostasis in the intercellular matrix regulation [30–34]. It is known that MCs may compose 10% of all immunocompetent cells in the skin [35]. Secretome functional characteristics allow MCs to act as inducers and regulators of the most important physiological and pathological processes—coordination of tissue microenvironment homeostasis, realization of innate and adaptive immunity, initiation and subsequent development of inflammatory and allergic reactions, angiogenesis, cell proliferation, and others among them [36]. At the same time, a number of secretome components, including proteoglycans, specific proteases, TGF-β, etc., take both direct and indirect part in changes occurring in the structure of the stroma of internal organs, endowing MCs with special potentials in terms of remodeling the intercellular matrix [37–39]. A frequent association of MC population indices with the intensity of fibrotic changes in the relevant organs suggested their active involvement in the induction of the biogenesis of collagen proteins and extracellular stages of fibrillogenesis. This determines the importance of developing and introducing into morphological practice methodological techniques that contribute to specifying the points of application of MC selective effects in the collagen fiber formation, as well as more objective assessment of their contribution to the development of pathological processes with progressive tissue fibrosis.

However, the use of certain histochemical techniques limits the interplay potential assessment between specific connective tissue cells and a fibrous component, which significantly reduces the informative content of visualized structures in relation to both understanding tissue histoarchitectonics and interpreting morphological aspects of the tissue microenvironment fibrous matrix remodeling.

With this study, we present the results of combinations of standard stainings of MC with the stainings of the intercellular matrix fibrous component of the skin connective tissue. These new combination stainings shed new light on the participation of mast cell secretion components in the molecular mechanisms of fiber formation and on the extracellular matrix structure remodeling.

The study of regeneration under altered gravity conditions, including the wound surface, is an acute issue for space biology. Histochemistry techniques are an effective tool for solving this problem.

5.1 Materials and Methods

This study investigated the skin [dorsal neck area at the level of 5–6 cervical vertebrae (CV–CVI)] of Wistar rats of the intact group ($n=6$) and animals with soft-tissue wounds simulated using excision of the skin and subcutaneous tissue (7 days after modeling the wound process, $n=6$). Skin biomaterial samples were fixed in buffered 4% formaldehyde and routinely embedded in paraffin. PBS was used for all washings and dilutions. Paraffin tissue sections (5 μm thick) were deparaffinized with xylene and rehydrated with graded ethanols. For pathohistological analysis, tissue sections were routinely stained with H&E. After dehydration and clearing in xylene, the sections were covered with permanent mounting medium and observed on ZEISS Axio Imager.A2, and Z2 research microscopes with the Camera Axiocam 506 and 712 color, or 712 mono.

Toluidine blue, brilliant green, neutral red, alcian blue stain, periodic acid Schiff reaction, and histochemical detection of chloroacetyl esterase were used to detect MCs (Table 1). In addition, methylene blue, azur I, and polychrome acid fuchsin with aniline blue were used, which are part of the May–Grünwald, Giemsa, and picro-Mallory solutions, respectively. Silver impregnation technique, picrofuchsin stain in isolation or using van Gieson protocol stain, aniline blue stain using picro-Mallory technique protocol, fuchsin–resorcinol stain using Weigert's method, and combined alcian blue stain and periodic acid Schiff reaction were used for simultaneous visualization of the intercellular matrix of the connective tissue with MCs [30, 40] (Buchwalow et al. 2018). Moreover, the identification of MCs is preferably performed after staining the structures of the extracellular matrix. Brilliant green (tetraethyl-4,4-diaminotriphenylmethane oxalate) for histochemical detection of MCs and related elements of the intercellular substance used at a concentration of 1% dye solution. For staining nuclei, Carazzi's hematoxylin was used before staining with brilliant green. Combinations of histochemical techniques used in this study are presented in Table 1. All reagents were purchased from BioVitrum, Moscow, 129626, Russia. Below, we provide more detailed protocols for the combination stainings with toluidine blue and brilliant green.

Table 1
Histochemical staining combinations

Combinations	Identification of intercellular matrix components	Mast cell detection
1	Silver ammonia solution impregnation protocol	Toluidine blue
2	Picrofuchsin	Toluidine blue
3	Van Gieson protocol stain (picrofuchsin)	Toluidine blue
4	Brilliant green*	Toluidine blue
5	Aniline blue	Toluidine blue
6	Acid fuchsin	Toluidine blue
7	Eosin	Toluidine blue
8	Van Gieson protocol stain (picrofuchsin)	Brilliant green
9	Picrofuchsin	Brilliant green
10	Aniline blue	Brilliant green
11	Eosin	Brilliant green
12	Weigert's method (fuchsin-resorcinol stain)	Neutral red
13	Aniline blue	Neutral red
14	Brilliant green*	Neutral red
15	Van Gieson protocol stain (picrofuchsin)	Neutral red
16	Alcian blue stain + periodic acid Schiff reaction	Alcian blue stain, periodic acid Schiff reaction
17	Gimsa solution (eosin) + May–Grunwald solution (methylene blue)	Giemsa solution (azur I) + May Grunwald solution (methylene blue)
18	Van Gieson protocol stain (picrofuchsin) + May–Grunwald solution (eosin)	May–Grunwald solution (methylene blue)
19	Picrofuchsin + May–Grunwald solution (eosin)	May–Grunwald solution (methylene blue)
20	Picro-Mallory staining protocol [polychrome acid fuchsin (acid fuchsin + orange G), aniline blue]	Picro-Mallory staining protocol [polychrome acid fuchsin (acid fuchsin + orange G), aniline blue]

*Brilliant green can be used to stain mast cells and fibrous elements of the connective tissue.

5.2 Combined Staining with Toluidine Blue and Silver Impregnation

5.2.1 Solutions

1. 0.25% aqueous potassium permanganate.
2. 1% oxalic acid.

3. 5% iron ammonium alum (ferric ammonium sulfate).
4. Ammonia silver: prepare immediately before use.
   - 2 mL of 10% aqueous silver nitrate.
   - 1 mL of 10% aqueous sodium hydroxide or potassium hydroxide.
   - 0.5 mL concentrated ammonia solution.
   - Mix well. Most sediment dissolves in 1–2 min.
5. Formalin solution:
   - 10 mL of concentrated aqueous formaldehyde.
   - 40 mL of distilled water.
   - Prepare before use.
6. 5% aqueous sodium thiosulfate.

5.2.2 Protocol

1. Dewax and bring sections to water. Rinse in distilled water.
2. Potassium permanganate—5 min. Rinse in distilled water for 1 min.
3. Decolorize with oxalic acid (3 min). Rinse with distilled water.
4. Treat with ammonium alum for 5 min. Rinse in four shifts of tap water for 3 min each.
5. Rinse the slides in distilled water for 2 × 3 min.
6. Ammonia silver for 5 min.
7. Rinse quickly in distilled water (one or two dips).
8. Formalin solution for 1 min.
9. Rinse in distilled water for 2 × 3 min.
10. Sodium thiosulfate for 1 min.
11. Rinse in distilled water.
12. Stain mast cells with toluidine blue (0.1% aqueous solution) for 5 min.
13. Rinse quickly in 99% isopropyl alcohol.
14. Coverslip with permanent mounting medium.

5.3 Combined Staining with Toluidine Blue and Picrofuchsin

5.3.1 Solutions

1. Picrofuchsin solution: 5–10 mL of a 1% aqueous solution of acid fuchsin is added to 100 mL of a saturated aqueous solution of picric acid.
2. 0.1% aqueous solution of toluidine blue.

5.3.2 Protocol

1. Dewaxing sections. Rinse in distilled water.
2. Staining with picrofuchsin solution for 5 min.
3. Rinse with tap water for 2 min.
4. Staining with 0.1% aqueous toluidine blue solution for 5 min.
5. Rinse quickly in 99% isopropyl alcohol.
6. Coverslip with permanent mounting medium.
7. Combined staining with brilliant green and van Gieson's stain.

5.3.3 Solutions

1. To prepare a 1% solution of brilliant green, dissolve 1 g brilliant green in 52 mL 95% ethyl alcohol, add distilled water to 100 mL volume, and filter.
2. Weigert's hematoxylin is prepared immediately before staining, mixing equal volumes of Weigert basic solutions (A and B), which are stored separately.
   - Weigert solution (A): 1 g hematoxylin is dissolved in 100 mL of 96% ethanol;
   - Weigert solution (B): 4 mL of a 29% aqueous solution of ferric chloride is mixed with 1 mL of concentrated hydrochloric acid and 95 mL of distilled water is added. When mixing the solutions, the second solution is poured into the first in a somewhat smaller amount, and then, the second solution is added to the resulting mixture dropwise from the pipette to an equal amount with the first solution.
3. Picrofuchsin solution: 5–10 mL of a 1% aqueous solution of acid fuchsin is added to 100 mL of a saturated aqueous solution of picric acid.

5.3.4 Protocol

1. Dewax sections and bring to the water.
2. Apply Wiegert's hematoxylin for 5 min.
3. Rinse with tap water for 5 min, then in distilled water for 5 min.
4. Stain with picrofuchsin solution for 5 min.
5. Apply four drops of 1% brilliant green solution to a slide for 40 s.
6. Immerse in 99% isopropyl alcohol, differentiate for 5–15 min under microscope control.
7. Coverslip with permanent mounting medium.

5.4 Combined Staining with Brilliant Green and Picrofuchsin

5.4.1 Solutions

1. 1% brilliant green solution (preparation see above).
2. Picrofuchsin solution (preparation see above).

5.4.2 Protocol

1. Dewaxing sections and bring to the water.
2. Apply four drops of 1% brilliant green solution on the slide for 40 s, then quickly immerse the slide in 99% isopropyl alcohol, and differentiate for 5–15 min under microscope control.
3. Staining with picrofuchsin solution for 5 min.
4. Rinse quickly in distilled water.
5. Quickly dehydrated in 99% isopropyl alcohol.
6. Coverslip with permanent mounting medium.

5.5 Results

5.5.1 Toluidine Blue

There are cases where isolated staining with toluidine blue allows differentiating mast cells and the surrounding stroma (Fig. 2f); however, the almost complete absence of its contrast makes it challenging to detect the fibrous component of the intercellular substance. The combination of toluidine blue staining with impregnation with silver has demonstrated the successful preservation of metachromatic staining of MCs (Figs. 1 and 5c, g). Identification of the fibrous component of the dermis is accompanied by selective contrasting of thin reticular or precollagen fibers stained in black, while thicker and more mature collagen fibers acquire different shades of brown–yellow color (Figs. 1 and 5c, g). The degree of effectiveness of using this method to assess the colocalization of collagen fibers with MCs depends on the glycosaminoglycans of their secretome, which determine the intensity of metachromasia. In particular, this combination appeared to be more successful in the study of laboratory animals with skin wound simulation compared to intact rats (Fig. 1). MC metachromasia allowed identifying their adherence to reticular fibers containing type III collagen.

The frequency of MC contacts with impregnated fibers was numerous and may represent a separate task of morphological analysis when studying the characteristics of the skin MC population. Part of the MC was freely located in the intercellular matrix, while the other lay adjacent to the impregnated fibers (Fig. 1f, g, k), sometimes forming a pronounced network (Figs. 1c–e, h, i and 5c). Attention was drawn to the fact that on some MCs, there are thin fibers that were in contact with the surface of the plasmalemma or which started at the microlocus of their adjacency to the cell and continued into the intercellular matrix of the tissue microenvironment (Figs. 1c, g, i, j, l and 5c, g). Similar patterns were observed when MCs of isolated multicellular groups were formed; radially and tangentially oriented thin impregnated fibers departed from the plasmalemma of these MCs (Fig. 1b). In addition, the MC degranulation can be observed through the exocytosis mechanism in the plasmalemma area aside from the contact with fibers.

MCs were sometimes identified, plainly mounted in the framework of the reticular fibers, covering the plasmalemma from all

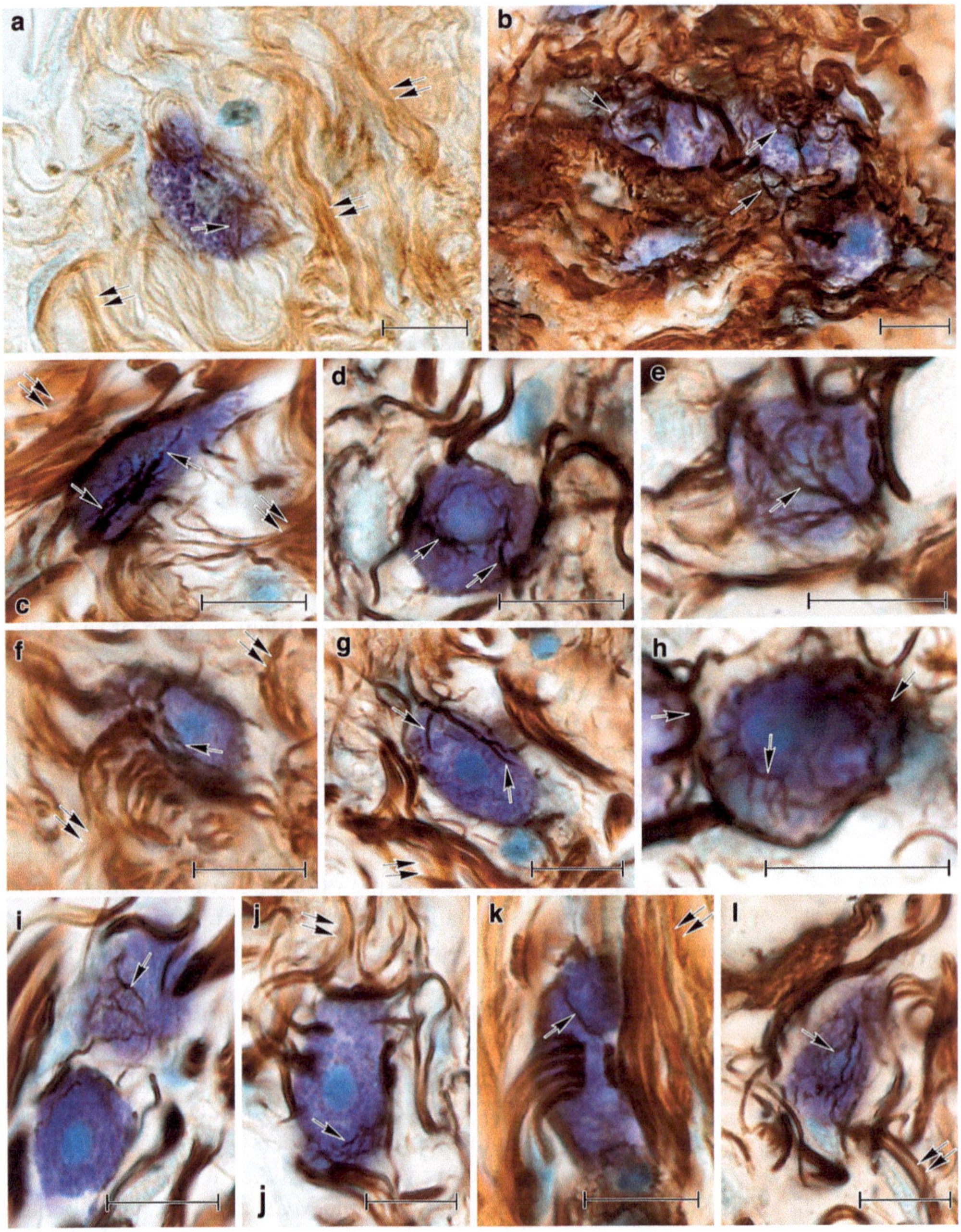

Fig. 1 Rat skin. A combination of silver impregnation with toluidine blue staining. (**a**) Normal, (**b–l**) wound healing process at various stages after skin damage. (**a**) The prevalence of mature collagen fiber bundles (indicated by a double arrow); the fiber formation low intensity in the field of extracellular matrix colocalization with the MC plasmalemma (indicated by an arrow). (**b**) High content of impregnated fibrous component around the MC (indicated by an arrow). (**c–l**) Variants of the reticular fiber localization of various caliber with MC (indicated by an arrow). The thinnest fibers extending from the MC and forming the network are located paracellularly. The paracellular foci of fiber formation in the form of an annular reticular network (**d, e, h**) are especially noteworthy. Reticular fibers can pass into mature thicker collagen fibers (indicated by a double arrow). Bar 10 μm. Reprinted with permission from [38]

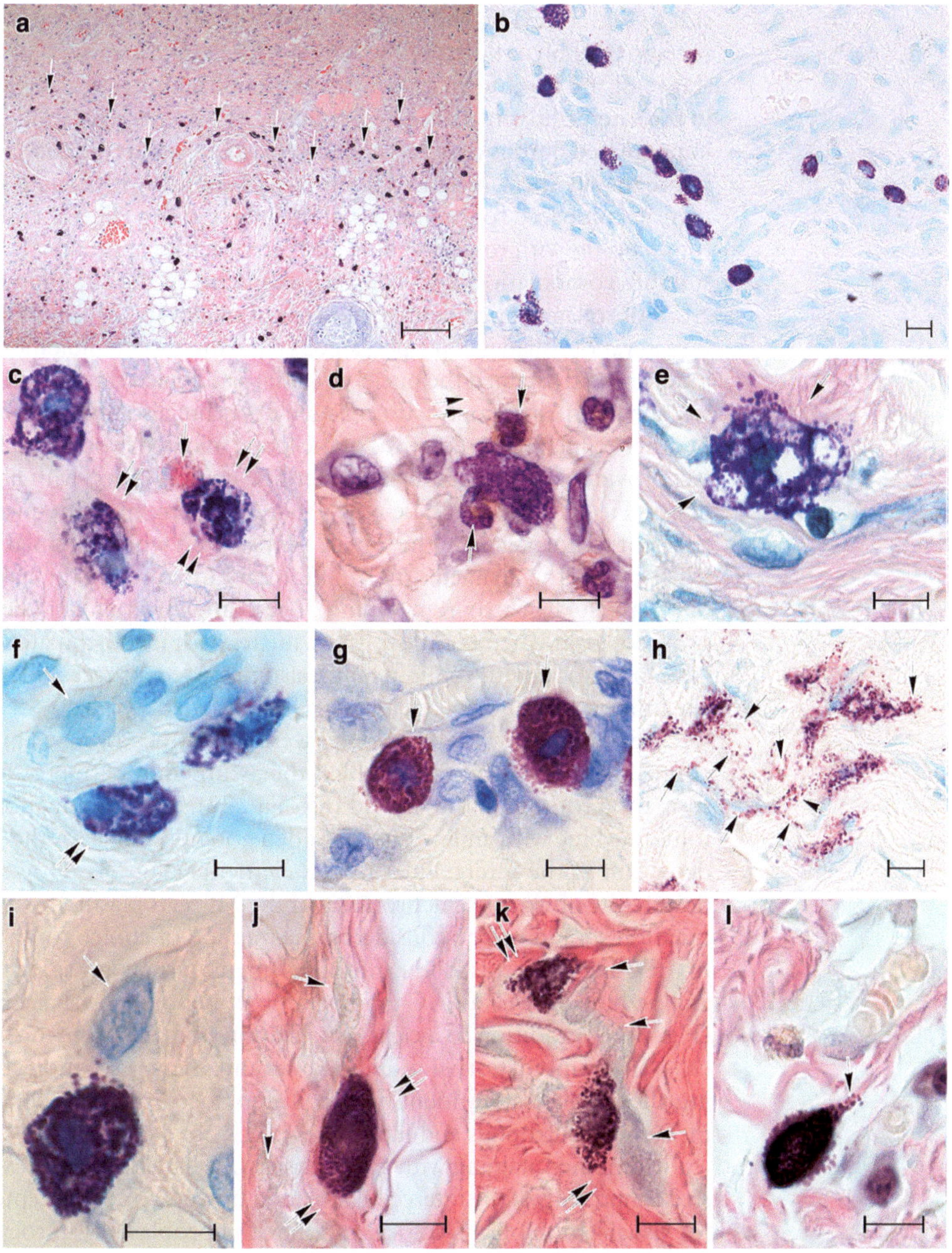

Fig. 2 Rat skin after modeling the wound process. (**a**–**f**) Staining with toluidine blue separately (**f**) or in combination with picrofuchsin (**a**), Giemsa stain (**b–d**) and van Gieson's stain (**e**), (**g–l**) staining with May–Grünwald solution in isolation (**g–i**) or in combination with van Gieson's stain (**j**), picrofuchsin (**k**), and Giemsa stain (**l**). (**a**) A general view of mast cells and the fibrous component of the intercellular matrix of the skin connective tissue. The border (arrow) of the dermis with a high content of MC (at the bottom, in the direction of the epidermis) and the zone of their absence is well visualized. Collagen fibers are colored red. (**b**) Predominant accumulation of MCs in a limited area of the dermis. (**c, d**) Contacts of MCs with neutrophils (indicated by arrows) surrounded by bundles of collagen fibers (indicated by double arrows). (**e**) Actively secreting MCs surrounded by collagen fibers (indicated by arrows). (**f**) MC localization with fibroblast (indicated by an arrow)

sides (Fig. 1k). The thinnest impregnated structures on the cell surface possibly reflect the morphological manifestations of initial stages of the fibril formation from tropocollagen macrocomplexes in the intercellular matrix. The frequency of MC colocalization with impregnated fibrous elements and fibroblasts, as well as the content of impregnated fibers, increased in the skin of animals with wound simulation compared to the animals of the intact group (Figs. 1 and 5).

Staining with toluidine blue and picrofuchsin appeared to be a suitable combination due to its achievementj of the simultaneous effect of collagen fiber identification in red shades and MC identification in various shades of purple (Fig. 2a). Furthermore, toluidine blue independently stained the nuclei of cells; this allowed simultaneously visualizing MCs not only with collagen fibers but also with the nuclei of other cellular representatives of the tissue microenvironment. This fact provides a means of evaluation for MC colocalization with fibroblasts (Fig. 5n). It should be noted that MCs were more frequently adjacent to fibroblasts in the skin of animals with wound simulation, which makes this characteristic significant for interpreting the activity of intercellular matrix remodeling. The completion of van Gieson protocol stain when contrasting the nuclei with iron hejjmatoxylin in combination with toluidine blue also provided a successful visualization of MC and collagen fibers (Fig. 2e).

During secretion, MC granules were sometimes localized in the connective tissue; this fact enabled the assumed initiation of fibril formation routing in the skin dermis (Fig. 2e). The combination of toluidine blue with aniline blue provided good potential for the MC detection in the intercellular matrix elements (Fig. 3l). However, despite the obvious metachromasia of the MC cytoplasm, this combination should not be considered the method of choice compared to the other protocols, due to fibrous elements poor contrasting with aniline blue, as well as MC granular component poor visualization (Fig. 3l). In addition to a clear visualization of metachromatic mast cells, the combination of toluidine blue and Giemsa solution staining allows successfully identifying collagen fibers, and granulocytes, including neutrophils and fibroblasts (Figs. 2b–d and 5e).

Combined staining with acid fuchsin was characterized by excessive contrasting of the intercellular matrix; however, it was acceptable for visualization of MC colocalization with fibroblasts or fibrocytes.

Fig. 2 (continued) and fibers (double arrow). (**g**) Adherence of MCs to the basal membrane of the capillary (indicated by arrows) and partial degranulation. (**h**) Exocytosis of secretory granules of MCs in localized dermis loci (indicated by arrows). (**i–k**) Options for MC contacting with fibroblasts (indicated by arrows) and collagen fibers (indicated by double arrows). (**l**) Secretion of MC granules in the region of collagen iber (indicated by an arrow) and paracellular space. Bar = 10 μm for the entire layout and 100 μm for the (**a**). Reprinted with permission from [38]

The least suitable was combined staining of the slides with toluidine blue and eosin. It might be recommended to apply Puebla-Osorio et al. [41] protocol to increase the effectiveness of this combination. Thus, the high content of glycosaminoglycans in mast cell granules preserves the metachromatic properties of the secretome when combining toluidine blue staining with other dyes that are selective for extracellular matrix fibers. However, the clearest visualization of mast cells with collagen fibers is achieved by simultaneous impregnation with silver or picrofuchsin staining. In view of identifying the initial stages of fibrillogenesis, the advantages of using the silver impregnation protocol become apparent.

5.5.2 Methylene Blue

The use of methylene blue as a part of May–Grünwald solution showed the MCs effective detection against the background of weakly colored pink collagen fibers; their contrasting is achieved partly due to the eosin presence in the stain composition (Figs. 2g–ig–i and 5o). Better detection of collagen fibers was provided by the combination of van Gieson's stain (Fig. 2j), picrofuchsin (Fig. 2k), and Giemsa stain (Fig. 2l). In addition, adjacent cells nuclei contrasting of a specific tissue microenvironment also allows from one degree to another concluding on MC colocalization with fibroblasts or fibrocytes (Figs. 2i–k and 5b, l). The combination with Giemsa stain together with adequate contrasting of the extracellular matrix may result in increased staining of mast cell secretion; this fact must be taken into account when staining (Fig. 2l). It is worth noting that the isolated Giemsa stain may also be applied to detect extracellular matrix with mast cells due to azur I being one of its components and to visualize the nuclei of other cells of the specific tissue microenvironment, including fibroblasts. Therefore, according to the results of the analysis of the effectiveness of using methylene blue to study mast cell colocalization with the fibrous component of connective tissue, it is effective only if it is used in combination with other solutions, e.g., May–Grunwald. This allows us to consider the use of methylene blue less convenient than toluidine blue when combined with fiber-selective dyes.

5.5.3 Brilliant Green

As for results of staining with brilliant green in combination with other histochemical protocols, it should be noted that it is considered to be highly effective for visualization of the skin MCs (Fig. 3a). When contrasting the nuclei with Mayer's hematoxylin, MCs are well visualized in turquoise blue shades, the number of which per unit of the tissue area is comparable to the results of staining with toluidine blue (Figs. 3c and 5a, d). The application of picrofuchsin with brilliant green combination allows successfully contrasting MCs against the background of collagen fibers stained red (Fig. 3b). Since, brilliant green has a weak affinity for nuclear structures and is quickly washed out during differentiation, it should be better combined with

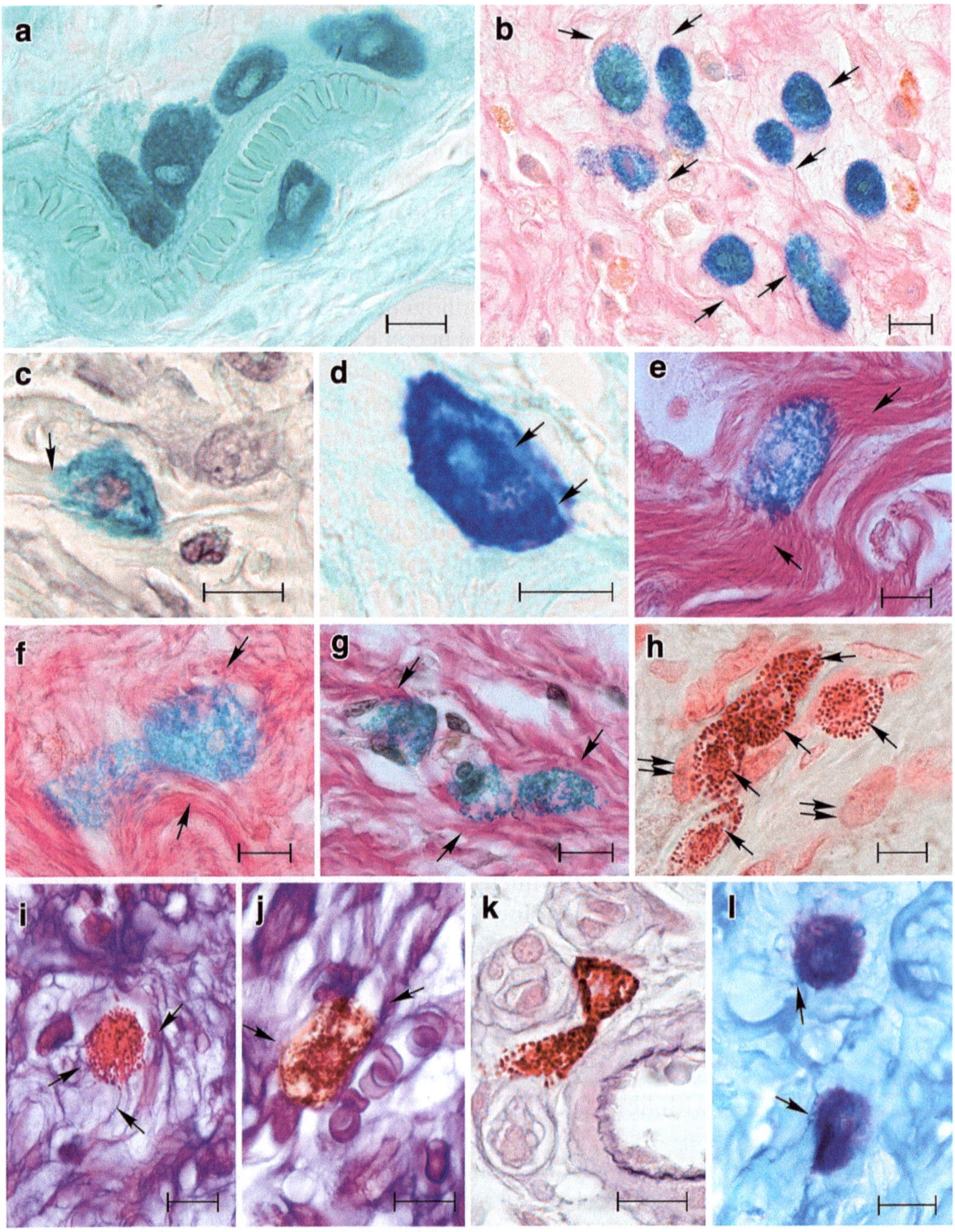

Fig. 3 Rat skin after modeling the wound process. (**a–g**) Staining with brilliant green in isolation (**a**) and in combination with staining with picrofuchsin and Mayer's hematoxylin (**b**), Mayer's hematoxylin (**c**), toluidine blue (**d**), picrofuchsin (**e, f**), van Gieson's stain (**g**), (**h–k**) staining with neutral red in isolation (**h**), in combination with aniline blue (**i, j**), Weigert's stain (express method) (**k**), and (**l**) combined staining with aniline blue and toluidine blue. (**a**) Perivasal location of the MC. (**b**) A group of MCs localized with thin collagen fibers (indicated by arrows). (**c**) Close contact of MC with a bundle of collagen fibers (indicated by an arrow). (**d**) Intracytoplasmic localization of collagen fiber in the MC (indicated by arrows). (**e–g**) MC localization with the fibrous component of the skin dermis (indicated by arrows). (**h**) MC (indicated by arrows) are localized near fibroblasts (indicated by double arrows), (**i, j**) Various options for the MC colocalization with collagen fibers (indicated by arrows).

classical van Gieson protocol stain, in which the nuclei are stained with Weigert's hematoxylin (Fig. 3g).

However, after combination with picrofuchsin nuclei contours, visualization of other cells becomes possible without counterstaining through microscopic examination (Fig. 3e, f). The tropism of brilliant green to the structures of the fibrous component of the intercellular matrix creates methodological conditions for its combination with toluidine blue and provides additional information when assessing the MC colocalization with fibers (Fig. 3d). In addition, this combination can provide information on the qualitative composition of the secretome, since there have been found various affinities of its components for stains. On the other hand, the aniline blue with brilliant green combination resulted in difficulties in detecting turquoise MCs on the blue-colored structures of the intercellular matrix; the fact prevents this protocol from being applicable in histochemical practice.

Similarly, staining with brilliant green was not effective in combination with eosin, causing nonselective masking diffuse staining of skin structures in pink shades.

Summarizing the above, it should be noted that for the study of mast cells and fibrillogenesis in the skin using brilliant green, it is most effective to combine this staining with the van Gieson protocol, in which not only the fibrous component of the extracellular matrix is selectively detected, but also the nuclei of the cells are contrasted.

5.5.4 Neutral Red

The use of neutral red as an independent reagent in MC detection is not convenient enough due to its ability to contrast cell nuclei, which can mask the visualization of granules stained in similar color shades. However, in some cases, this allows observing MC colocalization with fibroblasts (Fig. 3h). The combination with aniline blue stain has proved itself to be quite informative, allowing the evaluation of fiber colocalization with the skin MCs (Fig. 3i, j). It is advisable to avoid excessive staining of skin sections with aniline blue, since this results in the appearance of the difficult-to-differential background. Weigert's hematoxylin and neutral red stain combination allow successful visualizing of MCs and elastic fibers (Fig. 3k). The content of elastic fibers in the skin dermis is low compared to the content of collagen fibers; despite this fact, certain MCs come into contact with them. When completing an individual Weigert's hematoxylin staining, MCs are also visualized in dark violet shades; however, staining with neutral red can significantly improve the quality of their

Fig. 3 (continued) **(k)** Two contacting mast cells and an arteriole, in which the inner elastic membrane is clearly visible. **(l)** Mast cell colocalization with the fibrous component of the extracellular matrix of the dermis, exocytosis of granules with a parafibrillar arrangement (indicated by arrows). Bar 10 μm. Reprinted with permission from [38]

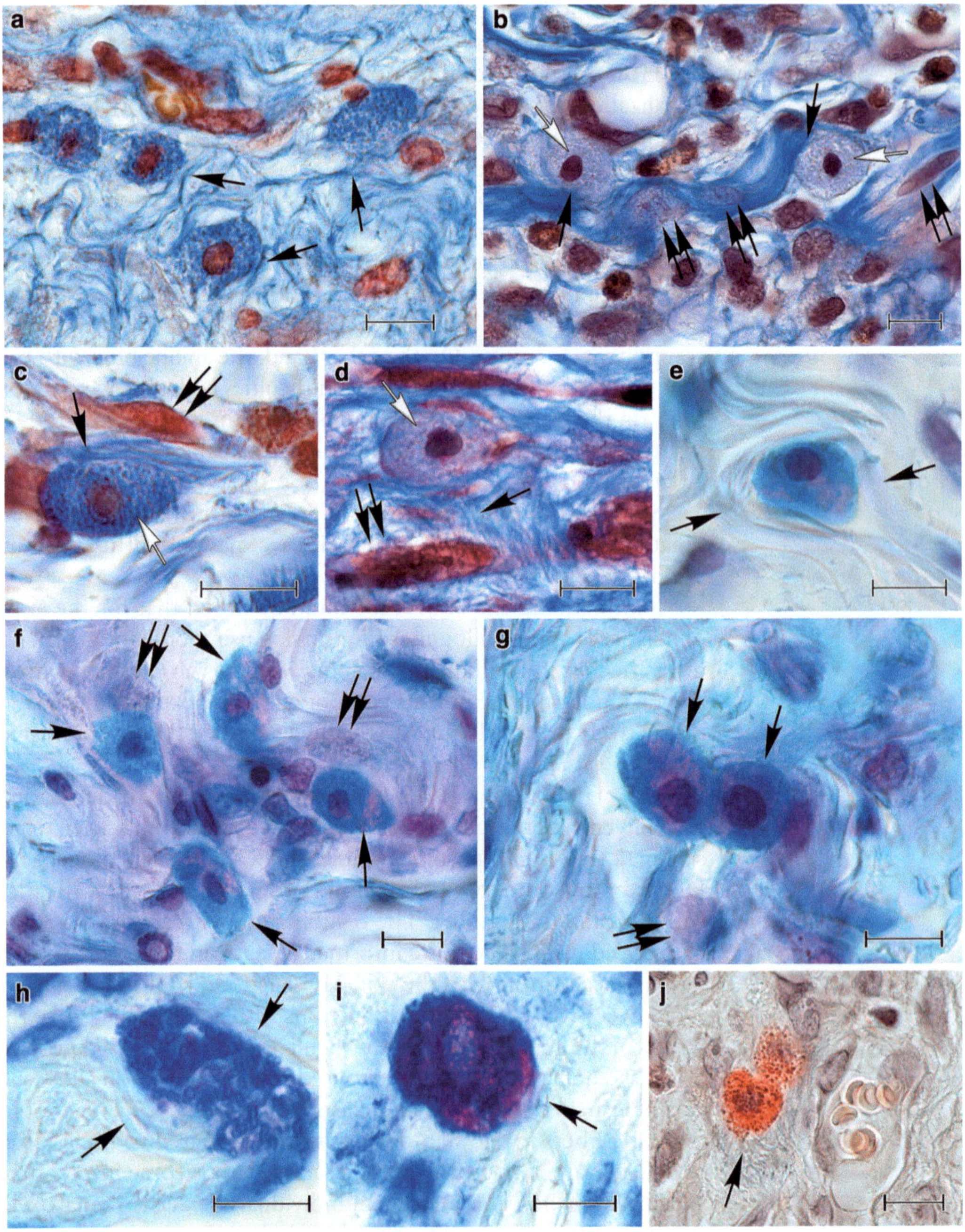

Fig. 4 Rat skin after modeling the wound process. (**a–d**) Picro-Mallory staining, (**e–g**) combined staining with alcian blue (pH 2.5) and periodic acid Schiff reaction, (**h, i**) staining with aniline blue and Giemsa solution, and (**j**) detection of chloroacetyl esterase activity. (**a**) MC localization at the site of active fiber formation. Collagen fibers that do not form bundles are determined (indicated by arrows). (**b**) MCs (white arrows) are localized with a large bundle of collagen fibers (indicated by black arrows) and fibroblasts (indicated by double black arrows) are determined. (**c, d**) Various options for the location of collagen fibers (indicated by an arrow) between MCs (arrows) and fibroblasts or myofibroblasts (indicated by a double arrow). (**e**) MC colocalization with a bundle of collagen fibers of the dermis (indicated by arrows). (**f–g**) The formation of the group of MCs (arrows) and

detection. The neutral red and brilliant green combination results in the dark staining of MCs due to the combined application of stains on the secretory material.

Thus, the most advantageous combination of neutral red for studying the participation of mast cells in fibrillogenesis is achieved by combining staining with aniline blue, which allows for advantage of the staining of cellular and fibrous structures in different colors.

5.5.5 Aniline Blue

The application of aniline blue in combination with polychrome acid fuchsin (the picro-Mallory staining protocol) led to a clear visualization of collagen fibers, and, when stained with MCs, their colocalization with extracellular matrix elements became clearly noticeable (Fig. 4a–d). At the same time, the detection of contractile proteins including smooth myocytes appears to be a beneficial advantage of this technique; this fact allows identifying myofibroblasts in the specific tissue microenvironment due to the red shades of the cytoplasm (Fig. 5j, k). This technique gives the opportunity to observe the beginning of fibroblast transformation into myofibroblast (Fig. 5i).

The application of aniline blue combination with a special Giemsa solution led to unexpected results. The fibrous component was stained diffusely, while Azur 1 from Giemsa's reagent resulted in the metachromatic detection of mast cells. As a result, mast cells could be identified against a bluish background containing fibers (Fig. 4i). However, mast cell granules looked unusual in some cases: the metachromatic core of the granule was not detected, and the outer granule contours became apparent, often turning into one another forming numerous intracellular passages (Fig. 4h).

Reasonably, the greatest advantage of the picro-Mallory protocol is the ability to indirectly judge the presence of myofibroblasts in the studied focus. At the same time, the successful detection of mast cells will obviously be achieved only if they have high functional activity and a high content of glycosaminoglycans in granules.

5.5.6 Alcian Blue

The application of alcian blue in combination with periodic acid Schiff reaction was quite successful in detecting MC colocalization with the connective tissue fibers (Fig. 4e–g). In addition to the unequal staining of the granular component of mast cells in blue or pink shades depending on the degree of biopolymer sulphation [30], nearby fibers became well-contoured (Fig. 4e–g).

Fig. 4 (continued) fibroblasts (indicated by double arrows) within a limited field of the dermis surrounded by collagen fibers, the interaction of MC (**g**). (**h–i**) Options for MC contacting with the fibrous component (indicated by arrows) of the extracellular matrix of the skin. (**j**) Collagen fibers (indicated by an arrow) close to MC localization. Bar 10 μm. Reprinted with permission from [42]

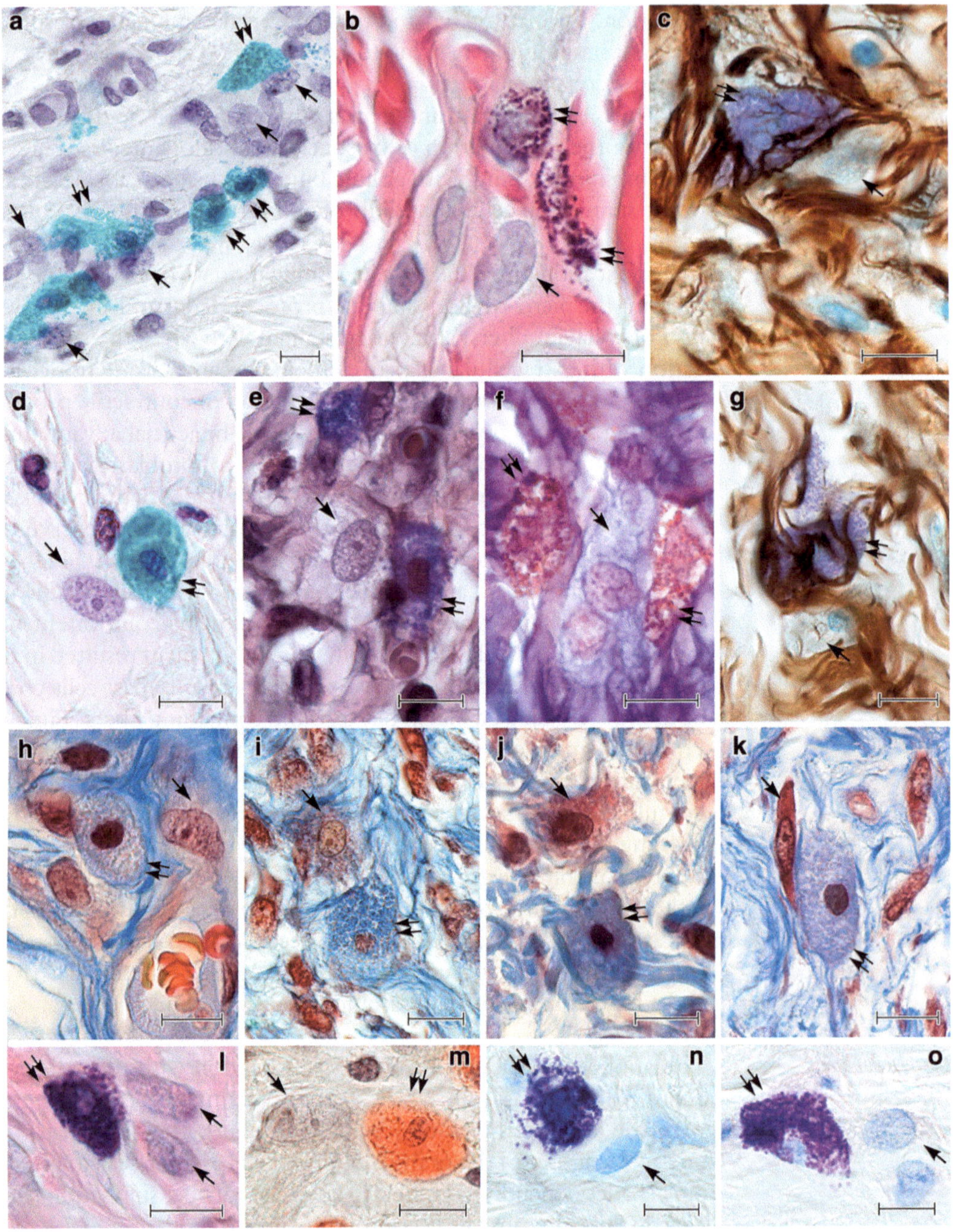

Fig. 5 Mast cell colocalization with fibroblasts in the rat skin after modeling the wound process. (**a, d**) Staining with brilliant green and Mayer's hematoxylin (**a, d**), (**b, l**) staining with May–Grünwald solution in combination with picrofuchsin, (**c, g**) impregnation with silver staining with toluidine blue, (**e**) staining with Giemsa solution with toluidine blue, (**f**) staining with aniline blue and neutral red, (**h–k**) staining according to picro-Mallory staining protocol, m identification of the activity of chloroacetyl esterase, (**n**) staining with toluidine blue, and (**o**) staining with May–Grünwald solution. Different variants of fibroblast colocalization (indicated by arrows) with MC (indicated by double arrows). (**a**) Group of fibroblasts is localized with several MCs forming a functional cluster of fibrillogenesis in the skin dermis. (**h–k**) Histochemical manifestations of the successive transformation of fibroblast (**h**) into myofibroblast (**i–k**) with the accumulation of contractile elements in the cytoplasm. Bar 10 μm. Reprinted with permission from [38]

6 Mast Cells, Elastic and Collagen Fibers: A Combination of Immunohistochemical and Histochemical Staining Protocols

The ability to simultaneously visualize specific cells of the immune and stromal landscape along with elastic fibers provides a perspective for developing a novel trend in space biology—assessing the tensometry of the tissue microenvironment. Considering mast cells as an example, we present a protocol for the simultaneous detection of mast cells using immunohistochemical staining of tryptase, and elastic or collagen fibers using a histochemical protocol. Thus, the presented technology is a combined protocol of histochemical and immunohistochemical staining, which can be conveniently used to study the adaptive mechanisms of the tissue microenvironment to zero gravity conditions. The results of spatial phenotyping of mast cells and elastic fibers on the entire section allow mapping the sample with an assessment of juxtacrine and paracrine colocalization.

For the simultaneous detection of tryptase-positive MCs and elastic fibers and collagen fibers, a combined protocol including standard immunohistochemical tryptase staining and elastic fibers staining according to Weigert's technique, or a protocol for simultaneous staining of elastic and collagen fibers according to the combined Weigert–Van Gieson method was used. The ratio of elastic and collagen fibers was estimated according to the manufacturer's protocol, after their simultaneous Weigert–Van Gieson stain (Fig. 6).

Triple labeling with simultaneous identification of collagen fibers, elastic network, and MCs with digital spatial profiling of the distribution density of profibrotic niches of the tissue microenvironment. The results of imaging and histoprofiling using artificial intelligence technology showed that space flight factors are accompanied by changes in the content of the fibrous component of the ECM (elastic and collagen framework, including reticular fibers) in the lungs, and the frequency of juxtacrine and paracrine MC effects on both collagen fibers and elastic fibers. These facts suggest the direct MC involvement in the formation of remodeling changes, partly owing to their active participation in collagen fibrillogenesis [42]. To our surprise, we found a close relationship between MCs and elastic fibers, which may reflect their physiological function in regulating the tensometry of the tissue microenvironment in zero gravity conditions. In particular, such activity was detected in the study of elastic fibers in a hereditary pathology—Marfan syndrome [43, 44].

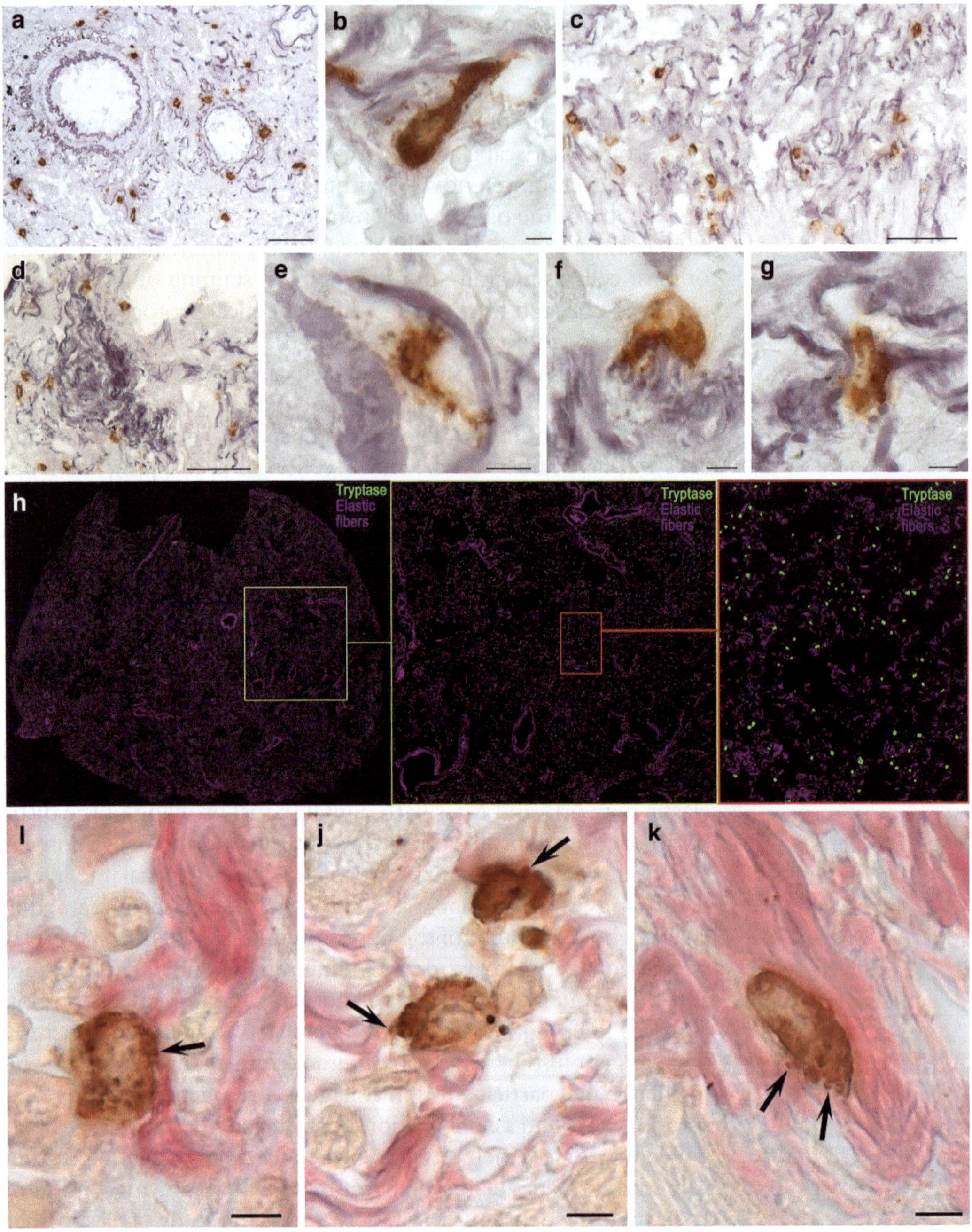

Fig. 6 MC interaction with pulmonary elastic fibers (**a–h**) and collagen fibers (i–k). Technique: combined elastic fiber staining according to the Weigert–van Gieson protocol and immunohistochemical detection of MC tryptase. (**a**) MC colocalization with elastic fibers mainly in the stroma of large vessels. (**b**) MCs in a state of active transport of secretory granules and a cytoplasmic outgrowth in contact with an elastic fiber. (**c**) A high content of elastic fibers with an unstructured histotopography in the respiratory region of the lung, in contact with MCs. (**d**) Focusing on the formation of excessive elastic fibers tightly adjacent to each other, with a predominant perifocal MC localization. (**e**) An actively secreting MC localized between elastic fibers, one of which has an abnormal thickness. (**f–g**) Different cytotopographic patterns of juxtacrine MC action on elastic fibers via targeted secretion of tryptase. (**h**) Spatial mapping of tryptase-positive MCs and elastic fibers. (**i–k**) Different patterns of MC colocalization with collagen fibers, with signs of juxtacrine tryptase action (arrowed). Scale: A, C, D—50 μm, the rest—5 μm

7 Combined Histochemical Approach in Assessing Tryptase Expression in the Mast Cell Population

To increase the efficiency of interpretation of mast cell's contribution to the state of a specific tissue microenvironment, it is necessary to detail the molecular composition of their secretome and analyze the pathways of degranulation. Developed method of combining immunomorphological and histochemical staining protocols contributes to the most objective detection of the integral level of tryptase expression in the intraorgan population of the skin mast cells. Novel technique for tryptase detection expands the possibilities of morphological analysis, provides researchers with additional data on the structure of the mast cell population and helps visualize the processing and cytological features and structural targets of tryptase during the development of adaptive and pathological reactions. Objective determination of the tryptase profile for organ-specific mast cell populations is in great demand in space biology to solve both fundamental mechanisms of adaptation to weightlessness and the development of new preventive measures to mitigate the negative consequences of the influence of space flight factors.

We studied the skin of rats in the area of the withers. The animals were divided into two experimental groups: intact (control) and those exposed to wounding (experimental). The biomaterial was fixed in neutral 10% formalin for 48 h, after that the skin samples were embedded in paraffin and histological sections 2 μ m thick were prepared. The staining of the microsections was carried out in two stages. At the first stage, immunohistochemical detection of tryptase was performed using the standard approach [2]. For this, the sections were dewaxed and incubated with mouse monoclonal antibodies (Abcam, #ab2378, dilution 1: 3000) overnight (Table 2). Secondary antibodies conjugated to HRP (Goat anti-Mouse IgG H&L, SDT GmbH, #AS-M1-HRP) were then applied and detected with ImmPACT™ DAB Peroxidase Substrate Kit (Vector Laboratories, #SK-4105) following the manufacturer's instructions for use. As a result, at the end of the first stage of the experiment, only tryptase-positive MCs, separately lying granules or fragments of the cytoplasm were visualized on skin microsections.

At the second stage, the sections were additionally stained with toluidine blue or May–Grunwald's solution following generally accepted protocols [26]. Since each of the histochemical dyes contrasted the cell nuclei to one degree or another, hematoxylin was not used. After the second stage of preparation, the sections were dehydrated and contained in a permanent mounting medium. The staining results are presented in Figs. 7–9.

Thus, the application of a new staining technique allows assessing tryptase profile in the skin MC population with more objective

Table 2
Antibodies and other reagents

Antibodies and other reagents	Source	Dilution	Vendor
Anti-Mast Cell Tryptase Mouse Monoclonal Antibody [AA1] (#ab2378)	Abcam, UK	1:3000	Abcam, Cambridge, UK
AmpliStain™ anti-Mouse 1-Step HRP (#AS-M1-HRP)	SDT GmbH, Germany	Ready-to-use	SDT GmbH, Baesweiler, Germany
ImmPACT DAB Peroxidase Substrate (#SK-4105)	Vector Laboratories, USA	Ready-to-use	Vector Laboratories, Burlingame, CA, USA
Toluidine blue for mast cells 2 × 250 × 250 mL (#07–002)	Ergoproduction, Russia	Ready-to-use	BioVitrum, Saint-Petersburg, Russia
May Grunwald's solution 1 L (#05–034)	Ergoproduction, Russia	Ready-to-use	BioVitrum, Saint-Petersburg, Russia

interpreting and characterizing MC involvement in adaptive or pathological processes not only in the field of space biology. These results increased objectivity in the investigation of tryptase biological activity in morphological studies relating to the connective tissue remodeling, regulation of inflammation, angio-, onco-, immunogenesis, etc. Visualization of biogenesis and secretory pathways of MC tryptase, cytotopography of granules, and their directed migration from the nucleus to the plasma membrane becomes possible. In terms of histotopography, it is useful to take into account the frequency of MC colocalization with granulocytes, lymphocytes, macrophages, and representatives of fibroblastic differon for further assessment and interpretation of aspects of intercellular signaling in a specific tissue microenvironment. This opens completely different ways of immunotopographic analysis in histology and may serve as a supporting procedure for studying the mechanisms of immunogenesis and remodeling the extracellular matrix of connective tissue in vivo.

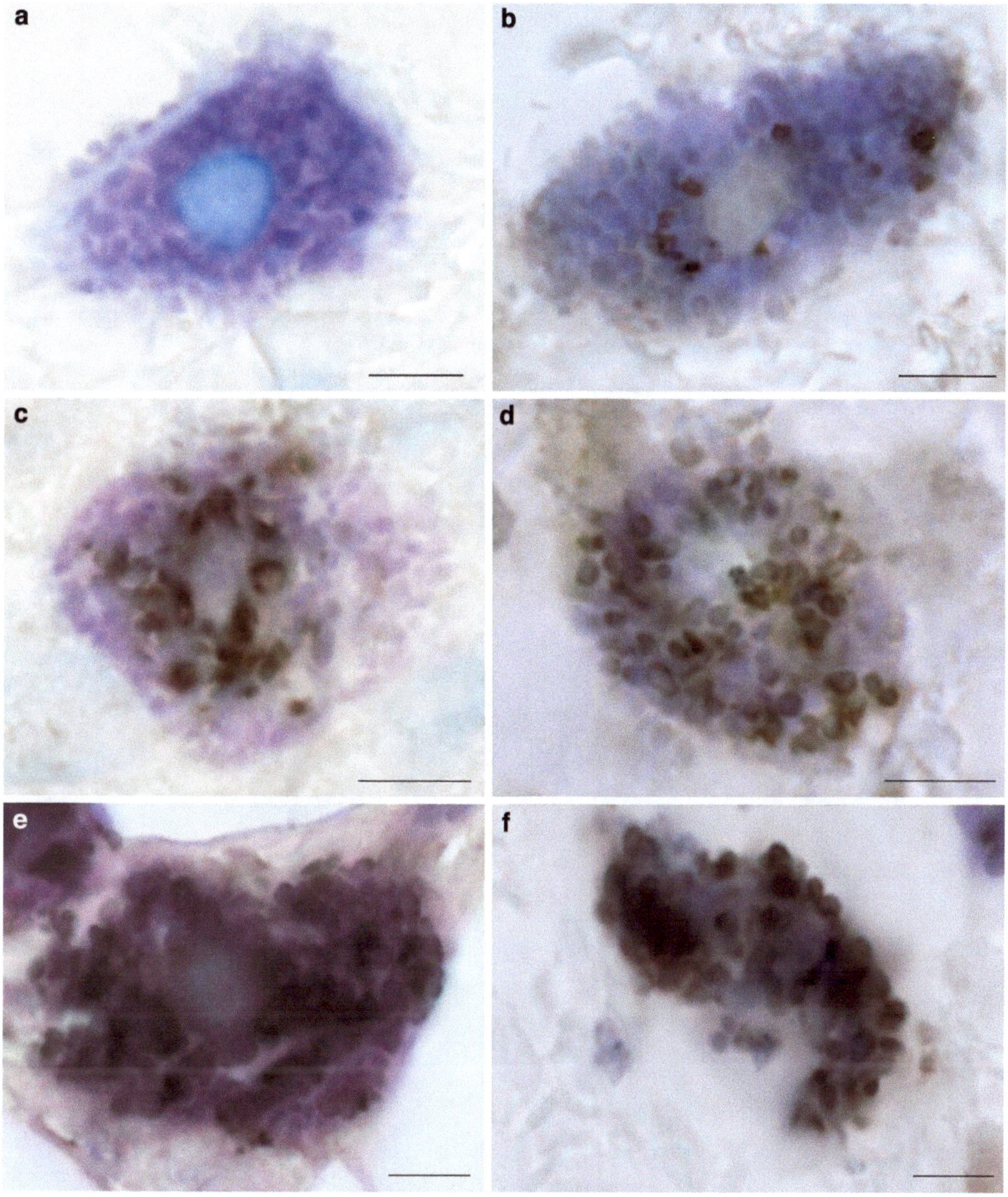

Fig. 7 Types of rat skin mast cells depending on the tryptase expression level. Fixation—10% neutral formalin. Immunohistochemical staining for MC tryptase with additional staining with May–Grunwald solution (**a, c, e**) or toluidine blue (**b, d, f**). **a–b**—control, **c–f**—in 7 days after wounding. MCs without tryptase (**a**), with single tryptase-positive granules (**b**), with tryptase-containing granules occupying 1/3 (**c**), 2/3 (**d**), and the entire MC cytoplasm (**e–f**). Scale = 5 μm. Reprinted with permission from [45]

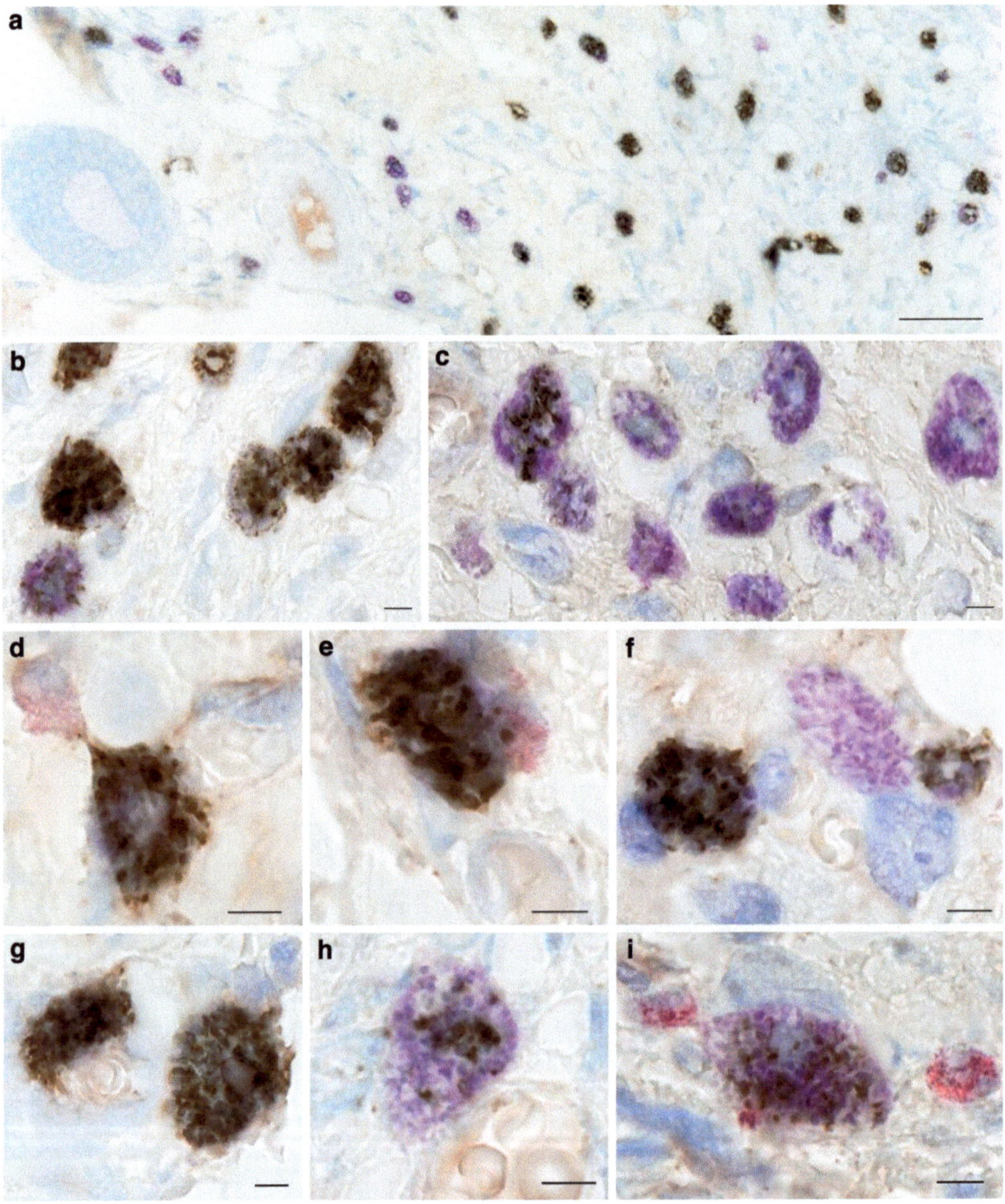

Fig. 8 Histotopography of rat skin mast cells. Fixation—10% neutral formalin. Immunohistochemical staining for MC tryptase with additional staining with May–Grunwald solution. **c, h**—control group, the rest—in 7 days after wounding. **a** and **b**—predominance of tryptase-positive MCs, some of which are in contact with each other (**b**). **c**—predominance of MC without tryptase expression. **d**—contacting with granular leukocyte. **e**—perivascular localization of tryptase-positive MC, contact with granulocyte. **f**—MCs with different levels of tryptase expression are located near the fibroblastic differon cells of the skin dermis. **g** and **h** Scale: a—50 μ m, others— 5 μm. Reprinted with permission from [45]

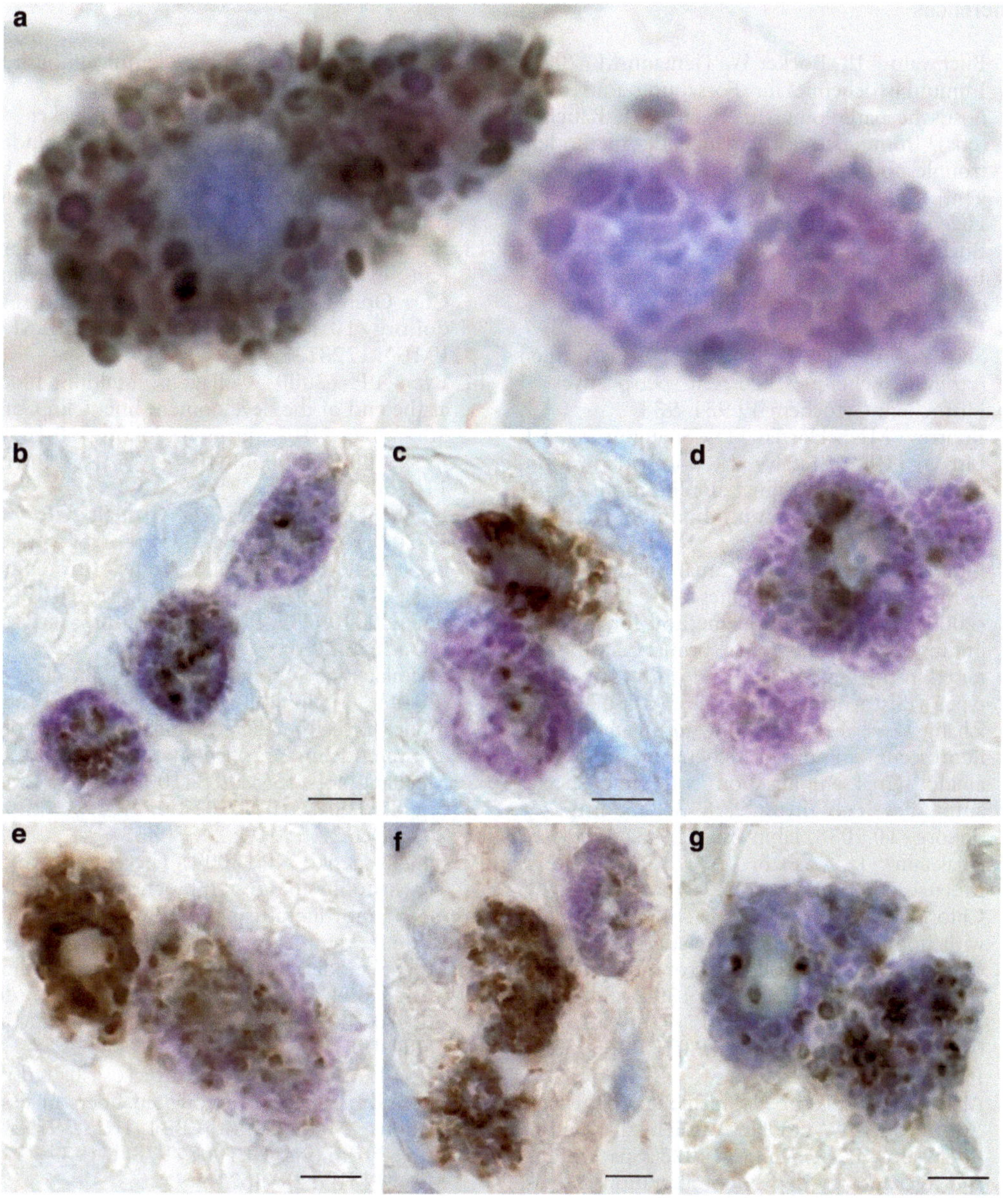

Fig. 9 Morphological equivalents of intercellular signaling in the skin mast cell population. Fixation—10% neutral formalin. Immunohistochemical staining for MC tryptase with additional staining with May–Grunwald solution (**a–f**) and toluidine blue (**g**). Control group (**b–d**) and an experimental group with wounding (**a, e–g**). (**a**)—MC contacting with different tryptase content. (**b–g**)—variants of colocalization of tryptase-positive MC. Scale = 5 μm. Reprinted with permission from [45]

References

1. Buchwalow IB, Böcker W, Tiemann M (2023) Immunhistochemie für Forschung und diagnostische pathologie. In: Raem AM, Rauch P (eds) Immunoassays: ergänzende methoden, troubleshooting, regulatorische anforderungen. Springer Berlin Heidelberg, Berlin, Heidelberg, pp 317–337
2. Buchwalow IB, Boecker W (2010) Immunohistochemistry: basics and methods, 1 edn. Springer, Heidelberg, Dordrecht, London, New York
3. Shi SR, Cote RJ, Taylor CR (2001) Antigen retrieval techniques: current perspectives. J Histochem Cytochem 49:931–938
4. Bobrow MN, Harris TD, Shaughnessy KJ, Litt GJ (1989) Catalyzed reporter deposition, a novel method of signal amplification—application to immunoassays. J Immunol Methods 125:279–285
5. Pirici D, Mogoanta L, Kumar-Singh S, Pirici I, Margaritescu C, Simionescu C, Stanescu R (2009) Antibody elution method for multiple immunohistochemistry on primary antibodies raised in the same species and of the same subtype. J Histochem Cytochem 57(6):567–575
6. Zhang W et al (2017) Fully automated 5-plex fluorescent immunohistochemistry with tyramide signal amplification and same species antibodies. Lab Invest 97:873–885. https://doi.org/10.1038/labinvest.2017.37
7. Atiakshin D, Kostin A, Buchwalow I, Morozow D, Samoilova V, Tiemann M (2023) Chloroacetate esterase reaction combined with immunohistochemical characterization of the specific tissue microenvironment of mast cells. Histochem Cell Biol 159(4):353–361. https://doi.org/10.1007/s00418-022-02174-1. Epub 2023 Jan 4PMID: 36598563; PMCID: PMC10081974
8. Atiakshin D, Kostin A, Shishkina V, Burtseva A, Buravleva A, Volodkin A, Elieh-Ali-Komi D, Buchwalow I, Tiemann M (2023) Space-flight- and microgravity-dependent alteration of mast cell population and protease expression in digestive organs of Mongolian gerbils. Int J Mol Sci 24(17):13604. https://doi.org/10.3390/ijms241713604. PMID: 37686410; PMCID: PMC10488096
9. Lojda Z, Gossrau R, Schiebler T (1976) Enzyme histochemistry. A laboratory manual. Springer, Berlin
10. Lojda Z, Gossrau R, Schiebler TH (1976) Enzyme cytochemical methods. Springer Verlag, Berlin, Heidelberg, New York
11. VanNoorden CJF, Frederiks WM (1992) Enzyme histochemistry. A laboratory manual of current methods. Bios Scientific Publishers, Oxford
12. Buchwalow I, Samoilova V, Boecker W, Tiemann M (2011) Non-specific binding of antibodies in immunohistochemistry: fallacies and facts. Sci Rep 1:28. https://doi.org/10.1038/srep00028
13. Bornstein P, Sage EH (2002) Matricellular proteins: extracellular modulators of cell function. Curr Opin Cell Biol 14(5):608–616. https://doi.org/10.1016/s0955-0674(02)00361-7. PMID: 12231357
14. Byers PH (2000) Collagens: building blocks at the end of the development line. Clin Genet 58(4):270–279. https://doi.org/10.1034/j.1399-0004.2000.580404.x. PMID: 11076051
15. Hitchcock AM, Yates KE, Costello CE, Zaia J (2008) Comparative glycomics of connective tissue glycosaminoglycans. Proteomics 8(7):1384–1397. https://doi.org/10.1002/pmic.200700787. PMID: 18318007; PMCID: PMC2577024
16. Loerakker S, Obbink-Huizer C, Baaijens FP (2014) A physically motivated constitutive model for cell-mediated compaction and collagen remodeling in soft tissues. Biomech Model Mechanobiol 13(5):985–1001. https://doi.org/10.1007/s10237-013-0549-1. Epub 2013 Dec 27. PMID: 24370853
17. Omel'yanenko NP (2009) The connective tissue (histophysiology and biochemistry). Izd-vo inostrannoy literatury, Moscow (In Russian)
18. Rittié L, Hutcheon AEK, Zieske JD (2017) Mouse models of corneal scarring. Methods Mol Biol 1627:117–122. https://doi.org/10.1007/978-1-4939-7113-8_8. PMID: 28836198
19. Rittié L (2017) Method for picrosirius red-polarization detection of collagen fibers in tissue sections. Methods Mol Biol 1627:395–407. https://doi.org/10.1007/978-1-4939-7113-8_26. PMID: 28836216
20. Rittié L, Type I (2017) Collagen purification from rat tail tendons. Methods Mol Biol 1627:287–308. https://doi.org/10.1007/978-1-4939-7113-8_19. PMID: 28836209
21. Buchwalow I, Boecker W, Tiemann M (2015) The contribution of Paul Ehrlich to histochemistry: a tribute on the occasion of the centenary of his death. Virchows Arch 466(1):111–116. https://doi.org/10.1007/s00428-014-1677-4. Epub 2014 Oct 31. PMID: 25359602
22. Dabbs DJ (2014) Diagnostic immunohistochemistry. Elsevier, Amsterdam

23. Ehrlich P (1878) Beiträge für theorie und praxis der histologischen färbung, vol Doktor. Leipzig University, Leipzig, p 65
24. Pearse A (1960) Histochemistry. Theoretical and applied. Little, Brown, Boston
25. Rath F-W (1981) Praktisch-diagnostische Enzymhistochemie. Anwendungsmöglichkeiten und Grenzen enzymhistochemischer Methoden in der klinischen Pathologie. VEB Gustav Fischer Verlag, Jena
26. Romeis B (2010) Mikroskopische technik. Spektrum Akademischer Verlag, Heidelberg
27. Bancelin S, Aime C, Gusachenko I, Kowalczuk L, Latour G, Coradin T, Schanne-Klein MC (2014a) Determination of collagen fibril size via absolute measurements of second-harmonic generation signals. Nat Commun 5:4920
28. Bancelin S, Decencière E, Machairas V, Albert C, Coradin T, Schanne-Klein MC, Aimé C (2014) Fibrillogenesis from nanosurfaces: multiphoton imaging and stereological analysis of collagen 3D self-assembly dynamics. Soft Matter 10(35):6651–7. https://doi.org/10.1039/c4sm00819g. PMID: 25058449
29. Ghazanfari S, Khademhosseini A, Smit TH (2016) Mechanisms of lamellar collagen formation in connective tissues. Biomaterials 97:74–84. https://doi.org/10.1016/j.biomaterials.2016.04.028. Epub 2016 Apr 27. PMID: 27162076
30. Atiakshin D, Samoilova V, Buchwalow I, Boecker W, Tiemann M (2017) Characterization of mast cell populations using different methods for their identification. Histochem Cell Biol 147:683–694
31. Elieh Ali Komi D, Wöhrl S, Bielory L (2020) Mast cell biology at molecular level: a comprehensive review. Clin Rev Allergy Immunol 58(3):342–365. https://doi.org/10.1007/s12016-019-08769-2. PMID: 31828527
32. Elieh-Ali-Komi D, Shafaghat F, Alipoor SD, Kazemi T, Atiakshin D, Pyatilova P, Maurer M (2025) Immunomodulatory significance of mast cell exosomes (MC-EXOs) in immune response coordination. Clin Rev Allergy Immunol 68(1):20. https://doi.org/10.1007/s12016-025-09033-6. PMID: 39976807; PMCID: PMC11842441
33. Frossi B, Mion F, Sibilano R, Danelli L, Pucillo CEM (2018) Is it time for a new classification of mast cells? What do we know about mast cell heterogeneity? Immunol Rev 282(1):35–46. https://doi.org/10.1111/imr.12636. PMID: 29431204
34. Heidarzadeh-Asl S, Maurer M, Kiani A, Atiakshin D, Stahl Skov P, Elieh-Ali-Komi D (2025) Novel insights on the biology and immunologic effects of histamine: a road map for allergists and mast cell biologists. J Allergy Clin Immunol 155(4):1095–1114. https://doi.org/10.1016/j.jaci.2024.12.1081. Epub 2024 Dec 27. PMID: 39734034
35. Dwyer DF, Barrett NA, Austen KF (2016) Immunological Genome Project Consortium. Expression profiling of constitutive mast cells reveals a unique identity within the immune system. Nat Immunol 17(7):878–887. https://doi.org/10.1038/ni.3445. Epub 2016 May 2PMID: 27135604; PMCID: PMC5045264
36. Wernersson S, Pejler G (2014) Mast cell secretory granules: armed for battle. Nat Rev Immunol 14(7):478–494. https://doi.org/10.1038/nri3690. Epub 2014 Jun 6. PMID: 24903914
37. Atiakshin D, Buchwalow I, Samoilova V, Tiemann M (2018) Tryptase as a polyfunctional component of mast cells. Histochem Cell Biol 149(5):461-477. https://doi.org/10.1007/s00418-018-1659-8. Epub 2018 Mar 12. PMID: 29532158
38. Atiakshin D, Buchwalow I, Tiemann M (2019) Mast cell chymase: morphofunctional characteristics. Histochem Cell Biol 152(4):253-269. https://doi.org/10.1007/s00418-019-01803-6. Epub 2019 Aug 8. PMID: 31392409
39. Rönnberg E, Melo FR, Pejler G (2012) Mast cell proteoglycans. J Histochem Cytochem 60(12):950–962. https://doi.org/10.1369/0022155412458927. Epub 2012 Aug 16. PMID: 22899859; PMCID: PMC3527880
40. Buchwalow I, Samoilova V, Boecker W, Tiemann M (2018) Multiple immunolabeling with antibodies from the same host species in combination with tyramide signal amplification. Acta Histochem 120(5):405-411. https://doi.org/10.1016/j.acthis.2018.05.002. Epub 2018 May 5. PMID: 29739626
41. Puebla-Osorio N, Sarchio SNE, Ullrich SE, Byrne SN (2017) Detection of Infiltrating Mast Cells Using a Modified Toluidine Blue Staining. Methods Mol Biol 1627:213-222. https://doi.org/10.1007/978-1-4939-7113-8_14. PMID: 28836204
42. Atiakshin D, Buchwalow I, Tiemann M (2020) Mast cells and collagen fibrillogenesis. Histochem Cell Biol 154(1):21–40. https://doi.org/10.1007/s00418-020-01875-9. Epub 2020 Mar 28. PMID: 32222902
43. Atiakshin D, Nikolaeva E, Gritsevskaya D, Semyachkina A, Kostin A, Volodkin A et al

(2024) Skin mast cells in Marfan syndrome: specific emphasis on connective tissue remodeling. Arch Dermatol Res 316(6):271
44. Atiakshin D, Nikolaeva E, Semyachkina A, Kostin A, Volodkin A, Morozov S et al (2024) The contribution of mast cells to the regulation of elastic fiber tensometry in the skin dermis of children with Marfan syndrome. Int J Mol Sci 25:9191
45. Atiakshin DA, Shishkina VV, Gerasimova OA, Meshkova VY, Samodurova NY, Samoilenko TV, Buchwalow IB, Samoilova VE, Tiemann M (2021) Combined histochemical approach in assessing tryptase expression in the mast cell population. Acta Histochem 123(4):151711. https://doi.org/10.1016/j.acthis.2021.151711. Epub 2021 Apr 7. PMID: 33838578

Check for updates

Chapter 11

Ovarian Reproductive and Endocrine Function Investigated in Microgravity Environment

Noemi Monti, Emanuele Galante, Alessandro Querqui, Giorgia Cucina, Antonio Angeloni and Mariano Bizzarri

Abstract

Long-term space missions expose astronauts to an altered gravitational field (microgravity) that significantly affects human physiology. Although it is critical to assess the effect of microgravity on female reproductive health, this aspect remains a poorly investigated one. To date, interest in the effects on the main ovary hormones (estrogens) has been in depth for its role in bone loss, while the impact of altered estrogen availability on fertility and ovarian function is unclear. The female reproductive apparatus depends on a finely tuned hormonal network involving the hypothalamic-pituitary-ovarian axis. At the heart of this system are the ovarian follicles, composed of granulosa and theca cells, which cooperate in the synthesis of steroid hormones in supporting oocyte maturation. Folliculogenesis and estrogen production are tightly regulated by gonadotropins—FSH and LH—which act on these two distinct cell types. Any disruption in this delicate endocrine cross talk may impair oocyte quality, ovulatory function, and overall reproductive health. New evidence suggests that microgravity may interfere with this delicate hormonal balance, potentially reducing estrogen synthesis and impairing follicular development mostly by downregulating aromatase activity. As the regulation of ovarian function is distributed across different tiers, investigation is particularly delicate. Nevertheless, molecular assessment of relevant parameters can be performed upon cells/tissues cultures growing in real and simulated weightlessness in order to collect data and drawing explicative models.

Keywords Microgravity, Steroidogenesis, Female fertility, Aromatase, Estrogens, Ovarian biology

1 Introduction

Humankind has always been fascinated by space, especially with a long-term vision of colonizing other planets. In recent decades, this aspiration gained momentum, as the new era of space exploration is characterized by the increasing frequency and duration of manned missions. The *Artemis Program*, spearheaded by NASA, is a prominent example of this new frontier [1, 2]. Artemis is a long-term program designed to return astronauts to the Moon, with the goal of establishing a permanent human presence on the lunar surface

Mariano Bizzarri (ed.), *Biomedicine and Biotechnology in Space*, Methods in Molecular Biology, vol. 3000,
https://doi.org/10.1007/978-1-0716-5174-2_11,

by the end of the 2020s. Additionally, the program includes plans to construct a lunar-orbiting space station, known as the Gateway, to serve as a support base for sustainable surface missions [3, 4]. Artemis marks a major step forward toward permanent lunar colonization and provides the groundwork for future crewed missions to Mars, which will require extended orbital stays exceeding 6 months. However, it is crucial to recognize that space is an inherently hostile environment for human physiology. Astronauts are exposed to multiple environmental stressors, including the absence of atmospheric pressure, extreme temperature variations, ionizing radiation, reduced air quality, altered pressure conditions, and gravitational changes such as microgravity—defined as a condition in which gravitational acceleration is lower than Earth's standard 1 g [5]. Exposure to this radically different environment significantly affects astronaut health and mission performance [6]. Although early studies considered microgravity to be biologically negligible [7], it is now well established that it profoundly affects human physiology. Documented effects include significant reductions in bone mineral density [8], atrophy and thinning of articular cartilage [9], decreases in muscle mass and strength [10], and alterations in cardiovascular and immune system function [11, 12]. However, the study of microgravity's impact on female reproductive function remains a largely underexplored area. To date, most research has focused on the endocrine regulation of sex hormones due to their broader systemic effects, including musculoskeletal impact [13]. Yet, in the context of stable settlements on the Moon or Mars, it is crucial to investigate how space environments may affect female reproductive health and fertility.

Although space agencies have prioritized the study of female reproductive health during and after spaceflight, reproducible data remain scarce due to the small sample size of female astronauts. Despite increasing participation, women still represent only 22.7% of astronauts who have reached low-Earth orbit [14]. Female astronauts often take pharmacological measures to protect their reproductive health during missions, such as suppressing the menstrual cycle using combined oral contraceptives containing progestins and estrogens [15, 16]. It is also worth noting that, compared to male astronauts, women tend to complete their first spaceflight at a later age (around 38 years), a factor that affects the timing of their first pregnancy. Data indicate that female astronauts increasingly postpone childbearing until after completing multiple missions, often resulting in delayed motherhood and greater reliance on assisted reproductive technologies (ART) [17, 18]. These trends complicate efforts to evaluate how microgravity influences female fertility, including possible effects on pregnancy outcomes, birth rates, and miscarriage risk. Additionally, studies show that 20%–30% of female astronauts experience menstrual irregularities, further complicating assessments of postmission reproductive function [19].

Due to these limitations, researchers have turned to ground-based analogs to simulate the effects of microgravity on female reproductive function. Two widely used models include the six-degree head-down tilt bed rest protocol [15] and the dry immersion model [20]. In one dry immersion study conducted from days 10 to 15 of the menstrual cycle, participants exhibited a significant increase in inhibin B (which inhibits FSH secretion), and a decrease in luteinizing hormone (LH) and progesterone levels. While menstrual cycle length, uterine size, and endometrial thickness remained unchanged, the dominant follicle showed a measurable increase in size. In contrast, an earlier 3-day dry immersion study in women with regular menstrual cycles found no significant effect of gravity absence on female reproductive function [21]. Likewise, results from bed-rest studies have been inconsistent. For example, Sandler and Winters observed only minor changes in progesterone secretion during the luteal phase, without identifying significant disruptions in ovarian function.

In 2023, the first in vitro study assessing the effects of simulated microgravity using a Random Positioning Machine (RPM) on metaphase II human oocytes was published [22]. Transmission electron microscopy (TEM) analysis revealed substantial structural alterations in organelles, including mitochondrial morphology, the endoplasmic reticulum, and cortical granules. These disruptions are likely related to cytoskeletal changes. The findings suggest that microgravity may negatively affect oocyte quality, maturation, and potentially fertilization and embryo development. In mouse models, exposure to microgravity has also been shown to impair oocyte maturation, particularly affecting meiotic spindle formation and inducing cytoplasmic anomalies such as blebbing, while significantly reducing the proportion of mature oocytes [23]. In these studies, oocyte maturation was halted at the first meiotic division, despite appropriate hormonal stimulation with FSH, LH, and 17β-estradiol. Additional findings using a rotating wall vessel system in mouse ovarian follicles showed decreased follicular survival, downregulation of proliferative markers such as PCNA, and reduced expression of GDF-9, a key factor in oocyte growth [24]. Interestingly, when follicles were embedded in a three-dimensional matrix prior to microgravity exposure, their survival and maturation rates were not significantly altered—highlighting the potential protective role of 3D culture systems in these conditions.

The female reproductive system is regulated by a complex cross talk between internal and external organs, modulated by hormonal, nutritional, and biophysical factors. This finely tuned interaction is essential for the production of sex hormones, the development of mature follicles, and the subsequent release of an oocyte competent for fertilization and zygote formation. Folliculogenesis and the synthesis of sex steroids in the ovary are tightly regulated by neuroendocrine factors, including gonadotropin-releasing hormone (GnRH) secreted by the hypothalamus [25]. GnRH, in turn, stimulates the synthesis and systemic release of two key gonadotropins: luteinizing hormone (LH) and follicle-stimulating hormone (FSH). These hormones act directly on the ovarian follicle, where they modulate the steroidogenesis pathway, driving the production of sex steroids. The follicle is considered the fundamental functional unit of the ovary and consists of two main somatic cell types that work cooperatively in steroid production and oocyte maturation: granulosa cells (GCs), which are in close contact with the oocyte, and theca cells (TCs), which form the outer follicular layer. Steroidogenesis begins in the theca cells, where cholesterol serves as the substrate for a cascade of enzymatic reactions leading to the production of androgens. These androgens then diffuse into the granulosa cells, where they are converted into estrogens. This conversion is catalyzed by the enzyme CYP19A1, also known as aromatase, which is responsible for the aromatization of androgens into estrogens, including estrone and estradiol. Each cell type within the follicle responds to a different gonadotropin: FSH acts on granulosa cells, while LH stimulates theca cells, in accordance with the two-cell, two-gonadotropin theory [26]. As a result, estrogen production occurs in a cyclic manner under the influence of LH and FSH and exerts negative feedback on GnRH release at the level of the hypothalamus. This delicate hormonal regulatory loop is essential for reproductive homeostasis and is vulnerable to disruption under spaceflight conditions. The ovary, therefore, plays a central role in maintaining female fertility, and its function may be particularly susceptible to environmental changes encountered during space travel. It is critical to examine also the functional cross talk between follicular cells, as disturbances in this interaction could impair steroidogenesis.

One of the primary obstacles in studying female reproduction under altered gravity conditions is the limited availability of appropriate biological models. Human ovarian tissue is difficult to access due to ethical concerns and the high degree of follicular heterogeneity resulting from cyclical hormonal changes [27]. Additionally, many immortalized ovarian cell lines fail to replicate the complex physiological behavior of native cells, limiting their utility in mechanistic or translational studies [5]. In this context, bovine ovaries offer a highly valuable and ethically acceptable alternative model for studying ovarian function [28]. Several studies have demonstrated that bovine ovarian physiology shares critical similarities with the human ovary, particularly in terms of molecular signaling pathways, hormonal regulation, and follicular dynamics [29, 30]. Bovine ovaries are also excellent models for investigating pathological conditions such as premature ovarian failure (POF) and polycystic ovary syndrome (PCOS), due to shared pathophysiological mechanisms with humans [31]. Importantly, cattle are monovular mammals, like humans, and exhibit comparable ovarian size, gestation duration, and follicular growth dynamics. These features make bovine models more physiologically relevant than traditional rodent models, which are often polytocous (i.e., giving birth to multiple offspring at once) and differ markedly in reproductive biology [32].

Moreover, bovine ovaries are readily available as by-products from routine slaughterhouse procedures, representing an ethically acceptable, cost-effective, and non-invasive model. Their accessibility allows for the detailed study of follicular development stages and enables both in vivo and in vitro experimental approaches.

Here, we describe a robust and reproducible protocol for the isolation and primary culture of bovine granulosa and theca cells, exposed to real (on board of the International Space Station) and simulated microgravity (by using a Random Positioning Machine, RPM [33]. This model will enable future research into the cellular and molecular mechanisms by which altered gravity affects ovarian cell function. Ultimately, this protocol would be instrumental in supporting the development of countermeasures to preserve female reproductive health during prolonged space missions and in preparation for future space colonization.

2 Materials

Work under strict sterile conditions in a laboratory to ensure the integrity and reproducibility of experiments.

2.1 Equipment

a. Horizontal Laminar Flow Hood: used to handle bovine ovaries and biological materials under sterile conditions during cell isolation.

b. Class II Biological Safety Cabinet: maintains a controlled, HEPA-filtered airflow to prevent contamination during cells seeding and reagents handling.

c. Phase-contrast microscope: allows regular monitoring of cell morphology, viability and confluence.

d. Centrifuge: used to process cell suspensions and collect pellets.

e. Orbital shaker: provides gentle agitation for enzymatic digestion.

f. CO_2 incubator (37°C, 5% CO_2): ensures a controlled environment for maintaining optimal temperature, humidity, and CO_2 levels essential for cell growth.

g. Water bath: gently warms reagents prior to the use.

h. Autoclave: sterilizes tools prior to the use to eliminate microbial contamination.

i. Refrigerators (+4°C/−20°C): used to store culture media and reagents, preserving stability and shelf-life.
j. Pipette and pipette aids: enable accurate measurement and transfer of small liquid volumes, ensuring precision in cell culture.
k. Cryopreservation system (liquid nitrogen tank or –80°C freezer): used to store cells long-term while maintaining viability and functional integrity.
l. Disinfectants (e.g., bleach, 70% alcohol): used to decontaminate working surfaces and liquid waste.
m. Cell counter (automated or non-automated): used to determine cells number.
n. Biohazard waste containers: safely dispose of contaminated materials such as pipettes, gloves and culture media.
o. Laboratory coat: provides personal protection and maintains sterility.
p. Disposable nitrile gloves: protect samples and the operator from contamination.
q. Safety goggles: shield eyes from splashes and hazardous materials, mainly during cell isolation.

2.2 Ovaries Collection and Transport

Laboratory supplies:

a. Sterile containers used for the collection and transport of bovine ovaries under aseptic conditions.
b. M2 medium at room temperature to preserve ovaries until manipulation.

2.3 Cells Isolation

a. Sterile tweezers.
b. 100 mm Petri dishes.
c. Sterile scalpels.
d. 15 mL Falcon™ conical tubes.
e. Sterile pipette tips (200–1000 μL).

Reagents and solutions:

f. Dulbecco's Modified Eagle's Medium supplemented (DMEM+) with
   i. L-Glutamine (1% v/v)
   ii. Penicillin-Streptomycin (1% v/v)
   iii. Amphotericin B (1% v/v)
   iv. Gentamicin (4% v/v)
g. Fetal Bovine Serum (FBS).
h. Dulbecco's Phosphate-Buffered Saline (PBS) w/o Ca^{2+}/Mg^{2+}.
i. Powder Collagenase type IV.

2.4 Primary Culture Maintenance

Laboratory supplies:

a. 100 mm Petri dishes.
b. 15 mL Falcon™ conical tubes.
c. Serological pipettes (5–10–25 mL).

Reagents and solutions:

d. DMEM+.
e. FBS.
f. Dulbecco's PBS w/o Ca^{2+}/Mg^{2+}.
g. Trypsin-EDTA solution.
h. Dimethyl sulfoxide (DMSO).

2.5 Materials for Simulated Microgravity Exposure

Laboratory supplies:

a. Random Positioning Machine (RPM; Yuri GmbH, Germany).
b. T25 Falcon™ culture flasks.
c. Parafilm™.
d. Fixing clamps.
e. Tape.

Reagents and solutions:

f. DMEM +.
g. FBS.
h. PBS (w/o Ca^{2+}/Mg^{2+}).
i. Trypsin-EDTA solution.

3 Methods

3.1 Collection of Ovaries

- Collect bovine ovaries by a licensed veterinarian at a local slaughterhouse (*see* **Note 1**) using a sterile scalpel.
- After excision, immediately place the ovaries in sterile M2 medium prewarmed to 37°C in sterile containers (e.g., standard urine collection cups) and transport them to the laboratory (*see* **Note 2**).

3.2 Isolation of Granulosa Cells

- Under a horizontal laminar flow hood, place 3–4 mL of prewarmed M2 medium into a sterile 100 mm Petri dish.
- Using sterile tweezer, transfer one ovary at a time into this Petri dish for processing. Puncture every ovary small to medium antral follicles (2–5 mm in diameter) using sterile scalpels to release the follicular fluid.

- After processing all collected ovaries, aspirate the follicular fluid with a sterile pipette and transfer it into a 15 mL falcon conical tube—add fresh M2 medium to ensure to aspirate all the follicular fluid in the Petri dish.
- Suspend the aspirate in the falcon in DMEM+ medium (w/o FBS) and centrifuge at 1200 rpm for 5 min at +4°C. Repeat the centrifugation step three times, adding fresh medium before each spin to ensure proper washing of the cells.
- Finally, resuspend the resulting cell pellet, containing bovine granulosa cells (bGC), in DMEM+ medium with 5% FBS (*see* **Note 3**).
- Seed bGCs into multiple sterile 100 mm Petri dishes and incubate at 37°C in a humidified atmosphere with 5% CO_2 (*see* **Note 4**).

3.3 Isolation of Theca Cells

- Under a horizontal laminar flow hood, use a sterile tweezer to transfer the ovaries into a new sterile 100 mm Petri dish containing 1–2 mL of M2 medium.
- Carefully dissect the follicular walls using a sterile scalpel from the same follicles previously used for granulosa cell isolation.
- Isolate the theca layer, mince the tissue pieces using the sterile scalpel and transfer them into sterile conical tubes containing 0.05% collagenase type IV solution in DMEM+ (without FBS), previously filtered through a 0.22 μm sterile filter (*see* **Note 5**). For optimal enzymatic activity, divide the tissue fragments into multiple 15 mL Falcon™ conical tubes, each containing 2–3 mL of the collagenase solution.
- Incubate the suspensions in a water bath with orbital shaker function (or in an orbital shaker placed inside an incubator) at 37°C with continuous agitation for 45–60 min. Every 15 min, briefly vortex the suspension to promote tissue dissociation.
- When the suspension appears cloudy, stop the enzymatic activity by adding an equal volume of DMEM+ medium containing FBS.
- Centrifuge the suspension at 1200 rpm for 5 min at +4°C, repeating the wash step three times and adding fresh DMEM+ before each spin to remove residual enzyme and debris.
- Finally, resuspend the cell pellet (bTC) in complete medium (with 5% FBS) and plate the suspension into multiple sterile 100 mm Petri dishes.
- Incubate bTC at 37°C in a humidified atmosphere containing 5% CO_2 (*see* **Note 6**) (Fig. 1).

3.4 Primary Cell Culture Maintenance

- On the following day, perform a complete medium change for bGC cultures. Carefully wash each 100 mm Petri dish twice with

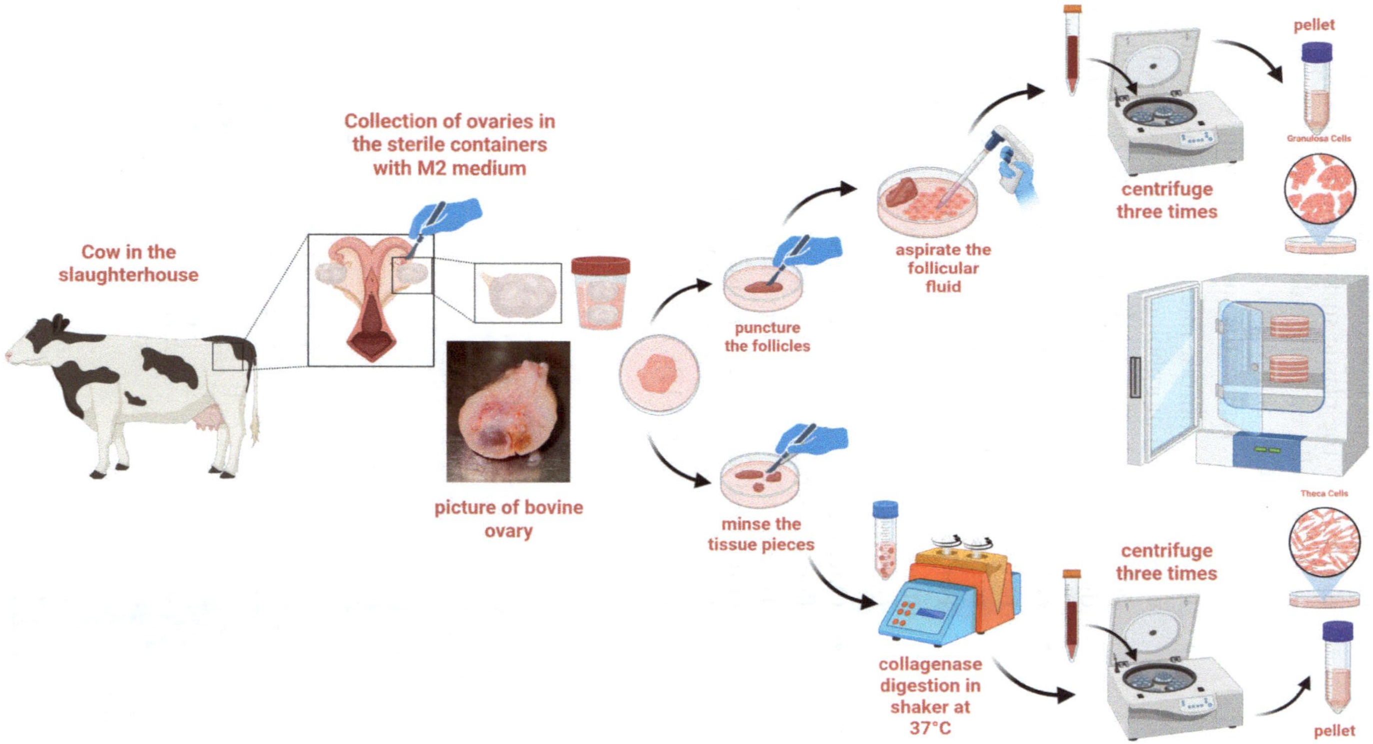

Fig. 1 The figure illustrates the schematic workflow for the collection and isolation of granulosa and theca cells from bovine ovaries. The picture was created with items in https://BioRender.com

prewarmed PBS (without Ca^{2+} and Mg^{2+}) and then add fresh complete medium.

- For bTC cultures, delay the first medium change until 48 h post-plating. This allows cells embedded within tissue fragments to attach to the plastic surface and begin spreading. As with granulosa cells, wash the cultures gently twice with prewarmed PBS (w/o Ca^{2+} and Mg^{2+}), then add fresh complete medium. Subsequently, replace the culture medium every 48–72 h.
- Under phase-contrast microscopy, granulosa and theca cells exhibit distinct morphologies: granulosa cells typically display a polygonal or hexagonal shape, whereas theca cells appear spindle-shaped with a fibroblast-like morphology (Fig. 2A).
- Assess expression of cell-type-specific marker through quantitative real-time PCR (qRT-PCR) to confirm the identity of the two cell populations and exclude potential cross-contamination. For bGC, Cyp19A1 (Aromatase) is recommended as a specific marker. While Cyp17A1 is suitable for identifying bTC (Fig. 2B) (*see* **Note 7**).
- Once bGC and bTC cultures reach approximately 80%–90% confluence, detach the cells using a 0.05% trypsin-EDTA solution for 10 min at 37°C. After detachment, the cell suspensions can be either used for further expansion (*see* **Note 8**) or cryopreserved. For freezing, resuspend the cells in a cryopreservation medium composed of 90% FBS and 10% DMSO.

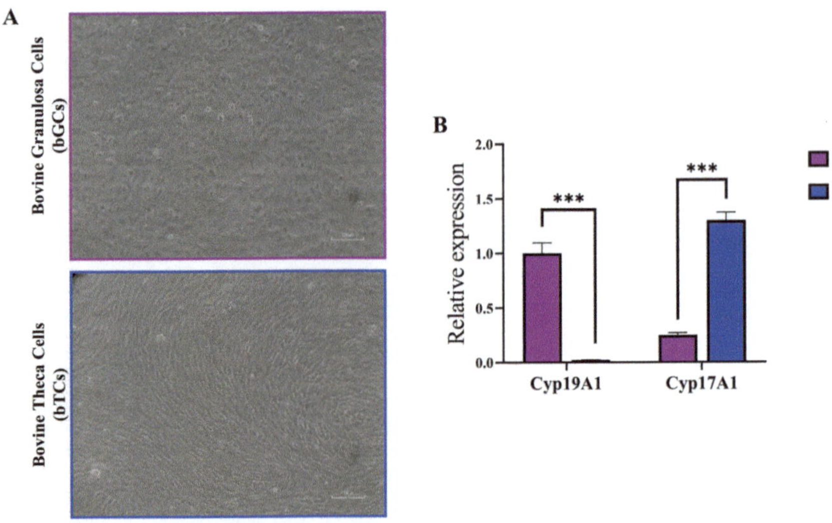

Fig. 2 Bovine Granulosa and Theca cells characterization. (**A**) Phase-contrast microphotographs showing bGCs and bTCs morphology. Scale bars: 100 μm. (**B**) mRNA expression of markers Cyp19A1 and Cyp17A1 evaluated by qRT-PCR. Data were normalized to RPS29 mRNA expression. Bars represent means ± SEM. ***$p < .001$

3.5 Simulated Microgravity Exposure

- To evaluate how simulated microgravity affects bGC and bTC behaviour and morphology, seed both cell types in parallel experimental and control conditions. We decided to investigate the effect of microgravity environment exposure for three different time points. When both cultures seeded in 100 mm Petri dishes reached approximately 90% confluence, *detach the cells and divide into T25 flasks* with DMEM+ and 5% FBS per cell type—plate the same number of culture flasks for both simulated microgravity (μg) and 1 g ground control conditions (*see* **Note 9**).
- On the following day, remove the medium from the flasks, wash with prewarmed PBS and completely fill every flask (μg and 1 g) with fresh DMEM+ medium with 5% FBS. Perform the refreshment of medium slowly to avoid the formation of any air bubbles. To prevent medium leakage tightly close the T25 flask and seal the caps with Parafilm™.
- Simulate microgravity conditions using a Desktop RPM, a specialized 3D clinostat developed by Yuri GmbH. Place the flasks at the center of the RPM platform. Firmly secure the flasks with adhesive tape and fixing clamps to prevent any movement and sliding during rotation.
- Operate the RPM in 3D random mode, using random motion and random direction, maintaining an average velocity of 60 deg/s to create a dynamic environment that prevents the application of a consistent unidirectional or omnidirectional 1 g force, rather than eliminating gravity entirely.
- Place the RPM inside a standard incubator set at 37°C with 5% CO2 and connect it to the control console using standard electrical cables. Simultaneously, place the corresponding 1 g control flasks in the same incubator, under identical temperature and CO2 conditions but without exposure to RPM movement (Fig. 3A).

3.6 bGC and bTC Co-Culture Microgravity Environment Exposure

- To investigate the effects of simulated microgravity on co-cultures of bGC and bTC cells, detach both cell types when cultures reached approximately 90% confluence in 100 mm Petri dishes. Under physiological conditions, the cellular composition of the ovarian follicle consists of approximately 60% granulosa and 40% theca cells. We suggest this ratio in co-culture experiments.
- Defining an appropriate seeding density is essential when establishing co-cultures, particularly in relation to the experimental timeline. We suggest a seeding density of 40% and a total of 3×10^5 cells per T25 flask.
- Remove the medium and wash each Petri dish twice with prewarmed PBS (without Ca^{2+} and Mg^{2+}). Detach the cells using 0.05% trypsin-EDTA for 10 min at 37°C. Halt trypsin activity by

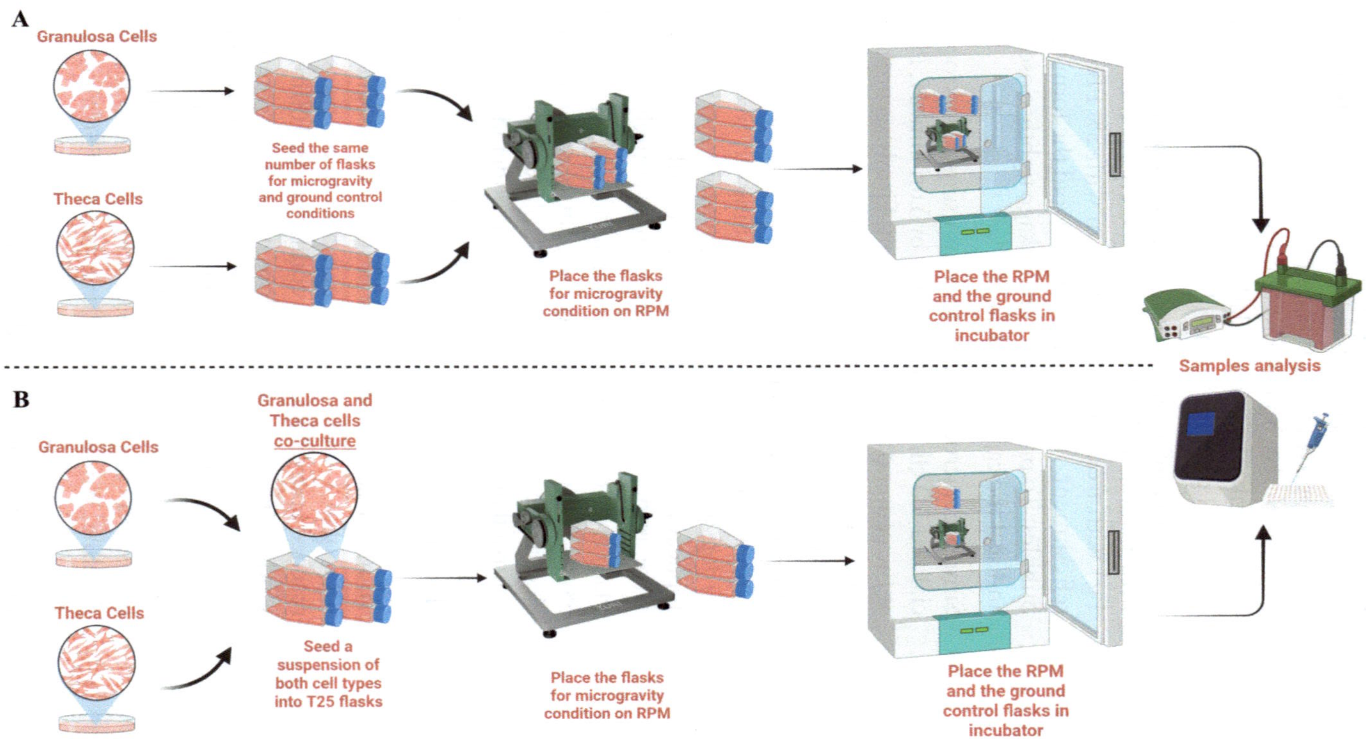

Fig. 3 The figure illustrates the schematic method to expose bGCs and bTCs monocoltures (**A**) and bGCs and bTCs co-culture (**B**) to microgravity environment through RPM device. The picture was created with items in https://BioRender.com

adding complete DMEM+ medium containing 5% FBS. Count the total number of viable cells from both cultures (*see* **Note 10**).

- To set up the co-culture, prepare a single suspension by combining the two cell types in the predetermined ratio and total cell number, resuspending them in DMEM+ supplemented with 5% FBS. Seed the suspension into T25 flasks and allow the cells to attach for 24 h before initiating simulated microgravity exposure (*see* **Note 11**).
- On the following day, remove the medium from each T25 flask and wash once with prewarmed PBS (without Ca^{2+} and Mg^{2+}). Then, completely fill both μg and 1 g control flasks with fresh DMEM+ medium with 5% FBS. Medium replacement should be performed slowly to avoid the formation of air bubbles, which can interfere with simulated microgravity conditions.
- Conduct the subsequent exposure to simulated microgravity following the same procedure described for monocultures (*see* **Note 12**) (Fig. 3B).

4 Notes

1. In some countries, the collection of ovaries from slaughterhouses may require prior authorization from the facility's management or veterinary authorities. Researchers are advised to comply with all applicable national and institutional regulations regarding the use of animal by-products for research purposes.
2. Ovaries can be stored in M2 medium either at 4°C or at 37°C prior to processing. Refrigeration at 4°C is recommended for overnight storage, as it helps preserve tissue integrity and minimizes bacterial or mycotic growth. Incubation at 37°C may be used if processing occurs within a few hours, but prolonged storage at this temperature is not advised.
3. Cell counting is not performed at this stage because the follicular fluid frequently contains a high number of red blood cells, which interfere with accurate quantification using manual or automated cell counters. Plating is therefore performed directly depending on the volume of cell pellet rather than cell number.
4. The number of Petri dishes used to seed cell suspensions depends on the volume and density of the resulting cell pellet. This applies to both granulosa and theca cell preparations and should be adjusted accordingly to avoid overcrowding or underseeding.

5. The optimal concentration of collagenase may vary depending on the enzymatic activity of the specific batch used. It is recommended to test and adjust enzyme concentration and digestion time based on preliminary observations of tissue dissociation efficiency to avoid overdigestion or cell damage.
6. After enzymatic digestion, the suspension may still contain small tissue fragments, not completely dissociated. These fragments contain viable theca cells and should not be discarded. It is recommended to seed the suspension, including tissue fragments, using sterile forceps to facilitate the transfer of tissue pieces into the culture dishes.
7. Additionally, for granulosa cells, FSH receptor and Anti-Müllerian Hormone (AMH) can also be used as specific markers; for theca cells, StAR and Cyp11A1 are recommended markers.
8. For all experiments, bGC and bTC should be used until passage 3.
9. In our setup, experiments were conducted at three different time points (7, 10, and 14 days), although the duration can be adapted depending on specific experimental goals. Plan to plate flasks specifically designated for downstream analyses such as RNA, DNA, or protein extraction, as well as for any other functional assays. These should be prepared in parallel for both μg and 1 g conditions to ensure consistent sampling and avoid repeated cell harvesting, which may affect experimental outcomes.
10. The number of viable cells could be then counted manually using a hemocytometer or with an automated cell counter.
11. It is recommended to gently pipette the mixed cell suspension up and down several times prior to seeding, to ensure a homogenous distribution of both granulosa and theca cells. Vigorous pipetting should be avoided to prevent mechanical stress or cell damage.
12. When analyzing the effects of simulated microgravity at multiple time points, it is crucial to stop the RPM as quickly as possible to prevent cells from readapting to Earth's gravity. Even brief exposure to normal gravity may influence the outcome, especially for gravity-sensitive processes. Additionally, plan flask placement carefully to avoid having to remove all culture flasks from the RPM platform at each time point, which could compromise the consistency of the experiment.

References

1. What is Artemis? https://doi.org/https://www.nasa.gov/what-is-artemis
2. Chavers G, Watson-Morgan L, Smith M, Suzuki N, Polsgrove T (2020) NASA's human landing system: the strategy for the 2024 mission and future sustainability. IEEE Aerospace Conference, Big Sky, MT, USA, pp 1–9
3. Johnson SK, Mortensen DJ, Chavez MA, Woodland CL (2021) Gateway – a communications platform for lunar exploration. In: 38th International communications satellite systems conference (ICSSC 2021), Hybrid Conference, Arlington, VA, USA, 2021, pp 9–16
4. Bizzarri M, Gaudenzi P, Angeloni A (2023) The biomedical challenge associated with the Artemis space program. Acta Astronaut 212:14–28
5. Mishra B, Luderer U (2019) Reproductive hazards of space travel in women and men. Nat Rev Endocrinol 15:713–730
6. Williams D, Kuipers A, Mukai C, Thirsk R (2009) Acclimation during space flight: effects on human physiology. CMAJ (Can Med Assoc J) 180:1317–1323
7. Montgomery PO Jr, Cook JE, Reynolds RC et al (1978) The response of single human cells to zero gravity. In Vitro 14(2):165–173
8. Sonawane R, Patil S, Rahaman J, Mukherjee D (2025) Effect of microgravity on bone tissue: mechanisms of osteodegeneration and advanced treatment modalities. Biochem Biophys Res Commun 771:152055
9. Nabavi N, Khandani A, Camirand A, Harrison RE (2011) Effects of microgravity on osteoclast bone resorption and osteoblast cytoskeletal organization and adhesion. Bone 49(5):965–974
10. Lee Satcher R, Fiedler B, Ghali A, Dirschl DR (2024) Effect of spaceflight and microgravity on the musculoskeletal system: a review. J Am Acad Orthop Surg 32(12):535–541
11. Mircea AA, Pistritu DV, Fortner A et al (2024) Space travel: the radiation and microgravity effects on the cardiovascular system. Int J Mol Sci 25(21):11812
12. Lv H, Yang H, Jiang C et al (2023) Microgravity and immune cells. JR Soc Interface 20(199):20220869
13. Ferrando AA, Lane HW, Stuart CA et al (1996) Prolonged bed rest decreases skeletal muscle and whole body protein synthesis. Am J Physiol-Endocrinol Metab 270:E627–E633
14. Cutigni M, Cucina G, Galante E et al (2025) Microgravity impairs endocrine signaling and reproductive health of women. A narrative review. Front Physiol 16:1558711
15. Tou J, Ronca A, Grindeland R, Wade C (2002) Models to study gravitational biology of mammalian reproduction. Biol Reprod 67:1681–1687
16. Jain V, Wotring VE (2016) Medically induced amenorrhea in female astronauts. NPJ Microgravity 2:16008
17. Ronca AE, Baker ES, Bavendam TG et al (2014) Effects of sex and gender on adaptations to space: reproductive health. J Womens Health 23:967–974
18. Jennings RT, Baker ES (2000) Gynecological and reproductive issues for women in space: a review. Obstet Gynecol Surv 55:109–116
19. Radowicka M, Pietrzak B, Wielgoś M (2013) Assessment of the occurrence of menstrual disorders in female flight attendants - preliminary report and literature review. Neuro Endocrinol Lett 34(8):809–813
20. Gorbacheva EY, Toniyan KA, Biriukova YA et al (2023) The state of the organs of the female reproductive system after a 5-day "dry" immersion. Int J Mol Sci 24(4):4160
21. Tomilovskaya E, Amirova L, Nosikova I et al (2021) The first female dry immersion (NAIAD-2020): design and specifics of a 3-day study. Front Physiol 12:661959
22. Miglietta S, Cristiano L, Espinola MSB et al (2023) Effects of simulated microgravity in vitro on human metaphase II oocytes: an electron microscopy-based study. Cells 12(10):1346
23. Wu C, Guo X, Wang F et al (2011) Simulated microgravity compromises mouse oocyte maturation by disrupting meiotic spindle organization and inducing cytoplasmic blebbing. PLoS One 6(7):e22214
24. Zhang S, Zheng D, Wu Y et al (2016) Simulated microgravity using a rotary culture system compromises the in vitro development of mouse preantral follicles. PLoS One 11(3):e0151062
25. Messinis IE, Messini CI, Dafopoulos K (2014) Novel aspects of the endocrinology of the menstrual cycle. Reprod Biomed Online 28(6):714–22
26. Armstrong DT, Goff AK, Dorrington JH (1979) Ovarian follicular development and function. In: Midgley AR, Sadler WA (eds) Raven Press, New York, pp 169–182
27. Rooda I, Méar L, Hassan J, Damdimopoulou P (2025) The adult ovary at single cell resolution: an expert review. Am J Obstet Gynecol 232(4S):S95.e1–S95.e16
28. Adams GP, Pierson RA (1995) Bovine model for study of ovarian follicular dynamics in humans. Theriogenology 43:113–120
29. Alvarez RH, Duarte KMR, Carvalho JBP et al (2023) Ovarian morphology and follicular dynamics associated with ovarian aging in Bos indicus beef cows. Anim Reprod Sci 254:107279
30. Mohammed BT, Donadeu FX (2018) Bovine granulosa cell culture. Methods Mol Biol (Clifton, N.J.) 1817:79–87

31. Malhi PS, Adams GP, Singh J (2005) Bovine model for the study of reproductive aging in women: follicular, luteal, and endocrine characteristics. Biol Reprod 73(1):45–53
32. Roberts JF, Jeff Huang CC (2022) Bovine models for human ovarian diseases. Prog Mol Biol Transl Sci 189(1):101–154
33. Roiati M, Raia T, Fedeli V et al (2023) The Ovo Space project: simulated microgravity affects epigenetic factors in bovine granulosa and theca cells. In: Frontiers in epigenetics and epigenomics - abstract book of the 1st epigenetics society international meeting "epigenetics of disease and development", 12–14 October 2023, Rome, Italy, p 290

Chapter 12

Defining Microbiome Impact on Host Physiology During Spaceflight Using *Caenorhabditis elegans*

Dana Blackburn, Bushra Rahman, Atiyya P. Saroyia, Audrey J. Parish, Monica Driscoll, Nathaniel J. Szewczyk, Siva A. Vanapalli and Buck S. Samuel

Abstract

Microbiome-integrated *Caenorhabditis elegans* cultivation methods enable investigation of host–microbiome interactions in the context of space-relevant stresses using three key innovations: introduction of live bacterial communities replacing chemically defined media, implementation of auxin-inducible degradation systems to prevent progeny production, and development of complementary hardware platforms. Polyethylene bags provide gas-permeable cultivation environments for large populations with complex microbiomes supporting downstream molecular analyses, while NemaCapsules with micropillar arrays and passive culturing chambers allow real-time phenotypic assessment through on-orbit imaging, transforming our ability to correlate molecular signatures with physiological outcomes in microgravity.

Keywords Microfluidics, Host–microbe interactions, Microgravity, Space health, Genetics

1 Introduction

Understanding the microbiome's impact on human health during spaceflight represents a frontier in space biology research. Despite growing recognition of the microbiome's importance, limited knowledge exists regarding microbial community structure and function in microgravity environments, largely due to constraints, such as small astronaut cohorts, individual genetic variability, and inherent differences among missions [1, 2]. *Caenorhabditis elegans* offers a powerful solution to these challenges as a genetically tractable model organism with conserved biological pathways and a short lifespan that facilitates analyses of the impacts of long-duration spaceflight on host physiology [3, 4]. Previous missions, including ICE-FIRST

Dana Blackburn and Bushra Rahman contributed equally to this work.

Mariano Bizzarri (ed.), *Biomedicine and Biotechnology in Space*, Methods in Molecular Biology, vol. 3000, https://doi.org/10.1007/978-1-0716-5174-2_12,

and CERISE, established the foundations for *C. elegans* research in microgravity, revealing alterations in gene expression related to muscle maintenance, metabolism, and longevity [5, 6].

The protocol that we document highlights methodological advancements that have significantly extended the utility of this robust system. First, transitioning from chemically defined media (CeMM) or metabolically inert food sources to physiologically relevant live bacterial communities allows for the integration of study of impacts of the microbiome on host physiology. While CeMM supported early experiments, it significantly altered *C. elegans* development compared to standard laboratory conditions [7]. Specifically, the CBIOMES (*C. elegans* Biological Investigation on Microbiome Effect in Space) mission introduces a diverse set of *C. elegans* strains to live *E. coli* OP50 and complex microbiomes (e.g., BIGbiome [8], a 63-member bacterial consortium with functional similarity to human microbiomes). Second, the application of auxin-inducible degradation (AID) systems to prevent progeny production [9] enables controlled long-duration studies in closed systems by eliminating resource competition from offspring and maintaining synchronized animal populations. These advances allow for the distinction of direct effects of microgravity from those mediated by altered microbiome composition.

In addition to the introduction of the microbiome, we present two optimized and modular hardware applications for molecular or real-time assessments of spaceflight impact on host physiology. First, polyethylene (PE) bags provide a versatile platform for *C. elegans* cultivation in spaceflight, particularly for host–microbiome interaction studies. These biocompatible systems maintain large nematode populations with live microbiomes over extended periods [10, 11]. The gas permeability of these systems ensures sufficient oxygen exchange while maintaining containment, critical for biosafety compliance aboard the ISS. Most importantly, PE bags accommodate sufficient nematode populations (numbering in the thousands) for comprehensive downstream molecular analyses, facilitating systems-level investigations of host–microbiome interactions in microgravity.

Complementing PE bags, integrated microfluidic platforms enable real-time phenotypic assessment. The NemaFlex-S device features micropillar arrays for measuring nematode muscle strength through deflection analysis [12]. NemaCapsules represent further advancement, incorporating Passive Nematode Culturing Chambers (PNCs) that maintain *C. elegans* cultures long-term without daily intervention [13]. These systems feature micropillar chambers for housing adult worms with sieve channels permitting bacterial media access from surrounding wells.

Here, we describe a modular pipeline that integrates molecular and microscopic analyses of microbiome impacts on host physiology using simple polyethylene containment and real-time imaging capabilities of NemaCapsules (Fig. 1). Combined, these flexible approaches represent robust methods to address many of the

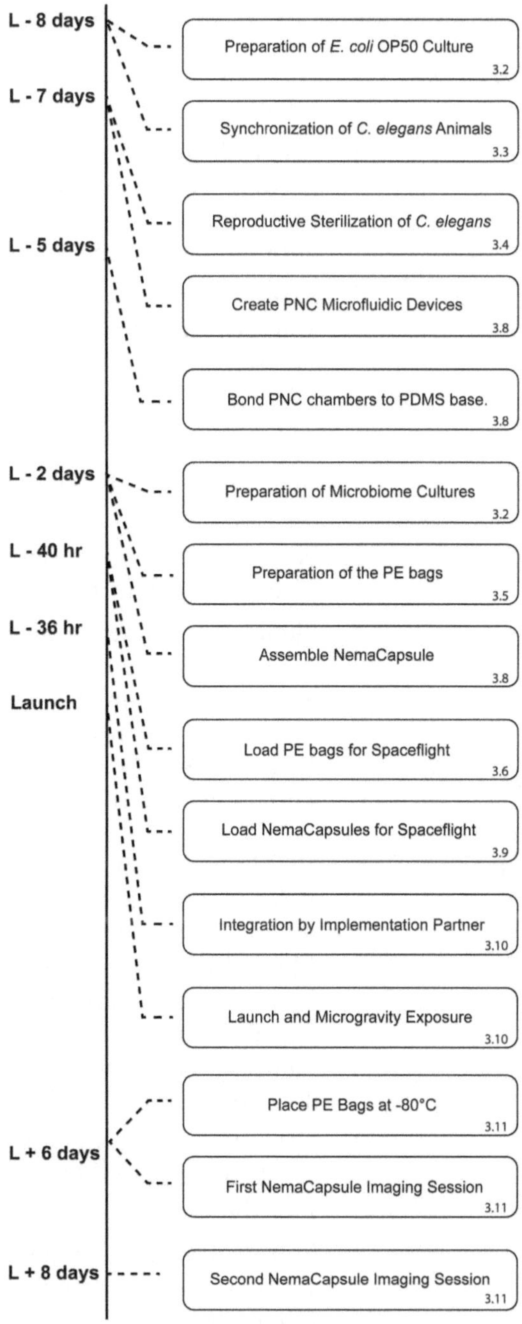

Fig. 1 Schematic of a potential launch schedule. The timepoints chosen for each experiment are determined by launch deadlines and desired collection times. Corresponding method section numbers are included for reference

challenges of studying the impact of space travel on the interaction of the gut microbiome and host physiology.

2 Materials

2.1 Bacterial Culture and Maintenance

1. Bacterial strains (Caenorhabditis Genetics Center): *E. coli* OP50 and natural *C. elegans* microbes [8, 14].
2. LB broth and LB agar plates: Autoclave 20 g/L of Lennox LB broth powder dissolved in water. For LB agar, add 15 g/L agar before autoclaving. After autoclaving, cool to 55°C and aliquot into Petri dishes (25 mL for a 100 mm Petri dish).
3. S basal: 5.85 g NaCl, 1 g K_2HPO_4, 6 g KH_2PO_4, and 1 mL cholesterol (5 mg/mL in ethanol) in 1 L of water. Autoclave to sterilize.
4. Conical tubes: Use sterile 50-mL volume tubes for growing and centrifuging bacterial cultures.
5. Salt solutions: Filter sterilize or autoclave 1 M $CaCl_2$ and 1 M $MgSO_4$.
6. Potassium Phosphate buffer (pH 6.0): 108.3 g KH_2PO_4 and 35.6 g K_2HPO_4 in 1 L of water.
7. Liquid Nematode Growth Medium (NGM): To 975 mL of water in a 1 L autoclavable bottle add 3.0 g NaCl and 2.5 g peptone. After autoclaving, cool to room temperature. Add the following sterile reagents while stirring: 0.5 mL of 1 M $CaCl_2$, 1.0 mL of $MgSO_4$, 25 mL of potassium phosphate buffer, and 1 mL of cholesterol (5 mg/mL in ethanol).

2.2 *C. elegans* Culture and Maintenance

1. NGM agar plates: Add 23 g of NGM powder to 975 mL of water in a 1 L autoclavable media bottle. After autoclaving, cool the mixture to 55°C and add the following sterile reagents while stirring: 0.5 mL of 1 M $CaCl_2$, 1 mL of 1 M $MgSO_4$, and 25 mL of potassium phosphate buffer (see Sect. 2.1). Aliquot into Petri dishes (10 mL in a 60 mm Petri dish or 25 mL in a 100 mm Petri dish) and allow plates to solidify in a laminar flow hood with lids off or partially ajar.
2. M9 buffer: Start with 700 mL of water in an autoclavable bottle. Add 6 g Na_2HPO_4, 3 g KH_2PO_4, 5 g NaCl, and 1 mL of 1 M $MgSO_4$. Add water to 1000 mL and autoclave to sterilize.
3. Culture flasks: For liquid growth and auxin treatment, use 12.5 cm^2 tissue culture flasks with vented caps.
4. Indole Acetic Acid (IAA) solution: 40 and 50 mM of 3-indoleacetic acid solution in sterile M9 buffer. Apply low heat and vortexing as needed to dissolve completely.

2.3 Bleach Synchronization of *C. elegans*

1. Bleach solution: Two parts of chlorine bleach with one part of 5 M NaOH. Prepare fresh each time.
2. Screw-cap tubes: Use 5 mL centrifuge tubes for harvesting eggs from gravid adult worms.

2.4 Mold Fabrication for PNC Microfluidics

1. Silicon wafers 4″ or 6″ (University Wafer): Serves as the base material for PNC mold fabrication.
2. SU-8 photoresist: Negative photoresist is used to create the high-resolution, high-aspect-ratio microstructures in the mold.
3. Isopropanol: For rinsing the wafer before adding the SU-8 photoresist and removing excess SU-8 photoresist from the wafer.
4. Dosimeter: For measuring UV lamp intensity.
5. 360 nm long pass (LP) filter: For blocking shorter wavelengths from reaching the SU-8-coated wafer during design imprinting.
6. UV safety glasses: For shielding against UV radiation emitted from the UV lamp during exposure.
7. SU-8 photodeveloper: For developing the wafer by removing unexposed photoresist and leaving behind the patterned features.
8. Wafer dipper: For anchoring the wafer in its fitted slot during development.
9. Trichloro(1H, 1H, 2H, 2H-perfluorooctyl)silane: For minimizing mechanical wear on molds after multiple PDMS castings.
10. Nunc storage vials with perforable lids: For contained storage of trichlorosilane during mold treatment.
11. 25-gauge needles and 1 mL syringe (Luer Lock): For aliquoting trichlorosilane into vials.

2.5 NemaCapsule Preparation

1. Polydimethylsiloxane (PDMS) resin: Sylgard 184 Silicone Elastomer Kit (Dow) is the standard elastomeric material for microfluidic devices. Mix A and B components in a 10:1 wt/wt ratio.
2. Biopsy punch (1 mm): For punching a hole into the circular impression in PNCs to create the worm loading port.
3. Razor blades: For making cuts at the edge of the sieve channels to allow the surrounding media to flow into the PNCs.
4. PDMS sheets: Used as the base substrate for the PNCs.
5. X-Acto® knife: For cutting the PDMS sheets into the shape of the housing hardware of the NemaCapsule unit (*see* **Note 1**).

6. 0.25–1 lb plates: For adding mechanical pressure on PNCs bonded to the PDMS base.

2.6 Loading Hardware for Ground and Spaceflight

1. Polyethylene (PE) tubing (2 mil, 1.5" flat): Used as primary culture hardware for large populations of *C. elegans* (*see* **Note 2**).
2. M9-T solution: Dilute a 10% Triton X-100 stock solution in M9 buffer to 0.01% Triton X-100 solution in M9 buffer.
3. Pluronic solution: 5% wt/vol solution of Pluronic F-127 in sterile deionized (DI).
4. Luer Lock syringes: Use 1 mL volume for injecting microbiomes into the PNCs. Use 10 mL volume for the pluronic treatment and filling the bulk NemaCapsule air space.
5. Syringe needles: 20-gauge Luer Lock needles for use in loading the pluronic solution and the microbiomes into the PNC loading port.
6. Loading syringe assembly: For loading the adult worms from the media suspension into the PNC loading port. The loading syringe consists of a connected assembly of a 1-mL syringe (Luer Lock), 20-gauge needle (Luer Lock), 6-in long Tygon capillary tubing (1/32″ ID × 3/32″ OD), and a stainless steel tubing coupler (15 mm, 20 gauge) (Fig. 2).
7. Stainless steel dowel pins (1 mm D × 5 mm L): To seal the PNC loading port after injecting the worms.

2.7 General Lab Equipment

1. Spin coater: For coating the silicon wafer with SU-8 photoresist during mold fabrication.
2. Mask aligner: To transfer patterns from the photomask onto the silicon wafer during mold fabrication.
3. Plasma cleaner: For preparation of the PNCs and the PDMS substrate for bonding.
4. Oven: For use in curing the PDMS to create PNCs and bonding PNCs to their PDMS base.
5. Vacuum desiccator: For use in degassing the PNCs.
6. Hot plates: For baking devices during PNC fabrication.

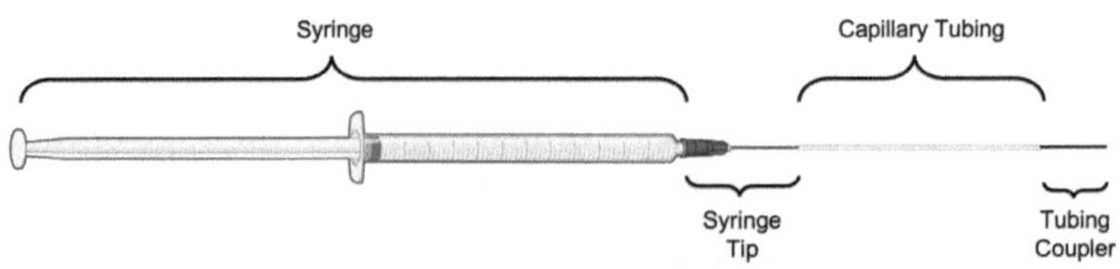

Fig. 2 Loading syringe schematic. Assembly consists of 1 mL syringe, 20-gauge tip, capillary tubing, and a tubing coupler to insert into loading port of PNC microfluidic device. Created with Biorender.com

7. Tube revolver rotator: For use in worm age synchronization to aerate during egg hatching.
8. Vacuum filtration flask: For use in aspirating supernatants in wash steps during worm and biome preparation.
9. Impulse heat sealer: For sealing PE bags and creating a closed system.
10. Benchtop centrifuge: For use in pelleting bacteria and worms in 5–50 mL conical tubes.
11. Platform rocker or orbital shaker: For use in growing worms to adulthood in tissue culture flasks.
12. Shaking and stationary incubators: For use in growing and maintaining bacterial strains. Temperatures needed in this assay range from 26°C to 37°C.
13. Worm incubator: For use in growing and maintaining the worm strains during ground and preflight experiments. Temperatures needed in this assay range from 15°C to 20°C.
14. Dissecting microscope: For use in visualizing disrupted worms during egg harvesting, counting worms, and loading NemaCapsules.
15. Autoclave: For sterilization of buffers and hardware components.

3 Methods

3.1 Maintenance of *C. elegans*

1. Inoculate LB broth (*see* **Note 3**) with a single colony of *E. coli* OP50. Incubate overnight at 37°C with shaking (150–200 rpm) (*see* **Note 4**).
2. Aliquot and centrifuge the *E. coli* OP50 culture in 50 mL conical tubes at 3260 × *g* for 12 min.
3. Aspirate the supernatant and resuspend the pellet in M9 buffer with one-tenth the volume of the original culture, resulting in a 10× concentration of *E. coli* OP50 (e.g., 10 mL for a 100 mL culture).
4. Add 200 μL of the *E. coli* OP50 culture to each 60-mm NGM agar plate (*see* **Note 5**).
5. Gently swirl each plate to spread the *E. coli* OP50 culture. Avoid spreading to the edges, as this will lead to more worms lost due to crawling on the dish walls and desiccation.
6. Allow plates to dry without the lid in the laminar flow hood. This takes approximately 1 h.
7. After at least 1 day at room temperature, use these plates to grow worm strains of interest (*see* **Note 6**) at 20°C until worms reach gravid adulthood.

8. To prevent worms from starving, passage worms by taking a slice of the agar with worms and inverting the slice onto a fresh NGM agar plate seeded with *E. coli* OP50. Worms can be propagated for ~1 month before new worms should be grown from freezer stocks or a starving stock plate (*see* **Note 7**).

3.2 Cultivation and Preparation of Microbiomes

1. From glycerol freezer stocks, streak for pure culture on an LB agar plate (*see* **Note 8**).
   a. For *E. coli* OP50, incubate overnight at 37°C.
   b. For *C. elegans* natural microbiome strains, incubate for 24–48 h at 25°C.
2. Inoculate LB with a single colony from each bacterial strain (*see* **Note 9**).
   a. For *E. coli* OP50 and simple microbiomes, media bottles can be used for large volumes (*see* **Note 10**).
   b. For complex microbiomes with large numbers of bacterial strains, use 10 mL of LB in separate 50 mL conical tubes (*see* **Note 11**).
3. Incubate cultures at the appropriate temperature overnight with shaking (150–200 rpm).
4. Measure the OD_{600} of each overnight culture (*see* **Note 12**). For *E. coli* OP50, move to Step 9, skipping Steps 5–8.
5. Use the lowest OD_{600} reading as the baseline to determine the amount of each strain needed for equal ratios. Multiply the baseline OD_{600} reading by the total volume of the bacterial culture (e.g., 10 mL of culture with $OD_{600} = 0.325$ would be $10 \times 0.325 = 3.25$).
6. Divide the number found in Step 5 by the OD_{600} reading for each of the other bacterial cultures. This gives the volume needed for each strain.
7. In a sterile Erlenmeyer flask or media bottle, add the volume of each strain as determined in Step 6. This mixture now has every strain in equal parts based on OD_{600}.
8. Measure the OD_{600} of the microbiome mixture.
9. Adjust the concentration to an $OD_{600} = 1.0$ (*see* **Note 13**) in S basal by centrifuging the bacteria in 50 mL conical tubes for 30 min at 3260 × *g* to pellet cells (*see* **Note 14**), aspirating the supernatant, and resuspending the pellet in S basal.
10. Take the final OD_{600} reading and adjust if needed (*see* **Note 15**).
11. To prepare mass-concentrated microbiomes for NemaCapsules start by weighing 50-mL conical tubes (m_1).

12. Add the microbiomes prepared above to the 50-mL conical tubes and centrifuge for 10–15 min at 3260×*g* to pellet.
13. Aspirate the supernatant and reweigh the tubes with the bacterial pellets (m_2).
14. Determine the weight of the bacterial pellet: pellet= $(m_2 - m_1)$ [g] or $(m_2 - m_1) \times 1000$ [mg]
15. Determine the media volume for a final concentration of 20 mg/mL: media volume [mL] = $\frac{(m_2 - m_1) \times 1000}{20\ \text{mg} / \text{mL}}$
16. Resuspend the bacterial pellet with the volume of liquid NGM determined in Step 15.
17. Prepare serial dilutions of each bacterial culture in sterile M9 buffer and drop 10 μL of each dilution on an LB agar plate. Tilt the agar plate to allow the droplet to run down the plate, stopping short of the edge. After overnight incubation at the appropriate temperature, examine plates for potential contamination.

3.3 Age Synchronization of *C. elegans* Cultures

1. Wash gravid adults off the *E. coli* OP50 maintenance plates with sterile M9 buffer. Add ~1 mL to a 60 mm plate and swirl the plate to lift the worms from the plate. Pipette the M9 buffer across the plate to loosen worms stuck in the bacteria.
2. Collect the worm microbe mixture in a sterile 5 mL screw-capped centrifuge tube and add M9 buffer to make a total of 3 mL.
3. Add 1 mL bleach solution to the tube.
4. Briefly vortex the tube to bring animals into suspension.
5. Slowly invert the tube back and forth constantly for approximately 3 min or until gravid hermaphrodites begin to break apart. Monitor visually by holding a tube to the light or using a dissecting microscope (*see* **Note 16**).
6. Wash away the bleach solution: (1) Centrifuge for 30 s at 1300×*g*; (2) Aspirate the supernatant to ~0.5 mL; (3) Add 3 mL of M9; (4) Shake to resuspend the pellet.
7. Repeat Step 6 twice more for a total of three washes.
8. Place tubes containing *C. elegans* eggs on a tube revolver rotator at room temperature.
9. Allow the eggs to hatch and synchronize at the L1 stage overnight (~16–18 h at 25°C). (Note: Animals are arrested at this developmental stage due to a lack of food in the media.)
10. Place a 10 μL aliquot of the solution on a microscope slide. Assess the number of L1 animals by bright field microscopy.
11. Adjust the solution concentration to 10 L1 worms per μL in sterile M9 buffer.

3.4 Reproductive Sterilization of *C. elegans*

3.4.1 Liquid Culture

1. Add 100 µL of 50 mM IAA to a 12.5 cm^2 tissue culture flask. The final concentration will be 1 mM.
2. Add 400 µL of synchronized L1 worms in M9 buffer from Sect. 3.3 for ~4000 total worms.
3. Add 4.5 mL of *E. coli* OP50 culture resuspended in S basal prepared in Sect. 3.2 to the tissue culture flask.
4. Place on a platform rocker (20 rpm) in an incubator at the desired temperature until adulthood is reached (*see* **Note 17**).

3.4.2 NGM Agar Plate

1. Producing IAA agar plates. Spread 200 µL of 40 mM IAA on a 10 mL 60 mm NGM agar plate for a final concentration of 1 mM per plate. Allow the IAA to soak into the plates with the lids off in a laminar flow hood until dry (~1 h) or store plates with the lids on overnight at room temperature in an opaque container.
2. After the NGM + IAA plates are dry, add 500 µL of *E. coli* OP50 culture resuspended in S basal prepared in Sect. 3.2 to each plate. Allow plates to dry without the lid in a laminar flow hood (~1 h) or overnight with the lid (longer if the relative humidity is >30%).
3. Add 10 µL of synchronized L1 worms in M9 buffer from Sect. 3.3 to each plate for ~200 worms per plate.
4. Store plates in an incubator without light, or in an opaque box, at the desired temperature until worms reach adulthood (*see* **Note 17**).

3.5 Preparation of PE Bags for Spaceflight

1. Unroll a section of 1.5″ width PE tubing and cut to 4.5″ in length using sterile scissors (Fig. 3a, *see* **Note 18**). Repeat until the desired number of PE bags is reached.
2. For each PE bag, create a seal 0.75″ from one end using an impulse heat sealer set to 2.5 to make a tab for sample labeling (Fig. 3b, *see* **Notes 19 and 20**). Allow the impulse sealer to cool for a few seconds in between each seal to avoid overheating.

3.6 Loading of PE Bags for Spaceflight

1. Pour liquid-cultured reproductively sterile adult worms from the tissue culture flasks (see Sect. 3.4) into 15 mL conical tubes.
2. Wash: Centrifuge for 30 s at 1000 rpm (1300 × *g*), aspirate the supernatant, and add 10 mL M9 buffer into the conical tube.
3. Repeat the Step 2 twice for three total washes to remove as much bacteria as possible.
4. After the final wash, adjust the volume for a final concentration of ~4 worms/µL (see Sect. 3.3, Step 10).

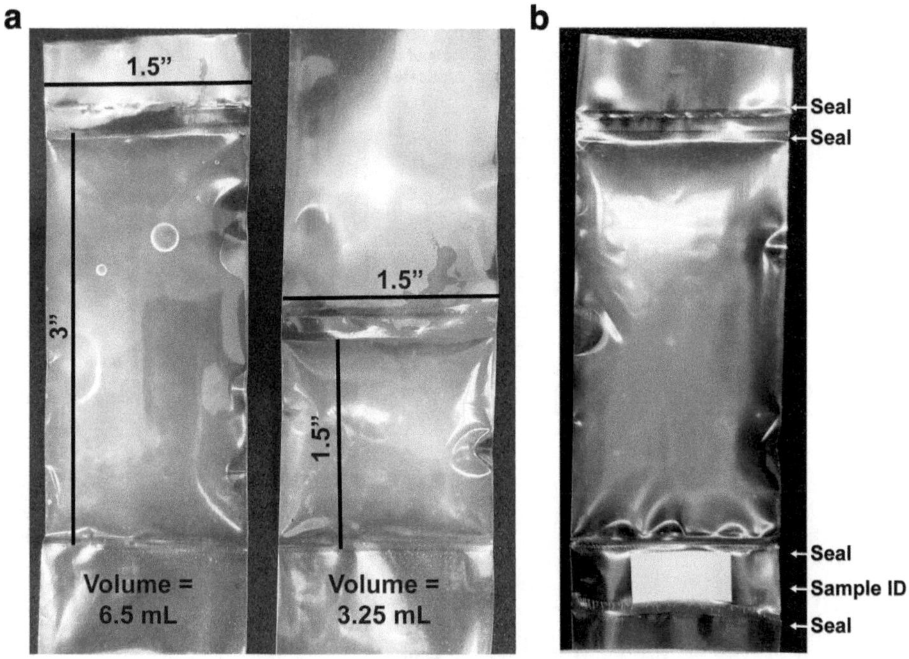

Fig. 3 PE bags loaded with worms and microbiome cultures. (**a**) PE bag dimensions with the corresponding liquid volumes that yield a permissible surface area to volume ratio (**b**) A PE bag featuring double seals that protectively integrate labels and prevent leaks

5. Using a P1000 pipette and filter tip, pipette sterile M9-T solution up and down to coat the inside of the pipette tip and limit worm adherence to the tip. Add 0.5 mL of the worms to a PE bag. For no worm controls, add 0.5 mL of sterile M9 buffer.
6. Add 6 mL of *E. coli* OP50 or a microbiome mixture in S basal prepared in Sect. 3.2 to the PE bag.
7. Push out excess air and use an impulse heat sealer to seal the open end of the PE bag 3" from the other seal (*see* **Note 21**). Add a second seal to prevent potential leaks (Fig. 2b).

3.7 Mold Fabrication of PNC Microfluidics for NemaCapsules

3.7.1 Photomask Design and Construction

1. Construct a 2-D AutoCAD rendering for the PNC microfluidics using the precise desired dimensions for the features of each PNC (Fig. 4a) and the spacing of all the replicates combined (Fig. 4b). For this assay, two types of PNCs are used: (1) crawling PNCs that are designed with a dense pillar array

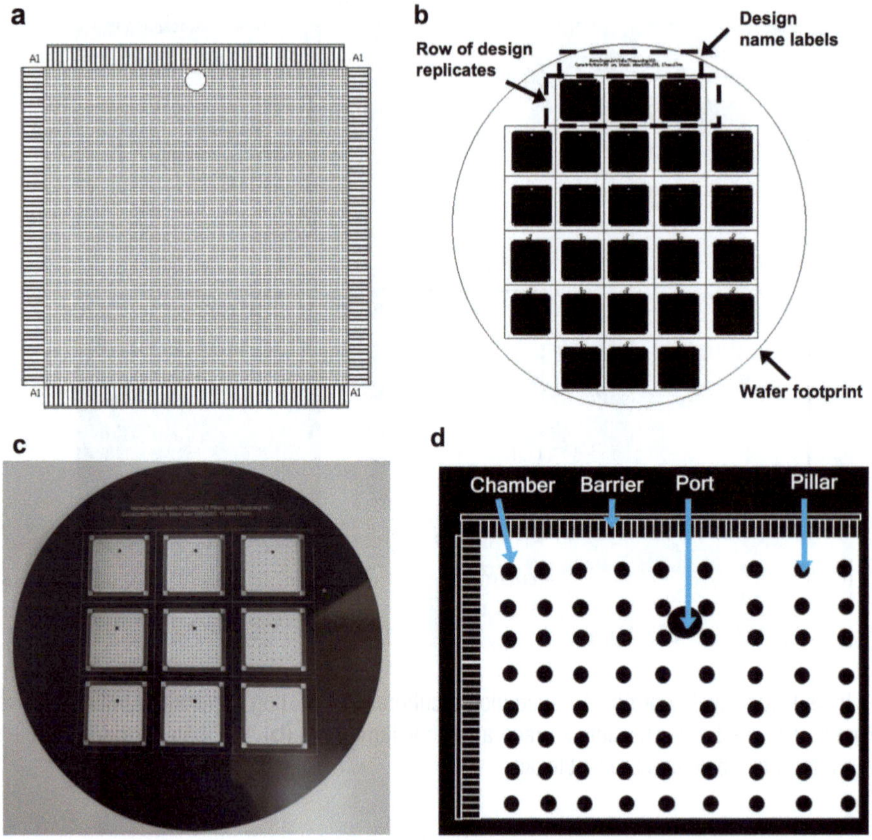

Fig. 4 CAD Design and Transparency. (**a**) AutoCAD drawing of single PNC (**b**) AutoCAD image showing layout of PNC replicates on wafer (**c**) printed design transparency from CADArt (**d**) black/white polarity designation for each feature

allowing worms to move in their natural crawling gait; and (2) swimming PNCs that are designed with a sparse pillar array to induce swimming, which provides a quick and easy visual assessment of the worms.

2. Obtain a photomask (Fig. 4c) of the above rendering from CAD/Art Services, Inc. with the following specifications: (1) emulsion side oriented down; (2) a photomask with transparent outside the surrounding circle (Fig. 4b); (3) black masks denote nonpolar features such as recesses or extruding features, while polar features are kept white (see Fig. 4d for example); and (4) the photomask outer diameter matches the desired silicon wafer size.

3.7.2 Photoresist Application

1. Open the valve to the gas cylinder connected to the spin coater with a working pressure of 60 kPa.
2. Turn on the connected vacuum pump and confirm a vacuum pressure reading of over 20 in Hg using vacuum mode on the spin coater. Exit vacuum mode.

Table 1
Spin coating input parameters

	Time (s)	Acceleration	Speed (rpm)
Step 1	15	260	15
Step 2	30	520	1200

3. Center the wafer on the stage and activate the vacuum to hold the wafer in place.
4. Create a program with the desired parameters to obtain ~100 μm depth of the SU-8 photoresist layer (Table 1) (*see* **Note 22**).
5. Clean the wafer by spraying isopropanol on top and spin by pressing "Run/Stop."
6. Pour SU-8 photoresist onto the center of the wafer in a slow, rotational motion until it forms a circle roughly ¼ of the wafer area (½ of the diameter).
7. Run the program for the wafer to be fully coated with SU-8 photoresist.
8. Remove the wafer by turning off "vacuum" mode on the unit and cover to protect it from light and debris. Take care not to touch the wafer or the SU-8 photoresist surface in the process.

3.7.3 Soft Baking

1. Place the wafer in the center of a hot plate preheated to 72°C for 8–10 min.
2. Move to a hot plate preheated to 110°C for 20–22 min.
3. Remove from the hot plate and allow it to cool to room temperature.

3.7.4 Mask Alignment and UV-Based Imprinting

1. Turn on the mask aligner and allow it to reach full power (50 V, 7 A).
2. Measure the lamp intensity by placing the metallic sensor side of the dosimeter onto the stage and aligning the stage under the lamp. Then turn off surrounding lights and expose the dosimeter side to UV light for 10 s (*see* **Note 23**).
3. Calculate the time required for UV exposure using the formula below, where exposure energy [mJ/cm^2] is obtained from the SU-8 photoresist manufacturer's manual, 100 μm is the SU-8 photoresist thickness, and the lamp intensity [mW/cm^2] is from Step 2. Add 3–5 s to the calculated time to enhance the cross-linking of the SU-8 photoresist, providing improved pattern fidelity and consistent development outcomes.

$$\text{Exposure time [s]} = \frac{1.4(3)\,(\text{Exposure Energy for} \sim 100\ \mu\text{m features})}{\text{Lamp intensity}}$$

4. Tape the photomask under a glass stage cover with the emulsion side facing down. Ensure that labels or writing on the photomask are readable from left to right.
5. Place the wafer on the stage with the SU-8-coated side facing up to contact the emulsion side of the photomask. Make sure to center the wafer for the design to superimpose it.
6. Use vacuum suction for maximum contact between the photomask and the wafer. Check for proper suction by ensuring the glass stage cover cannot be lifted manually.
7. To eliminate the overexposure of the SU-8 photoresist and enhance feature profiles and dimensions, cover the photomask and wafer area with a 360 nm long-pass (LP) filter. **This step is vital.**
8. Align the stage with the UV lamp and expose the photomask-wafer assembly for the time calculated in Step 3.
9. Postexposure Bake: After imprinting, repeat Sect. 3.7.3 for a postexposure bake to harden the photoresist.

3.7.5 Wafer Development

1. Fill a container larger than the wafer with the SU-8 photo developer under a fume hood.
2. Using the wafer dipper, gently submerge the wafer in the photo developer bath and incubate for 10 min. Make sure to minimize vibrations, as excessive motion can distort barriers and sieve channels.
3. Remove the wafer from the bath and gently pour isopropanol over the mold features with a squeeze bottle. If a milky white fluid appears, there is still excess SU-8 photoresist remaining, and more rinsing with the photo developer is needed.
4. Repeat Steps 2 and 3 in 2–3-min increments until only clear fluid appears after the isopropanol test.
5. Allow the developed wafer to dry at room temperature (typically 5–10 min).
6. Hard Bake: After development, bake the dried wafer on a hot plate preheated to 150°C for 10–30 min. The purpose of this is to stabilize the SU-8 structures and enhance the effectiveness of subsequent silanization processes.

3.7.6 Silanization

1. Trichlorosilane treatment ensures mold features stay intact during PNC preparation. Using a syringe with a 25-gauge needle, inject 1 mL of trichlorosilane across four storage vials

through the perforable lids in a fume hood. To avoid touching the trichlorosilane, leave the needle in the lid of the last vial.

2. Tape the vials to different walls inside a vacuum desiccator in a fume hood. Change the silane every few months, depending on the frequency of usage.
3. Place the mold in the desiccator and run the vacuum pump for 2 h. After turning off the pump, leave the mold in the vacuum-tightened desiccator for 24–48 h.

3.8 NemaCapsule Hardware Preparation

3.8.1 PNC and PDMS Base Preparation

1. Pour 0.8 g of PDMS resin per chamber into the molds created in Sect. 3.7 to generate 1.5–2 mm-thick PNC microfluidic devices.
2. Degas in a vacuum desiccator for 15 min, then place the molds in a laminar flow hood until the air bubbles have dissipated.
3. Place molds in an oven at 70°C for at least 2 h to cure (*see* **Note 24**).
4. Using a biopsy punch, create a 1-mm diameter hole in the large circular impression in each PNC to create the worm loading port. Then make cuts at the edges of the sieve channels with a razor blade to allow passive flow of the bacterial culture into the device (Fig. 5a).
5. To create a PNC base, cut a PDMS sheet (do not remove plastic from the sheet) with an X-Acto knife to the dimensions of the PNC housing hardware (*see* **Note 25**).

3.8.2 PNC Bonding to PDMS Base

1. Use a plasma cleaner to prepare the PNCs and PDMS base for bonding two chambers per PDMS base, one crawling and one swimming (Fig. 5a, b).
2. Turn on the power for the unit and pressure readings.
3. Remove the stage and position the PNCs next to the PDMS base with the surfaces being bonded facing up. For the PNCs, the side with the imprinted features should face up.
4. Insert the stage back into the plasma cleaner. Generate a tightly closed unit by activating the pump and turning the valve (mounted on the front door) clockwise to evacuate atmospheric pressure.
5. Wait for the pressure readings to equilibrate to 0.5–0.6 Torr.
6. Once the pressure equilibrates, turn the radio frequency (RF) power to "high" for 90 s.
7. Once complete, turn the RF power to "off," turn off the vacuum, and turn the pressure valve counterclockwise to let atmospheric pressure back in to open the unit door.

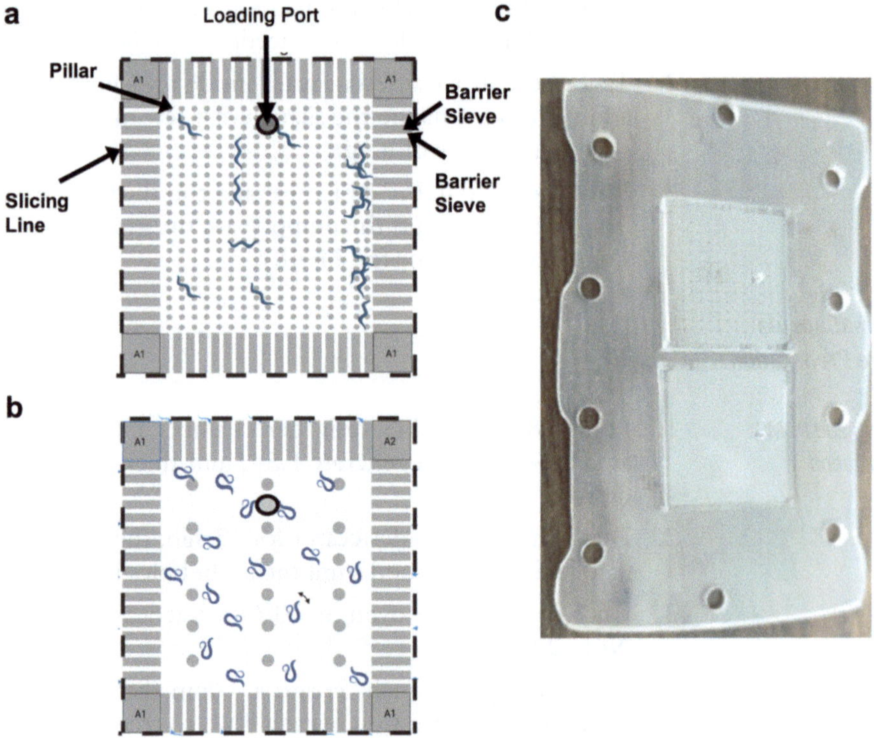

Fig. 5 PNC microfluidic devices. (**a**) schematic of a PNC chamber designed for crawling (**b**) swimming chamber schematic (**c**) both devices bonded to shaped PDMS substrate

8. Remove the PNCs and PDMS base. Flip the features side of the PNCs to come into flush contact with the top surface of the PDMS base and apply pressure with fingers for a few seconds.
9. Place a Kimwipe over the bonded PNC assembly with ~1 lb of weights on top. Keep in an oven at 70°C for at least 3 days (*see* **Note 26**).
10. After the PNCs are completed (Fig. 5c), assemble the NemaCapsules in a sterile environment 1–3 days before worm loading.
11. The PDMS sheet serves as the NemaCapsule floor, while the opposite side remains open until after the adult worms are loaded in Sect. 3.9. Store open NemaCapsules inside an ethanol-sterilized container at room temperature.

3.9 Loading of NemaCapsules for Spaceflight

1. Fill a 10-mL syringe with a 20-gauge needle with the pluronic solution and gently inject the solution into the loading port of each chamber until liquid drops come out of the sieve channels on each side of the device. This ensures the smooth passage of worms suspended in the chambers.

Surround devices with sterile deionized water and leave for at least 15 min.

2. Collect adult worms from the NGM + IAA plates by adding ~1–2 mL liquid NGM per plate, gently swirling, and pouring the worm suspension into a shallow container (e.g., 60-mm Petri dish lid). More media can be used to flush worms from plates in small increments to increase worm density as needed.
3. Fill a loading syringe assembly (Fig. 3) with 1 mL liquid NGM and expel air bubbles by decreasing the volume to 0.5 mL.
4. Visualize the adult worms collected in Step 2 with a dissecting microscope. Aspirate 45 adult worms into the loading syringe (*see* **Notes 27 and 28**).
5. Remove the worm suspension from the microscope stage and place the open NemaCapsule in a Petri dish onto the stage. To ensure sterility, close the Petri dish when not actively loading the NemaCapsule.
6. Connect the loading syringe's hollow pin into the loading port up to half the depth (*see* **Note 29**) and inject the worms into the arena.
7. After loading the worms, add 1 mL liquid NGM to the loading port with the loading syringe. The liquid will be expelled through the sieve channels to remove excess *E. coli* OP50 from Step 2 and any potential *C. elegans* offspring.
8. Discard the excess liquid by dumping it into a waste container.
9. Using a 1 mL syringe with a 20-gauge needle, inject 0.1 mL of the desired microbiome (prepared in Sect. 3.2.11) into the loading port. To prevent contamination, use one syringe per NemaCapsule.
10. Seal each loading port with a dowel pin (*see* **Note 30**).
11. After the NemaCapsules have been loaded with worms, attach an oxygen-permeable membrane to the open side of the NemaCapsule.
12. Fill the remaining air space inside with the appropriate microbiome as determined by the implementation partner (IP) (Fig. 6).

3.10 Integration and Launch

1. Place the prepared PE bags and/or NemaCapsules in their designated secondary containment (*see* **Note 31**) and incubate at the desired temperature (typically 15°C–20°C) until handover.
2. After handover, secondary containment is examined and documented by the IP to ensure compliance before depositing into flight equipment (*see* **Note 32**).

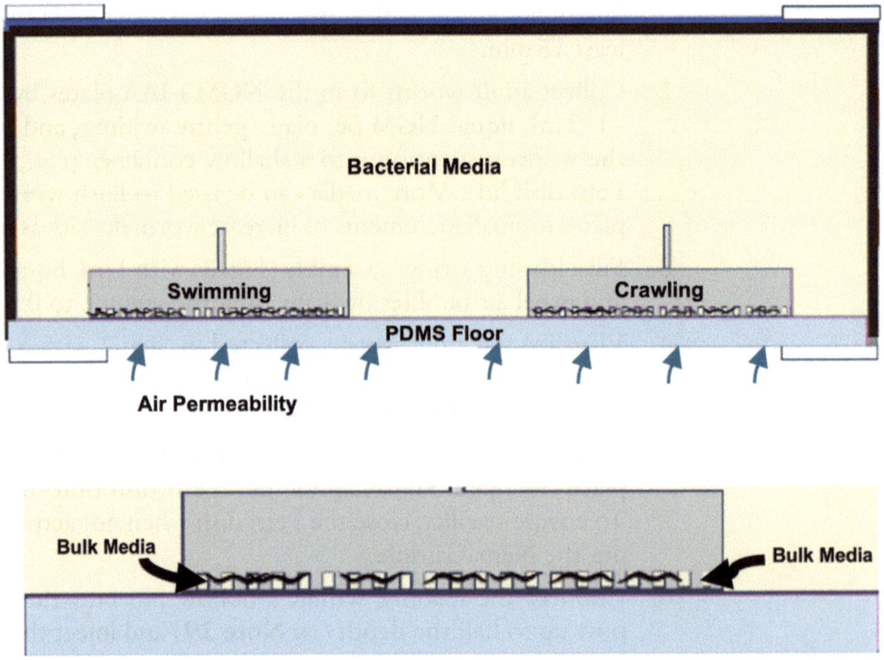

Fig. 6 NemaCapsules loaded with worms and microbiomes. Cross-sectional schematic shows the microfluidic chambers supported by the PDMS floor within the housing provided by IP and immersed in bacterial media (top), as well as a closeup of the worms within the chambers (bottom)

3.11 On-orbit Operations

3.11.1 PE Bag Operations

1. PE bag samples can be placed directly into a −80°C freezer when the desired *C. elegans* age and time in microgravity is reached (e.g., 7–9 days postloading at 15°C for gut microbiome analysis).

3.11.2 NemaCapsule Imaging

1. NemaCapsules are imaged on the same day that the PE bags are placed in the freezer. With adequate crew time, it is optimal to repeat the imaging 2 days later.
2. Remove samples from secondary containment and equilibrate to room temperature for 1.5–3 h before imaging.
3. While the samples are adjusting to room temperature, setup the camera with the following specifications (Fig. 7a): (1) Place the camera on a stand that elevates it 12.5 cm from the bench. Use a stand that minimizes vibrations, such as a styrofoam box; (2) A holder (created by the IP) for one NemaCapsule that allows light in from the sides sits 8.5 to 11.5 cm from the camera lens. The holder should be at a height of 14 cm to allow one PNC to be in the field of view (FOV); (3) Place LED lights on either side of the NemaCapsule holder at a distance of 8 cm and a height of 14 cm.

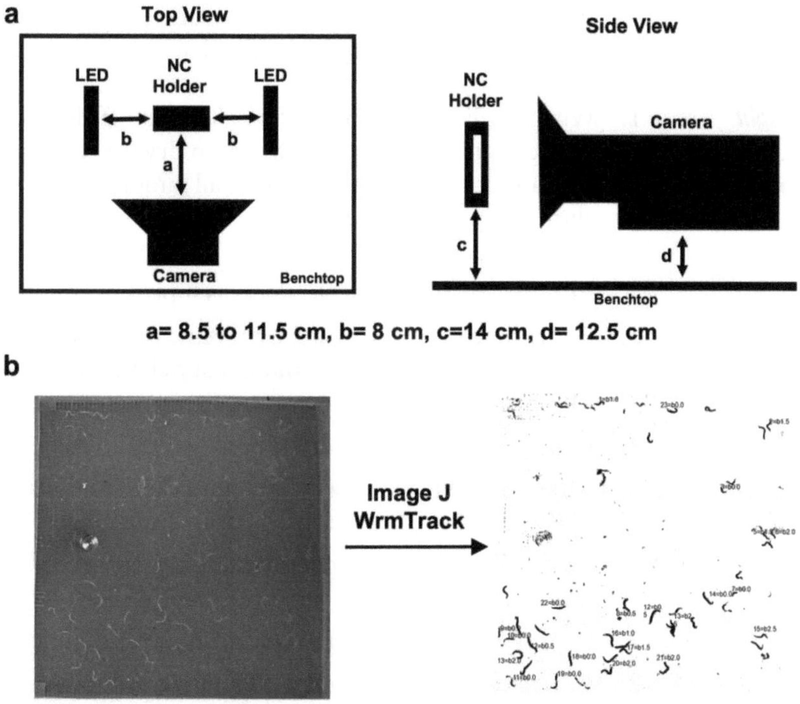

Fig. 7 Video acquisition and processing. (**a**) Schematic showing flight-like imaging setup from top (left) and side view (right) (**b**) Image of crawling animals from flight-like Canon XF705 (left) and binarized animals thresholded through Image J and run through WrmTrck plugin for speed values (right)

4. The camera should have the following settings: (1) Turn the image stabilization off; (2) Zoom to 15×; (3) Use 24 frames per second (fps). Other fps values may be used, but testing before launch is required to assess whether image processing will impact speed calculations.
5. Once the samples have equilibrated to room temperature and the imaging equipment is assembled, remove only one NemaCapsule from the secondary containment.
6. Place the NemaCapsule in the holder. Adjust the camera distance to ensure one entire PNC is in the FOV.
7. Adjust the camera focus for a clear contrast of the worms (Fig. 7b).
8. Image the PNC for 2 min.
9. After imaging one PNC, remove the NemaCapsule from the holder, rotate it 180°, and reinsert it into the holder so that the other PNC is in the camera FOV.
10. Image the second PNC for 2 min.
11. Return the NemaCapsule to the secondary containment and repeat Steps 3–8 for the rest of the NemaCapsules.

12. After all the NemaCapsules have been imaged, return the secondary containment to the incubator.

3.12 Post-Flight Assays of Host Physiology in PE Bags

1. With sterilized scissors, cut small slices off the PE bags into a sterile Petri dish (60 mm), while on dry ice to keep the remaining samples frozen. Allow the small sample slices to thaw before aliquoting for physiological assays. Return frozen samples to a −80°C freezer.
2. Add worms to a 2 mL 96-deep well plate.
3. Centrifuge at 300 × *g* for 1 min, remove liquid to 100 μL using an aspiration manifold or multichannel pipettor. Add 600 μL of M9-T solution.
4. Repeat Step 3 three more times.
5. After the final wash, aspirate the liquid to a final volume of 100 μL.
6. Surface-sterilize the worms by adding 100 μL of a 10% bleach solution for 5 min.
7. Repeat Step 3 two more times to remove the bleach solution.
8. *C. elegans* can be imaged with an inverted microscope to identify phenotypic differences (*see* **Note 33**). Transfer samples to a black 96-well plate with a clear bottom and image wells according to the microscope manufacturer's instructions.
9. Additionally, populations or individual worms may be processed for molecular characterization, such as microbiome profiling or RNA sequencing, following established protocols [8, 15].

3.13 Post-Flight Analyses of Swimming and Crawling Behavior in NemaCapsules

1. The locomotory swimming and crawling behavior of *C. elegans* can be analyzed through the image processing of the NemaCapsule videos.
2. Swimming behavior is quantified by manually counting the thrashes of individual worms. A single thrash is a complete bending movement of the worm's body from one side to the other.
3. Choose three time points for each video, approximately 40–45 s apart.
4. At each time point, count the number of thrashes per worm for 15 s.
5. Repeat Step 4 for 15 worms per time point.
6. Crawling behavior is quantified through speeds obtained with the WrmTrck plugin on Image J [16].
7. Download and install the WrmTrck plugin from the open-source page http://www.phage.dk/plugins/wrmtrck.html.

8. Convert the video file (obtained in Sect. 3.11) from an MXF format to an AVI format.
9. Open the AVI videos in Image J. Check the *Grayscale* box in the window that appears for an 8-bit grayscale video.
10. Using the rectangle tool, select only the PNC and hit *Ctrl-X* to crop out the excess area around the PNC.
11. To set the distance-based scale, draw a line across the width of the PNC. Go to *Analyze>Set Scale* and enter the measurement of the width of a PNC in the *Known distance* box. Then, check the box next to *Global.*
12. Removing background noise yields only worms against a dark background (Fig. 7b). Create the background image without the worms: Go to *Image> Stacks>Z-Project.* Under *Projection type,* select *Min Intensity.* This image is labeled as MIN [file name] by default.
13. Subtract the background in Step 1, yielding only worms: Go to *Process>Image Calculator.* For *Image 1,* select the original file name, and for *Image 2,* select the MIN file, and for the *Operation* select *Subtract.* Ensure *Create new window* is selected and hit *OK.*
14. Threshold the background-subtracted image to produce binarized images of worms (Fig. 7b).
15. Select the newly background-subtracted video from Step VI and go to *Image>Adjust>Threshold.*
16. Ensure *Dark background* and *Don't reset range* are selected.
17. From the left drop-down menu, select *Triangle* and hit *Apply.*
18. The only box selected is *Calculate threshold for each image* and hit *OK.*
19. Under *Plugins,* select the *WrmTrck* plugin and input the correct parameters (*see* **Note 34**). Hit *OK* to run the calculations.
20. Save the raw output results as a *.csv* or *.txt* file and all the binarized videos, containing "labels" in their filenames, as new *.avi* files. In the final compilation of speed values, make sure to only use the ones associated with the ID tagging a moving worm.

4 Notes

1. NemaCapsule is the name for the fully integrated unit of two PNCs and their housing hardware. The IP determines the housing hardware based on its available resources. One example is a BioCell (BioServe Space Technologies, Boulder, CO, USA).

2. With each new roll of PE tubing, perform a biocompatibility test to ensure that any batch effects of manufacturing do not interfere with the growth and development of the worms.
3. The volume of culture is based on the number of plates needed. For example, 10 of the 60 mm plates would need 20 mL of LB, whereas 10 of the 100 mm plates would need 100 mL of LB.
4. Sects. 3.1, 3.3–3.6, and 3.9 are repeated for ground control samples with a 48–72 h offset (determined by launch cadence). Prepared microbiomes (Sect. 3.2) are used for flight and ground control samples.
5. To scale up the assay, 100 mm NGM agar plates can be seeded with 1 mL of *E. coli* OP50 per plate. Each 100 mm plate can support the growth of up to 3–4× the number of *C. elegans* animals on a 60 mm plate.
6. This assay uses worm strains with P*pie-1*::TIR1::mRuby and *spe-44*::degron inserted onto the chromosome to suppress progeny production using the auxin-inducible degradation (AID) system. If non-AID strains are used, Sect. 3.4 can be skipped. Experimental timing and microbiome concentrations may need to be adjusted as offspring compete for resources.
7. After *C. elegans* strains are taken from the freezer, create a stock plate by allowing animals to reach starvation on an NGM + *E. coli* OP50 plate and store at 15°C for 2–3 months.
8. Bacterial culture plates can be placed at 4°C for storage and used for up to 3 weeks before new cultures should be grown from the freezer stocks.
9. *E. coli* OP50 cultures are prepared twice—1 day before Sect. 3.4 and 1 day before Sects. 3.6 and 3.9. *C. elegans* natural microbiomes are prepared once—1 day before Sects. 3.6 and 3.9.
10. Determine the final volume based on the final number of PE bags that will be made and double the initial LB volume for inoculation as a starting point.
11. If large quantities of the complex biomes are needed, inoculate a second tube of each bacterium rather than add more volume to one tube. Keeping it to 10 mL in a 50 mL tube results in a high overnight concentration of each bacterium.
12. A plate reader can be used to measure all of the strains at once in a 96-well plate.
13. Use $C_1V_1 = C_2V_2$ to find the volume of the original culture needed to concentrate to get the final volume to get an $OD_{600} = 1.0$. It is best to start with extra as it is easier to dilute than to concentrate the culture.

14. Larger volumes of bacteria can be centrifuged if Nalgene bottles and a larger centrifuge are available.
15. The OD_{600} does not need to be exactly 1.0. Aim to keep them all within 0.1 of each other. Biomes can be stored at 15°C or 20°C in a screw-cap bottle for up to one week for use with flight and ground control samples.
16. Visualize the worms with a dissecting microscope every 30 s starting at the 2-min mark to determine when worms begin to break.
17. The time for a given worm strain to reach adulthood will vary based on its genetic background, rearing temperature, and bacterial food source. For example, *C. elegans* wild-type AID strain N2 [9] requires 72 h at 20°C and 132 h at 15°C, while a *daf-16* mutant AID strain takes 84 h at 20°C and 144 h at 15°C.
18. The size of the PE bags can be altered. For 6.5 mL of liquid culture, the area between the seals should be 1.5″ W × 3″ L. Other measurements can also be used if the surface-to-volume ratio remains stable [10].
19. The impulse heat sealer fluctuates in temperature. Test the temperature settings on an extra piece of PE tubing before loading samples. Start with an initial setting of 2.5 and adjust for a complete seal that does not melt the PE tubing.
20. A second seal can be made outside the first seal for labeling. Print labels and insert them into the PE bags between the two seals to preserve sample identification.
21. A small clip (Amazon B09ZTV1MBR) can also help with sealing. Place the clip as close to the top of the liquid as possible and push out the excess air before sealing.
22. To adjust the depth of the SU-8 photoresist, start with the specifications from the manufacturer, including the volume (1 mL/in. of the wafer diameter) of the SU-8 photoresist and the final rotation speed. Further modifications can be made through trial and error to achieve the optimal depth of the PNCs.
23. Wear UV-protective goggles during exposure and do not make direct eye contact with the UV light.
24. To ensure the features of the mold have the correct dimensions, a PNC can be imaged with an inverted microscope and measured with the microscope software. Aerial dimensions are assessed with the features face down on the microscope slide. Chamber and feature depths are imaged with a cross-section of the PNC placed on its side on the microscope slide.

25. A semicured 20:1 (A:B wt/wt) PDMS resin can be used as a base for PNCs placed in Petri dishes or well plates.
26. Make 2–3× the required number of PNCs to allow for errors and defects.
27. When aspirating worms, avoid progeny, as there can be low numbers even after auxin treatment.
28. Minimize air bubbles while loading worms into the chambers by keeping the tip of the hollow pin in the liquid until sufficient worms are collected.
29. When the hollow pin is pushed fully into the port, back pressure is generated from the PNC and requires forceful expulsion of the worms. This results in worms being paralyzed or injured and introduces a large volume of air bubbles.
30. To prevent worms from drying out while loading the other NemaCapsules, the PNCs can be surrounded with 2 mL of the appropriate microbiome culture.
31. The type of secondary containment is based on the available resources for flight, which might include a sealed box or bag with filtered air holes to ensure adequate oxygen flow for the survival of *C. elegans* animals. PE bags should be placed in a single layer with minimal overlap for sufficient oxygen exchange.
32. Temperature may be maintained by insulated containment or powered equipment for launch. This is determined based on experiment requirements and cold stowage limitations.
33. Use fluorescence microscopy to visualize gut microbes if fluorescently tagged microbes are used in the microbiomes. Additionally, fluorescent reporter worm strains can be imaged with fluorescence microscopy to identify altered gene expression in different microbiomes.
34. The tutorials on the plugin webpage will provide a starting point for the worm input parameters (https://www.phage.dk/plugins/wrmtrck.html). Calculate the pixel area range through manual Image J calculations of the worm footage. Ensure the input video fps matches the frame rate of the video obtained from its properties metadata.

References

1. Voorhies AA, Mark Ott C, Mehta S et al (2019) Study of the impact of long-duration space missions at the International Space Station on the astronaut microbiome. Sci Rep 9:9911
2. Morrison MD, Thissen JB, Karouia F et al (2021) Investigation of spaceflight induced changes to astronaut microbiomes. Front Microbiol 12:659179

3. Higashibata A, Hashizume T, Nemoto K et al (2016) Microgravity elicits reproducible alterations in cytoskeletal and metabolic gene and protein expression in space-flown *Caenorhabditis elegans*. NPJ Microgravity 2:1–8
4. Szewczyk NJ, Tillman J, Conley CA et al (2008) Description of international *Caenorhabditis elegans* experiment first flight (ICE-FIRST). Adv Space Res 42:1072–1079
5. Honda Y, Higashibata A, Matsunaga Y et al (2012) Genes down-regulated in spaceflight are involved in the control of longevity in *Caenorhabditis elegans*. Sci Rep 2:487
6. Willis CRG, Szewczyk NJ, Costes SV et al (2020) Comparative transcriptomics identifies neuronal and metabolic adaptations to hypergravity and microgravity in *Caenorhabditis elegans*. iScience 23:101734
7. Szewczyk NJ, Udranszky IA, Kozak E et al (2006) Delayed development and lifespan extension as features of metabolic lifestyle alteration in C. elegans under dietary restriction. J Exp Biol 209:4129–4139
8. Zhang F, Weckhorst JL, Assié A et al (2021) Natural genetic variation drives microbiome selection in the *Caenorhabditis elegans* gut. Curr Biol 31:2603–2618
9. Kasimatis KR, Moerdyk-Schauwecker MJ, Phillips PC (2018) Auxin-mediated sterility induction system for longevity and mating studies in *Caenorhabditis elegans*. G3 (Bethesda) 8:2655–2662
10. Pollard AK, Gaffney CJ, Deane CS et al (2020) Molecular muscle experiment: hardware and operational lessons for future astrobiology space experiments. Astrobiology 20:935–943
11. Soni P, Anupom T, Lesanpezeshki L et al (2022) Microfluidics-integrated spaceflight hardware for measuring muscle strength of *Caenorhabditis elegans* on the International Space Station. NPJ Microgravity 8:1–12
12. Rahman M, Hewitt JE, Van-Bussel F et al (2018) NemaFlex: a microfluidics-based technology for standardized measurement of muscular strength of C. elegans. Lab Chip 18:2187–2201
13. Banse SA, Blue BW, Robinson KJ et al (2019) The stress-chip: a microfluidic platform for stress analysis in *Caenorhabditis elegans*. PLOS ONE 14:e0216283
14. Dirksen P, Assié A, Zimmermann J et al (2020) CeMbio – the *Caenorhabditis elegans* microbiome resource. G3 (Bethesda) 10:3025–3039
15. Zhang F, Weckhorst JL, Assié A et al (2020) High-throughput assessment of changes in the *Caenorhabditis elegans* gut microbiome. In: Curran SP (ed) Aging: methods and protocols. Springer US, New York, NY, pp 131–144
16. Nussbaum-Krammer CI, Neto MF, Brielmann RM et al (2015) Investigating the spreading and toxicity of prion-like proteins using the metazoan model organism *C. elegans*. J Vis Exp. 2015 95:52321

Part V

Technological Devices

Chapter 13

3D Bioprinted Tumor Constructs for Photothermal Therapy: Emerging Opportunities in Microgravity Research

Maria Laura Sforza, Maila Chirivì, Claudia Bearzi, Giovanni Perotto, Paolo Rosa, Francesca Petronella, Roberto Rizzi and Luciano De Sio

Abstract

Thanks to advancements in innovative technologies, it is now possible to conduct testing for new drugs and medical treatments without using animals. This significant development not only saves time and costs but also eliminates the need for animal sacrifice. In this context, we present a 3D-printed construct made from glioblastoma U87-MG cells combined with biomimetic keratin-coated gold nanoparticles (Ker-AuNPs). By utilizing the thermoplasmonic properties of these Ker-AuNPs, we can test photothermal therapy (PTT)—a minimally invasive treatment designed to destroy cancer cells—using a more realistic representation of the tumor environment. Additionally, our research represents a significant turning point for potential applications in microgravity conditions, enabling us to speculate on the efficacy of PTT in a controlled gravity environment.

Keywords Bioprinting, Human tissues, Glioblastoma, Photo-thermal therapy, Plasmonics, Cancer, Optics, Gold nanoparticles, Biomimicry, 3D model

1 Introduction

Among the minimally invasive cancer treatments is photothermal therapy (PTT), which is achieved by combining a dedicated light source and photoresponsive nanomaterials. The latter, when illuminated, can generate heat locally, allowing a targeted ablation of tumor cells [1]. Among the various nanomaterials used for PTT applications, plasmonic nanoparticles, such as gold (AuNPs) or silver (AgNPs), are among the best choices due to their localized surface plasmon resonance (LSPR) properties. Briefly, when irradiated with visible or near-infrared light sources, electrons located at the metal/dielectric interface begin to oscillate, causing a local increase in temperature [2].

AuNPs are biocompatible, stable, and exhibit high photothermal efficiency; however, their capping agents can still produce toxicity [3, 4]. Therefore, it is possible to coat them with materials that preserve

Mariano Bizzarri (ed.), *Biomedicine and Biotechnology in Space*, Methods in Molecular Biology, vol. 3000,
https://doi.org/10.1007/978-1-0716-5174-2_13,

the photothermal properties while improving biocompatibility, such as keratin [5]. It has already been demonstrated that keratin-coated gold nanoparticles (Ker-AuNPs) possess high biocompatibility, as evidenced in experiments performed measuring the cellular viability of human glioblastoma U87-MG cells in 2D cultures in the presence of Ker-AuNPs [6]. However, the complexity of the tumor microenvironment cannot be reproduced in cultures [7].

Help comes from 3D bioprinting (3DB), a technique that uses bioinks based on living cells and appropriate materials to reconstruct the complexity of tissues, tumors, and organs, mimicking the complex architecture and the various cellular interactions [8, 9].

Due to its promising potential, 3DB has become the subject of studies conducted in microgravity (μg) or the absence of gravity. These conditions significantly alter physical, chemical, and biological processes, paving the way for new pathways in understanding diseases and therapies [10], particularly in oncology [11].

We report here a protocol, including characterizations and control experiments, developed by Chirivì et al. [12] to combine Ker-AuNPs, U87-MG cells, 3DB, and PTT. The developed protocol can be implemented to create a new toolbox for experiments in controlled gravity, thereby opening the door to innovative tumor treatments. A summary scheme of the technique is shown in Fig. 1.

The protocol reported in this work is optimized in normal gravity conditions. However, it can be easily adapted for performing similar experiments in a controlled gravity environment. Indeed, under μg

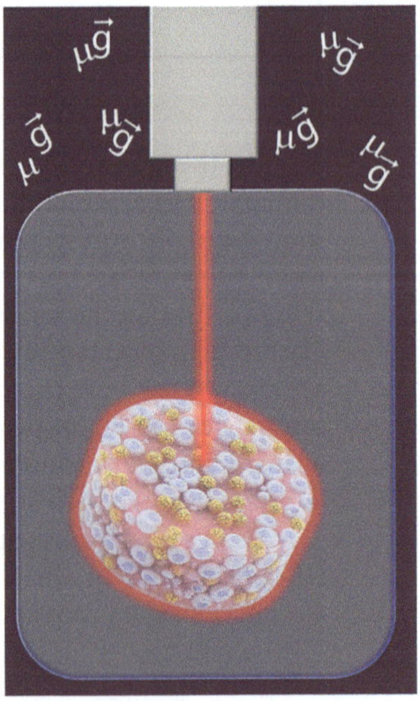

Fig. 1 3DB construct with glioblastoma U87-MG cells and Ker-AuNPs are subjected to laser irradiation under μg conditions

conditions, tumor cells exhibit varied behaviors. In some cases, an increase in aggressiveness due to immune system suppression has been found, while in others, it has been shown that μg hinders tumor proliferation by promoting apoptosis. In particular, suppression of vascular endothelial growth factor (VEGF) has been demonstrated, reducing tumor angiogenesis, as well as a reduction in nuclear localization of the transcriptional regulator YAP1, which limits proliferation and thus invasiveness [11]. Proliferation is also limited from a mechanical perspective, as the adhesion forces are lower, leading to a reduction in colony formation [11]. The reorganization of the cytoskeleton, along with the reduction of YAP1 and adhesion forces, resulted in a decrease in stiffness [10, 13]. The reduction of stiffness favors the penetration of oncological therapies and treatments [14], including the PTT of the reported protocol.

Under μg conditions, glioblastoma cells exhibited reduced proliferation and increased apoptosis, resulting from a decrease in thapsigargin-induced store-operated calcium entry (SOCE) and downregulation of ORAI1 protein expression [15]. When 3DB in μg with vector gravity rotation is used to simulate space, glioblastoma cells reconfirmed the decrease in invasiveness, highlighting a 55% reduction in YAP1 after 48 h of reduced gravity. Moreover, losses of aggregation and adhesion are observed, also due to the decrease in the vinculin protein [16].

In light of the reported findings, the 3D bioprinting of a construct made from U87-MG cells to biomimetic Ker-AuNPs in reduced gravity is feasible, potentially enhancing PTT compared to terrestrial conditions.

2 Materials

1. Sodium citrate
2. Keratin
3. MilliQ® water
4. FluoReporter FITC Protein Labeling Kit
5. $NaHCO_3$ 1 M
6. DMSO
7. Human U87-MG glioblastoma cells
8. Dulbecco's modified Eagle's medium (DMEM)
9. Heat-inactivated FBS 10%
10. FBS 5%
11. Penicillin 100 U/mL
12. Streptomycin 100 mg/mL
13. Trypsin–EDTA

14. Gelatin type A3 from porcine skin
15. Alginate 6% w/v; HEPES 25 mM
16. Irgacure 2959
17. Calcium Chloride 0.3 M
18. EDTA 5 mM
19. Collagenase II 337 U/mg
20. Collagenase D 0.29 U/mg
21. HBSS
22. PBS 1×
23. LIVE/DEAD Fixable Aqua Dead Cell Stain Kit-Pacific Orange 1:1000
24. Lipophilic tracer DiI 1:5000
25. Hoechst 1:1000
26. PFA 4%.

3 Methods

After preparing the fluorescent-labeled Ker-AuNPs and growing the U87-MG Cells, it is possible to make the bioink and perform the 3DB. Before the PPT experiments, it's crucial to perform a viability assessment and an immunofluorescence assay to demonstrate the effectiveness of the 3DB process.

3.1 Fluorescent Labeled Ker-AuNPs

1. Use the Turkevich method [17] to synthesize spherical AuNPs with a diameter of approximately 25 nm, using sodium citrate as the reducing agent.
2. After the synthesis, mix a solution of AuNPs (1.5 mM) with keratin to achieve a 1:100 weight ratio of Au to keratin (*see* **Note 1**).
3. Gently shake the solution overnight, then centrifuge the AuNPs twice at 14,300 ×*g* and resuspend them in MilliQ® water to remove unbound keratin.
4. Check the quality of the Ker-AuNPs using UV-visible absorption spectroscopy (range: 400–800 nm). A peak around 530 nm should be visible (*see* **Note 2**).
5. Label a fresh Ker-AuNPs solution using the FluoReporter FITC Protein Labeling Kit: prepare a buffer solution of $NaHCO_3$ 1 M, pH 9. Mix 1955 μL of buffer and 45 μL of Ker-AuNPs solution 44 mg/mL to obtain 2 mL of solution with a final concentration of keratin of about 1 mg/mL. Dissolve the FITC powder in 10 μL of DMSO. Add the solubilized FITC to the buffered Ker-AuNPs solution and mix using a stirrer for 1 h under dark conditions at room temperature.

6. Dialyze the Ker-AuNPs-FITC solution in a dialysis cassette with a molecular weight cut-off (MWCO) of 3500 kDa for 3 days against MilliQ® water.
7. Check the fluorescence of FITC-labeled Ker-AuNPs using fluorescence spectroscopy, exciting the samples at 491 nm and analyzing the emission spectrum around 542 nm (*see* **Note 3**).

3.2 U87-MG Cell Culture

1. Grow human U87-MG glioblastoma cells as already reported in Tagliaferro et al. [18]. Briefly, maintain the cells in Dulbecco's modified Eagle's medium (DMEM) supplemented with 10% heat-inactivated fetal bovine serum (FBS), 100 U/mL penicillin, and 100 mg/mL streptomycin in a humified atmosphere with 5% CO_2 at 37°C.
2. Change the medium every 3 days and detach the cells using trypsin–EDTA only when they are confluent and fully expanded (*see* **Notes 4** and **5**).

3.3 Bioink

1. Prepare gelatin (type A3 from porcine skin) methacryloyl (GelMa) following the protocol reported in Maiullari et al. [19].
2. Dissolve 6% w/v low molecular weight alginate and 6% w/v of GelMa in 1 mL of HEPES 25 mM.
3. Filter (0.22 μm) the obtained solution to ensure sterility.
4. Trypsinize, centrifuge, and resuspend the cells in the freshly prepared bioink stock solution to a final concentration of 12×10^6 cells/mL.
5. Add the bioink with the fluorescent-labeled Ker-AuNPs at a final concentration of 75 μM and 1 mg/mL of Irgacure 2959 photoinitiator (*see* **Note 6**).

3.4 3DB of Constructs

1. Load the preformulated bioink and a calcium chloride solution (0.3 M) in 1 mL sterile glass syringes placed on microfluidic pumps.
2. Generate 10 layers thick constructs, with layers being perpendicular to each other and a 50 μm distance between the hydrogel stripes in the X–Y plane. The final result should be a bioprinted scaffold with dimensions of $8 \times 8 \times 8 \times 1$ mm³.
3. Collect all scaffolds, place them in a 60 mm dish and cross-link them with UV light at low intensity (365 nm, 4–5 mW/cm²) for 5 min.
4. Wash the samples with 25 mM HEPES buffer containing 5 mM EDTA for 3 min, to remove as much ALG as possible, and culture them in a growth medium.
5. Check the growth of the cells in the construct by acquiring bright-field micrographs at 0, 24, and 56 h after the bioprinting process (*see* **Note 7**).

3.5 Viability Assessment

1. Extract the cells from the bioprinted constructs using a digestion solution composed of 300 U/mL collagenase II (337 U/mg) and 0.65 U/mL collagenase D (0.29 U/mg) in HBSS buffer for 10 min at 37°C.
2. Block digestion using a stop solution containing 1× PBS enriched with 5% FBS.
3. Centrifuge the cells at 1200 rpm for 5 min at RT and resuspended them in LIVE/DEAD Fixable Aqua Dead Cell Stain Kit-Pacific Orange (1:1000).
4. After 30 min at room temperature, wash the cell suspension with 1× PBS and analyze it using flow cytometry (*see* **Note 8**).

3.6 Immunofluorescence Assay

1. Label the U87-MG cells with the lipophilic tracer DiI (1:5000) for 20 min at 37 °C, and counterstain the nuclei with Hoechst (1:1000) for 15 min.
2. Wash the U87-MG cells twice with PBS, trypsinize them, and centrifuge them. Then, encapsulate them in the bioink stock solution at 12×10^6 cells/mL.
3. Resuspend the fluorescent Ker-AuNPs in the bioink.
4. After 0, 24, and 56 h of culture, wash the obtained 3DB constructs with 1× PBS for 10 min and fix them in 4% PFA for 1 h.
5. Use a laser scanning confocal microscope to acquire 3DB samples (*see* **Note 9**).

3.7 PPT Experiments

1. Illuminate the samples (24 and 56 h post-bioprinting, with and without Ker-AuNPs) for 160 s at normal incidence from the side. Use a CW laser operating at 532 nm (highest absorption band of Ker-AuNPs) with $I = 9.1\ W/cm^2$.
2. Collect thermal images during illumination after 120 s using a high-resolution thermal camera equipped with a close-up IR lens (2.9× magnifying factor, working distance 15 cm), from the same side of the laser, with a slight angle to the normal incidence direction.
3. Depict the time–temperature dependence of samples upon laser beam illumination and compare it also with the thermal images.

4 Notes

1. Use an amount of keratin three orders of magnitude higher than what is expected to be conjugated to guarantee that all the AuNP binding sites are saturated.
2. It is preferable to carry out a TEM analysis to verify the sizes of Ker-AuNPs.

3. It is preferable to exploit also fluorescence microscopy to see if FITC-labeled Ker-AuNPs are well dispersed.
4. It is essential to ensure that there is an adequate amount of cells.
5. Also, create bioink without adding Ker-AuNPs, so you can repeat the experiments with only cells in the constructs as a control.
6. It is preferable to incubate some U87-MG cells with Ker-AuNPs at 75 μM for 24 h for ultrastructural TEM analysis.
7. Follow the steps also using the bioink without Ker-AuNPs, to verify how the latter influences cell growth.
8. The steps should also be repeated for constructs without Ker-AuNPs and 2D cultures with and without Ker-AuNPs.
9. It is preferable to use volumetric 3D reconstruction for enhanced in-depth visualization.

References

1. Zou L, Wang H, He B, Zeng L, Tan T, Cao H, He X, Zhang Z, Guo S, Li Y (2016) Current approaches of photothermal therapy in teating cancer metastasis with nanotherapeutics. Theranostics 6:762–772.
2. Huang X, El-Sayed MA (2010) Gold nanoparticles: optical properties and implementations in cancer diagnosis and photothermal therapy. J Adv Res 1:13–28.
3. Javed R, Zia M, Naz S, Aisida SO, Ain NU, Ao Q (2020) Role of capping agents in the application of nanoparticles in biomedicine and environmental remediation: recent trends and future prospects. J Nanobiotechnol 18:172. https://doi.org/10.1186/s12951-020-00704-4
4. Al-Jawad SMH, Taha AA, Al-Halbosiy MMF, Al-Barram LFA (2018) Synthesis and characterization of small-sized gold nanoparticles coated by bovine serum albumin (BSA) for cancer photothermal therapy. Photodiagnosis Photodyn Ther 21:201–210.
5. Annesi F, Pane A, Pezzi L, Pagliusi P, Losso MA, Stamile B, Qualtieri A, Desiderio G, Contardi M, Athanassiou A, Perotto G, De Sio L (2021) Biocompatible and biomimetic keratin capped Au nanoparticles enable the inactivation of mesophilic bacteria via photo-thermal therapy. Colloids Surf A Physicochem Eng Asp 625:126950.
6. Guglielmelli A, Rosa P, Contardi M, Prato M, Mangino G, Miglietta S, Petrozza V, Pani R, Calogero A, Athanassiou A, Perotto G, De Sio L (2021) Biomimetic keratin gold nanoparticle-mediated in vitro photothermal therapy on glioblastoma multiforme. Nanomedicine J 16:121–138.
7. Liu Y, Bhattarai P, Dai Z, Chen X (2019) Photothermal therapy and photoacoustic imaging via nanotheranostics in fighting cancer. Chem Soc Rev 48:2053–2108.
8. Maiullari F, Costantini M, Milan M, Pace V, Chirivì M, Maiullari S, Rainer A, Baci D, Marei HE-S, Seliktar D, Gargioli C, Bearzi C, Rizzi R (2018) A multi-cellular 3D bioprinting approach for vascularized heart tissue engineering based on HUVECs and iPSC-derived cardiomyocytes. Sci Rep 8:13532. https://doi.org/10.1038/s41598-018-31848-x
9. Dey M, Ozbolat IT (2020) 3D bioprinting of cells, tissues and organs. Sci Rep 10:14023. https://doi.org/10.1038/s41598-020-70086-y
10. Silvani G, Bradbury P, Basirun C, Mehner C, Zalli D, Poole K, Chou J (2022) Testing 3D printed biological platform for advancing simulated microgravity and space mechanobiology research. NPJ Microgravity 8(1):19. https://doi.org/10.1038/s41526-022-00207-6
11. Akbarialiabad H, Jamshidi P, Aminzade Z, Azizi N, Taha SR, Sadeghian N, Varghese LJK, Kouhanjani MF, Niknam N, Babocs D, El-Assaad F, Russomano T, Murrell DF, Paydar S, Bunick CG, Christiansen R, Mark Melin M (2025) Leveraging space innovations for cancer breakthroughs on Earth. Trends Cancer 11(5):433–440.
12. Chirivì M, Bearzi C, Rosa P, Miglietta S, Petronella F, De Falco E, Calogero A, Pani R, Petrozza V, Perotto G, Rizzi R, De Sio L (2022) Biomimetic keratin-coated gold nanoparticles for photo-thermal therapy in a 3D bioprinted glioblastoma tumor model. Int J

Mol Sci 23:9528. https://doi.org/10.3390/ijms23179528

13. López Garzón NA, Pinzón-Fernández MV, Saavedra T JS, Nati-Castillo HA, Arias-Intriago M, Salazar-Santoliva C, Izquierdo-Condoy JS (2025) Microgravity and cellular biology: insights into cellular responses and implications for human health. Int J of Mol Sci 26(7):3058. https://doi.org/10.3390/ijms26073058
14. Chirivì M, Maiullari F, Milan M, Presutti D, Cordiglieri C, Crosti M, Sarnicola ML, Soluri A, Volpi M, Święszkowski W, Prati D, Rizzi M, Costantini M, Seliktar D, Parisi C, Bearzi C, Rizzi R (2021) Tumor extracellular matrix stiffness promptly modulates the phenotype and gene expression of infiltrating T lymphocytes. Int J of Mol Sci 22(11):5862. https://doi.org/10.3390/ijms22115862
15. Graf J, Schulz H, Wehland M, Corydon TJ, Sahana J, Abdelfattah F, Wuest SL, Egli M, Krüger M, Kraus A, Wise PM, Infanger M, Grimm D (2024) Omics studies of tumor cells under microgravity conditions. Int J of Mol Sci 25(2):926. https://doi.org/10.3390/ijms25020926
16. Silvani G, Basirun C, Wu H, Mehner C, Poole K, Bradbury P, Chou J (2021) A 3D-bioprinted vascularized glioblastoma-on-a-chip for studying the impact of simulated microgravity as a novel pre-clinical approach in brain tumor therapy. Adv Ther 4(11):2100106. https://doi.org/10.1002/adtp.202100106
17. Kimling J, Maier M, Okenve B, Kotaidis V, Ballot H, Plech A (2006) Turkevich method for gold nanoparticle synthesis revisited. J Phys Chem B 110:15700–15707
18. Tagliaferro M, Rosa P, Bellenchi GC, Bastianelli D, Trotta R, Tito C, Fazi F, Calogero A, Ponti D (2022) Nucleolar localization of the ErbB3 receptor as a new target in glioblastoma. BMC Mol Cell Biol 23:13. https://doi.org/10.1186/s12860-022-00411-y
19. Maiullari F, Chirivì M, Costantini M, Ferretti AM, Recchia S, Maiullari S, Milan M, Presutti D, Pace V, Raspa M, Scavizzi F, Massetti M, Petrella L, Fanelli M, Rizzi M, Fortunato O, Moretti F, Caradonna E, Bearzi C, Rizzi R (2021) In vivo organized neovascularization induced by 3D bioprinted endothelial-derived extracellular vesicles. Biofabrication 13:035014. https://doi.org/10.1088/1758-5090/abdacf

Check for updates

Chapter 14

Regenerative Medicine in Space: Advancing Bone Repair with Adipose-Derived Stem Cells and Bone-on-Chip Technology

Paola Pontecorvi and Cinzia Marchese

Abstract

Regenerative medicine is vital for counteracting astronaut bone loss in microgravity, where reduced mechanical loading accelerates osteopenia and impairs fracture healing. Current countermeasures like exercise and drugs help, but advanced regenerative solutions are needed. Bone-on-chip (BOC) technology, a microfluidic system mimicking bone physiology, enables precise studies of bone formation and degeneration in space. Adipose-derived stem cells (ASCs) committed to osteogenic differentiation (osASCs) offer a promising alternative to primary osteoblasts due to their ease of isolation, expansion potential, and strong regenerative properties. Integrating osASCs into BOC models facilitates research on bone loss mechanisms and novel osteoporosis treatments. This chapter details ASC isolation, osteogenic differentiation, and application in microgravity-simulating platforms, with the aim of contributing to progress in both space medicine and bone regeneration on Earth.

Keywords Regenerative medicine, Osteopenia, Adipose-derived stem cells, Osteogenic differentiation, Bone-on-chip

1 Introduction

Regenerative medicine in space is becoming essential to address the physiological challenges astronauts face during long-term missions, such as muscle loss, osteopenia, and cardiovascular alterations due to microgravity [1–3]. On Earth, osteoporosis involves a gradual loss of bone density, while in space, astronauts experience a much more accelerated form of bone loss due to the lack of mechanical loading, known as microgravity-induced osteopenia [4, 5]. In microgravity, bone tissue homeostasis is disrupted, making fracture healing a critical challenge due to unloading-induced bone mineral density (BMD) loss and accidental injuries. The absence of mechanical loading reduces osteoblast activity and increases osteoclast-mediated bone resorption, leading to a net loss of BMD at a rate of approximately 1%–1.5% per month in weight-bearing bones [6]. Impaired bone healing in space results from suppressed osteogenesis, altered

Mariano Bizzarri (ed.), *Biomedicine and Biotechnology in Space*, Methods in Molecular Biology, vol. 3000, https://doi.org/10.1007/978-1-0716-5174-2_14,

signaling pathways, such as Wnt/β-catenin and TGF-β, fluid shifts affecting inflammation, and deficient vascularization due to impaired endothelial cell function [7, 8]. These factors slow bone regeneration, raising concerns for astronauts performing high-risk operational tasks, particularly during extravehicular activities (EVAs). Current countermeasures include resistance exercise (e.g., Advanced Resistive Exercise Device - ARED), pharmacological interventions like bisphosphonates, and dietary strategies, but the development of in-space treatments such as bioprinted scaffolds, mechanical stabilization, or stem cell therapies is crucial for long-duration space missions, where returning to Earth for the treatment of bone injuries may not be a viable option [9]. Addressing these challenges is essential for astronaut health and mission success. Pioneering studies on tissue and bone regeneration in microgravity have opened new paths in this field [9]. Despite ongoing research, significant knowledge gaps remain regarding how human bone tissue regenerates in microgravity, both at the molecular and cellular levels (specifically in terms of cell-to-cell communication) and at the anatomical and morphological levels. Gaining a deeper understanding of these processes in microgravity could be highly beneficial for developing targeted countermeasures, including pharmacological and nonpharmacological interventions, that can be applied to both astronauts and patients suffering from osteoporosis.

Bone tissue experimental models are constantly advancing to better replicate specific aspects of bone physiology, disease mechanisms, and therapeutic responses. These models fall into two main categories: ex vivo models, where bone tissue is cultured outside the body, and in vitro models, where isolated cells are grown in either 2D or 3D environments [10]. In vitro bone models are more accessible than living tissue explants and allow for the controlled culture of human cells outside a living organism. Additionally, they enable the testing of multiple parameters within a single experiment. However, conventional 2D in vitro models are not well-suited for long-term studies and often lack clinical relevance due to the absence of key in vivo factors, despite their high efficiency and scalability help offset this drawback. Meanwhile, large-scale 3D in vitro models face challenges such as the need for a high number of cells and the risk of necrosis in the core of large scaffolds. These limitations have driven the emergence of organ-on-chip (OOC) technologies, which integrate microfabrication techniques with tissue engineering and microfluidics to create sophisticated microenvironments that mimic organ-specific functions, while minimizing cell requirements [11].

Microfluidics involves the manipulation of small fluid volumes within channels that typically range from tens to hundreds of micrometers in size. Microfluidic devices offer several advantages, including a high surface-to-volume ratio, the presence of shear stress, and the ability to precisely control chemical and physical gradients—features that are difficult to achieve in macroscale cell studies and challenging to regulate in in vivo experiments. By enabling the coculture of different cell types and fine-tuning spatiotemporal gradients and mechanical properties, microfluidic systems facilitate the development of biomimetic environments, forming the basis of OOC platforms [12]. A common design consists of two vertically stacked channels separated by a semipermeable membrane. Before seeding cells, this membrane is coated with an appropriate extracellular matrix (ECM) to create a more physiologically relevant microenvironment [13]. Observations of how bone and stem cells behave in microgravity reveal significant changes in morphology, cytoskeleton structure, and mechanotransduction - the process by which cells respond to mechanical stimuli [14]. Utilizing OOC in space or simulated microgravity allows for the rapid study of processes that might take years on Earth.

The bone-on-chip (BOC) models simulate bone tissue physiology, enabling researchers to study how bone cells react to the absence of mechanical stimuli - an essential factor in tackling osteopenia in astronauts [15], providing a controlled environment to analyze the bone formation and resorption [16], and a platform for testing new drugs and personalized treatments [17]. In this context, stem cells offer significant advantages over primary osteoblasts derived from patient tissue. Extracting primary osteoblasts from a patient's bone typically requires invasive procedures, such as bone biopsies, which are not only uncomfortable but also pose potential risks to the patient [18]. Furthermore, osteoblasts isolated from bone tissue rapidly change their phenotype once removed from their natural environment, limiting their ability to accurately represent bone biology over extended culture periods [19].

Stem cells hold significant promises for regenerative medicine, tissue engineering, and cell therapies due to their ability to self-renew and differentiate into various cell types, offering therapeutic potential for many diseases. Their capacity for expansion and long-term culture further supports their use in BOC systems, making them a more sustainable and reproducible alternative to primary osteoblasts. Mesenchymal stem cells (MSCs), originally found in bone marrow (BMSCs), are adult stem cells that can be isolated from various connective tissues, including adipose tissue, trabecular bone, skin, skeletal muscle, and umbilical cord. However, isolating MSCs typically requires large amounts of tissue, making tissue availability and donor site morbidity important considerations when selecting MSC sources. Adipose-derived stem cells (ASCs) are considered an optimal choice because they overcome these limitations [20]. ASCs exhibit multipotency, meaning they can differentiate into various cell types like adipocytes, chondrocytes, myoblasts, and osteoblasts in vitro. Studies have shown that ASCs can undergo osteogenic, adipogenic, and chondrogenic differentiation when cultured with specific media

[21]. While BMSCs have demonstrated the highest osteogenic potential, ASCs also exhibit osteogenic differentiation within 14–21 days in 2D culture conditions [22]. Moreover, ASCs from osteoporotic donors demonstrated better proliferation, differentiation, and bone regeneration capabilities than BMSCs, emphasizing their importance in maintaining bone homeostasis [23].

Furthermore, ASCs offer significant advantages over induced pluripotent stem cells (iPSCs) in constructing bone models, particularly for personalized medicine and research on astronauts' bone health. Derived from a patient's adipose tissue, ASCs inherently reflect the individual's genetic and epigenetic profile, making them ideal for creating tailored models. These stem cells can easily differentiate into osteoblasts, which deposit the bone ECM and facilitate matrix mineralization [22]. In contrast, iPSCs require reprogramming of somatic cells from skin biopsies [24], a complex process that introduces potential variability in their genetic and epigenetic makeup, making them less precise in replicating a patient's specific bone characteristics. Furthermore, harvesting ASCs is minimally invasive and more straightforward compared to the process of creating iPSCs, which is time-consuming and expensive. Additionally, ASCs have a strong regenerative potential and offer a more direct and potentially safer approach for bone regeneration compared to iPSCs [25]. When used for bone modeling, especially in the context of space research or personalized therapies, ASCs provide a much more relevant and efficient solution, enabling models that are directly applicable to the specific biological conditions of the patient or astronaut. Thus, ASCs stand out as a superior choice for BOC models, offering both practicality and precision in comparison to other MSCs and iPSCs [26].

Furthermore, the use of stem cell secretome and stem cell potentiation strategies [27, 28] shows potential in developing more effective treatments for osteoporosis and other bone-related conditions.

In conclusion, BOC models with osteoblasts derived from ASCs could provide valuable insights into bone loss in space and represent a cutting-edge platform for developing regenerative therapies. These strategies, initially aimed at preventing bone degeneration in astronauts, could be adapted to treat patients with osteoporosis and other degenerative bone diseases on Earth.

This chapter details the process of isolating and expanding ASCs from patient lipoaspirates or biopsies in 2D cultures, followed by their differentiation into osteoblasts. To simplify the concept of BOC, these ASC-derived osteoblasts (osASCs) are then seeded onto ibidi µ-Slide III 3D Perfusion platforms (ibidi GmbH, Gräfelfing, Germany) ready to be exposed to microgravity conditions using the Random Positioning Machine (RPM). This approach enables the investigation of microgravity effects on bone biology in a patient-specific context.

2 Materials

Work under strict sterile conditions in a laboratory equipped with essential cell culture technologies to ensure the integrity and reproducibility of experiments:

1. Biosafety Cabinet Level B2: to handle biological materials and prevent contamination by maintaining a controlled, HEPA-filtered airflow.
2. 37°C incubator with 5% CO_2: to provide optimal conditions for cell growth and differentiation.
3. Water bath (37°C): to gently warm reagents, media, and cell suspensions to physiological temperatures.
4. Pipettes and pipette aids: to accurately measure and transfer small volumes of liquids, ensuring precision in cell culture experiments.
5. Autoclave: to sterilize all reusable equipment to eliminate potential contaminants.
6. Centrifuge: to process cell suspensions, separate cellular components, and collect pellets, while ensuring that centrifugation parameters are appropriately set to maintain cell viability.
7. Phase-contrast microscope: to regularly monitor cell morphology, viability, and differentiation, ensuring proper documentation of results.
8. Cell counter (manual or automated): to quantify cell density and viability for accurate experimental setup.
9. Cryopreservation system (liquid nitrogen or −80°C freezer): to store cells long-term while maintaining their viability.
10. Fridge and freezer (+4°C/−20°C): to store reagents, media, and biological samples at refrigerated temperatures, preserving their stability and preventing degradation.
11. Biohazard waste containers: to dispose of contaminated materials, including media, pipettes, and gloves. Follow all biosafety and waste disposal regulations when handling and discarding biological and chemical waste.
12. Chemical fume hood: to handle toxic reagents safely.
13. Appropriate disinfectants (bleach, 70% alcohol): to decontaminate liquid waste using before disposal and clean working surfaces. Properly segregate hazardous chemical waste and follow institutional guidelines for its disposal to ensure environmental and personnel safety. Chemical fume hood: to handle toxic reagents safely.

2.1 Cell Isolation, Culture, and Characterization

1. Curved tweezers, 10 cm length, full curve, 0.4 mm tip width.
2. Microdissecting scissors, 10 cm length, 25 mm angled blade.
3. Autoclave bags and indicator tape.
4. 0.5, 1.5, 2, 15, and 50 mL sterile tubes.
5. Tissue culture flasks (T75, T25).
6. 100 mm Petri dishes.
7. 0.22 μm filters.
8. Cell scrapers.
9. 5, 10, 25 mL sterile serological pipettes.
10. 10, 20, 200, 1000 μL sterile filtered tips.
11. 100 μm cell strainer.
12. 6-well plates.
13. Phosphate Buffered Saline (PBS).
14. Powdered Collagenase Type I.
15. Penicillin/Streptomycin (PS).
16. DMEM/F12 medium.
17. Fetal Bovine Serum (FBS).
18. Penicillin–Streptomycin-L-Glutamine (PSG) solution.
19. Type IV collagen.
20. 0.5 mM EDTA/0.05% trypsin.
21. Trypan blue.
22. Primary antibodies against surface markers (CD34, CD73, CD90, CD45, and CD105).
23. 7-AAD staining solution.
24. Multichromatic flow cytometer system (we use CytoFlex, Beckman Coulter, Brea, CA, USA).

2.2 Osteogenic Differentiation

1. 6-well plates.
2. 0.5, 1.5, 2, 15, and 50 mL sterile tubes.
3. 5, 10, 25 mL sterile serological pipettes.
4. 10, 20, 200, 1000 μL sterile filtered tips.
5. PBS.
6. DMEM/F12 culture medium.
7. FBS.
8. PSG solution.
9. Type IV collagen.
10. 0.5 mM EDTA/0.05% trypsin.
11. StemXVivo® Osteogenic/Adipogenic Base Media (R&D Systems, Bio-Techne, USA).

12. StemXVivo® Human Osteogenic Supplement (R&D Systems, Bio-Techne, USA).
13. Alizarin Red.
14. Paraformaldehyde (PFA) 8%.

2.3 Cell Culture on Microfluidic Chips Under Microgravity Conditions

1. μ-Slide III 3D Perfusion system (ibidi GmbH, Gräfelfing, Germany).
2. 0.5, 1.5, 2, 15, and 50 mL sterile tubes.
3. 5, 10, 25 mL sterile serological pipettes.
4. 10, 20, 200, 1000 μL sterile filtered tips.
5. PBS.
6. DMEM/F12 culture medium.
7. FBS.
8. PSG solution.
9. Type IV collagen.
10. Desktop Random Positioning Machine (RPM) (Dutch Space - Leiden, The Netherlands).

3 Methods

3.1 ASC Isolation, Culture, and Characterization

1. *Preparation of 0.075% collagenase solution in PBS:* mix 40 mL of PBS with 0.03 g of powdered collagenase type I and store it at +4°C. Add 0.8 mL of penicillin/streptomycin (PS) to the solution, then filter it using a 0.22 μm filter and transfer it into a 50 mL tube for storage until use.
2. *Coating of tissue culture flasks with 10% collagen IV solution in PBS:* prepare a total of 4 mL of type IV collagen solution by mixing 3.6 mL of PBS with 0.4 mL of 250 μg/mL collagen IV stock solution, previously stored at −20°C. Spread the solution evenly inside the tissue culture flask using a cell scraper, applying 800 μL for a T75 flask and 400 μL for a T25 flask. Place the coated flasks in a 37°C incubator for 20 min to allow binding, then transfer them to +4°C for storage until use (*see* **Note 1**). Before using the flask, incubate it at 37°C for 10 min and leave the residual coating solution inside.
3. *Sterilization of instruments:* all necessary instruments that are not single-use, such as tweezers and scissors, must be autoclaved in the required quantity to ensure sterility during the ASC isolation procedure. Use autoclave bags and apply indicator tape. Set the autoclave to 1 bar and sterilize for 30 min in saturated steam at 121 ± 0.5°C.
4. *Adipose tissue collection:* obtain lipoaspirates or biopsies from hospitalized healthy donors or specific patients who

underwent elective liposuction surgery or abdominal wall resections for plastic surgery, after providing informed consent. The tissue collection must follow the ethical guidelines of the Declaration of Helsinki, as revised in 2008, and must be performed under a study approved by the Institutional Review Board of the departments involved. The biopsy must be delivered fully immersed in DMEM/F12 medium supplemented with 20% fetal FBS and 1% PSG (*see* **Note 2**).

5. *Preparation of washing solutions:* prepare three 50 mL tubes, each containing 10 mL of PBS with 5% PS (0.5 mL PS per tube).
6. *Adipose tissue sample cleaning:* unscrew the syringe cap and carefully transfer the lipoaspirate/biopsy into the first tube without allowing the syringe to come into contact with the tube. Gently agitate the tube to wash the sample. Allow the mixture to settle, enabling the fat to separate and float above the blood and other residues. Using a 25 mL pipette, carefully transfer the floating fat layer to the second tube, avoiding sudden aspiration or contact between the fat and the tube walls. Once the majority of the fat has been removed, use a 10 mL pipette for more precise collection. Repeat this washing process for the remaining two tubes to eliminate debris and red blood cells.
7. *Adipose tissue sample processing:* after the final wash, allow the fat to settle in the last tube. Prepare two 100 mm Petri dishes: one for placing sterile scissors and tweezers, and the other for transferring the adipose tissue using a 25 mL pipette. Remove connective tissue (identified as white fibrous strands) using tweezers, then finely mince the adipose tissue with scissors.
8. *Enzymatic digestion:* add 15 mL of 0.075% collagenase solution in PBS to the adipose tissue in the Petri dish and resuspend thoroughly using a 10 mL pipette. The volume of collagenase should be adjusted based on the biopsy size/lipoaspirate volume. Incubate the sample at 37°C for 1 h, pipetting after 30 min to enhance digestion, then incubating for an additional 30 min.
9. *Cell isolation and filtration:* prepare a tube containing DMEM/F12 medium supplemented with 20% FBS. To halt the digestion process, add 5 mL of this 20% FBS medium into 50 mL tube and filter the digested fat through a 100 μm cell strainer to remove residual debris. Separate the stromal-vascular fraction (SVF), which contains ASCs, by centrifugating the filtered suspension at 2000 rpm for 5 min.
10. *SVF pellet washing:* prepare 20 mL of PBS supplemented with 2% PSG (0.4 mL PSG per 20 mL). After discarding the supernatant, wash the pellet with this PBS-PSG solution and centrifuge again at 2000 rpm for 5 min.

11. *Cell seeding:* prepare a collagen IV-coated tissue culture flask that has been preincubated, labeling it with the biopsy identification number according to the primary culture record. Discard the supernatant and resuspend the pellet in 10 mL DMEM/F-12 medium (1:1 vol/vol) supplemented with 20% FBS, 100 U/mL penicillin, 100 mg/mL streptomycin, and 2 mM L-glutamine (1% PSG). Transfer the suspension into the coated flask and move it to a humidified incubator at 37°C with 5% CO_2 (*see* **Note 3**).
12. *Cell culture maintenance:* the following day, perform a complete medium change, washing the flask thoroughly with PBS before adding fresh medium. The SVF contained a mixed population of stromal cells, including ASCs, along with circulating cell types, hematopoietic stem cells, and endothelial progenitor cells. Over successive passages in culture, ASCs were selectively enriched due to their adherence to the plastic culture ware. Change the medium twice per week. Once the cells reached 80%–90% confluence, subculture ASCs using 0.5 mM EDTA/0.05% trypsin for 5 min at 37°C and reseed for further expansion (*see* **Note 4**).
Assess cell viability using the Trypan blue dye exclusion assay and observe morphology through phase-contrast microscopy: ASCs in 2D culture display fibroblast-like morphology (Fig. 1).
13. *Cell culture characterization:* Confirm the homogeneity and purity of ASC population by analyzing surface marker expression with multichromatic flow cytometry in a cell sample (*see* **Note 5**). Incubate the cell samples with the specific antibodies for each panel in the dark at 4°C for 30 min. Ensure that the samples are thoroughly mixed for optimal binding of the antibodies to the surface markers. After incubation, centrifuge the samples at 1200 rpm for 7 min at 4°C to pellet the cells. Carefully resuspend the cell pellets in PBS containing 10% FBS to maintain cell viability and prepare for flow cytometry analysis. Stain the cells with 7-AAD to differentiate live cells from dead cells. This dye will intercalate with DNA in dead cells, enabling the distinction between viable and nonviable populations. Analyze the stained samples using the flow cytometer to assess the expression of surface markers. ASC populations are defined by the expression of mesenchymal markers (CD105, CD90, CD73) and the absence of hematopoietic markers (CD45, CD34).

3.2 Osteogenic Differentiation of ASCs

1. Seed 6×10^4 ASCs per well in a six-well plate. Once the cells reach 50%–60% confluence, culture them in 2 mL StemXVivo® Osteogenic/Adipogenic Base Media plus 1% PS per well. After 24 h, add the StemXVivo® Human Osteogenic Supplement (1:20) to induce osteogenesis, following the manufacturer's instructions.

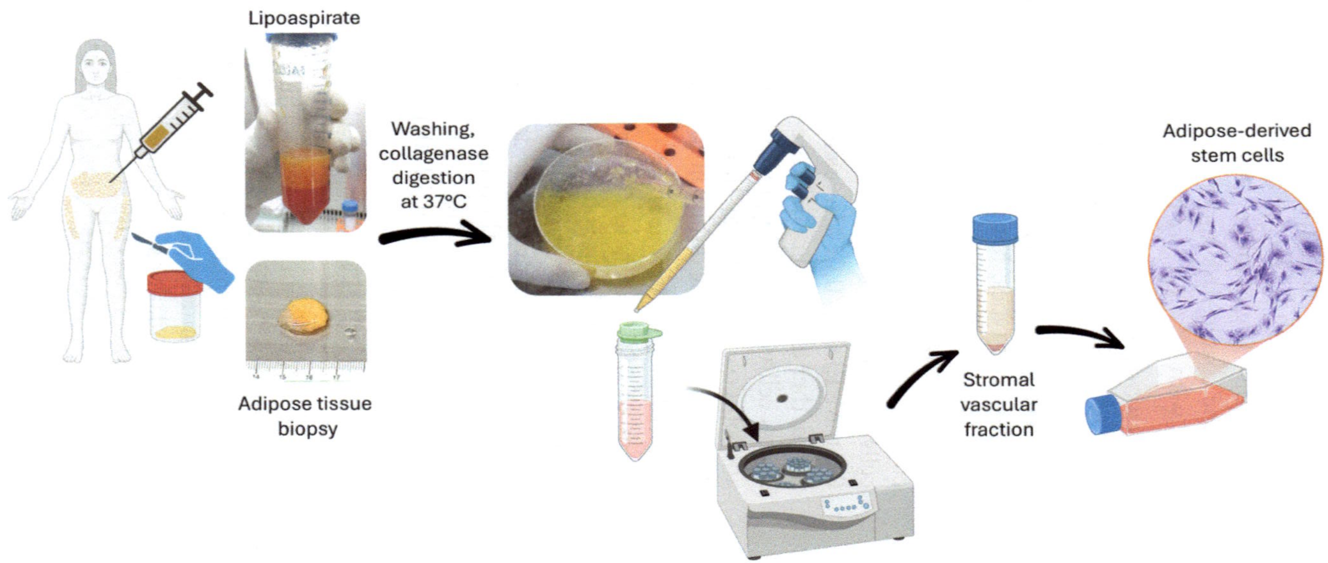

Fig. 1 ASC isolation from human adipose tissue. The picture was created with items from Biorender.com

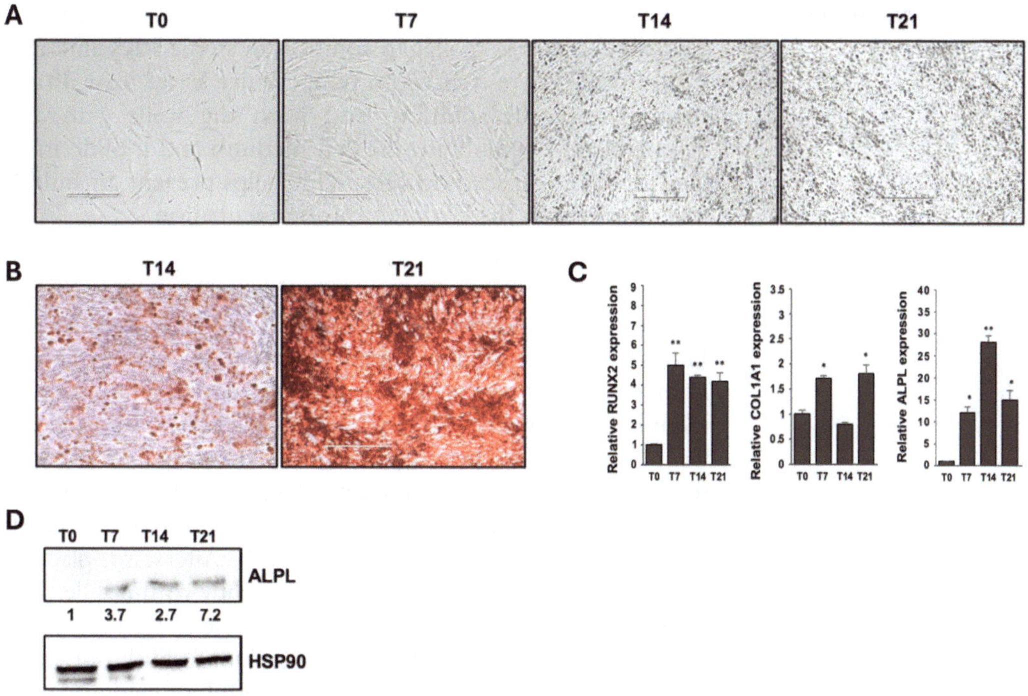

Fig. 2 ASC osteogenic differentiation. (**A**) Phase-contrast microphotographs showing ASCs at days 0, 7, 14, and 21 after osteogenic induction. Scale bars: 200 μm. (**B**) Phase-contrast microphotographs depicting mineralization (intense red clusters) in ASCs stained with Alizarin Red at 14- and 21-day time points of osteogenic induction. Scale bars: 400 μm. (**C**) mRNA expression of osteogenic markers RUNX2, ALPL, and COL1A1 was evaluated by qRT-PCR. Data were normalized to GAPDH mRNA expression. Bars represent means ± SD of three ($n = 3$) independent experiments, each performed in technical triplicate. *$p < 0.05$ and **$p < 0.01$ vs. T0 control. (**D**) Western blot analysis of ALPL protein expression in osASCs. Hsp90 was used as an internal control. Densitometric analysis of ALPL expression is shown as relative expression compared to the T0 control sample

2. Assess the morphological changes of ASCs induced to osteogenic differentiation at 7, 14, and 21 days with phase-contrast microscopy and the mineral matrix deposition in fully osteogenically committed ASC cultures, by using Alizarin Red staining. Prepare an appropriate volume (1.5 mL/well) of 4% PFA solution by diluting 8% PFA in PBS. Fix the ASCs treated with osteogenic medium for 14 and 21 days in 4% PFA for 15 min at room temperature, under the fume hood. Stain them with 2% Alizarin Red for 15 min at room temperature. Capture images of four random fields per treatment at 20× and 40× magnifications using an imaging system and analyze the images using an appropriate image analysis software (we used ImageJ 1.5).
3. Through Western Blot and qRT-PCR analyses, assess the expression of key genes of osteogenesis at the molecular level, such as RUNX2, ALPL, and COL1A1 in osASCs (Fig. 2).

3.3 Bone-on-Chip in Microgravity

1. To coat the wells of the μ-Slide III 3D Perfusion system, prepare a 10% type IV collagen solution in sterile PBS and apply 30 μL/well. Leave it at room temperature for at least 30 min, then aspirate the solution, and wash the wells with PBS. Remember to equilibrate the cell medium and μ-Slide in the incubator before seeding cells. This helps prevent air bubbles from forming in the channels during incubation.
2. For seeding osASCs, prepare a suspension with the desired concentration in complete osteogenic medium (StemXVivo® Osteogenic/Adipogenic Base Media plus 1:20 StemXVivo® Human Osteogenic Supplement plus 1% PS). Determine the number of cells to be plated based on the duration of the experiment in microgravity. We suggest 1.5×10^4 cells in 30 μL in each well. When seeding, be careful not to shake the slide, as this could result in uneven cell distribution. Once seeded, cover the slide with the supplied lid and incubate at 37°C with 5% CO_2. osASCs should attach to the slide within 24–48 h. Afterward, place the polymer coverslip, ensuring it adheres tightly to the slide. Press gently between the wells to ensure a good seal. Then, slowly fill the channels with 70 μL of cell-free medium to flush out any air bubbles, then fill the Luer adapters with 60 μL of medium.
3. Simulate microgravity conditions using a Desktop RPM, a specialized 3D clinostat developed by Dutch Space. Add fresh complete osteogenic medium before starting the experiment for optimal cell growth. To do this, carefully aspirate the old medium and replace it with 130 μL of fresh medium per channel. Seal the lid of the μ-slide with sealing film, ensuring a secure closure to prevent medium leakage, and position the slide at the center of the RPM platform. Firmly secure the slide with ample adhesive tape to prevent any movement. Operate the RPM in 3D random mode, using random motion and random direction, maintaining an average velocity of 60 deg/s to create a dynamic environment that prevents the application of a consistent unidirectional or omnidirectional 1 g force. Place the RPM inside a standard incubator set at 37°C with 5% CO_2 and connect it to the control console using standard electrical cables.

4 Notes

1. Collagen IV-coated flasks could be stored at +4°C with protective film on the cap for up to 1 month.
2. Lipoaspirates or adipose tissue biopsies could be stored in DMEM/F12 medium supplemented with 20% FBS and 1% PSG at +4°C for a maximum of 24 h.

3. For SVF seeding immediately after cell isolation, use a T25 flask or smaller, depending on the size of the SVF pellet, as cell proliferation is enhanced when cells are in close contact with each other.
4. For all experiments, ASCs should be used between passages 3 and 7 from the isolation.
5. Organize the surface markers CD34, CD73, CD90, CD45, and CD105 into two distinct panels for optimal simultaneous analysis with flow cytometer.

References

1. Bonanni R, Cariati I, Marini M et al (2023) Microgravity and musculoskeletal health: what strategies should be used for a great challenge? Life 13:1423
2. Grimm D, Grosse J, Wehland M et al (2016) The impact of microgravity on bone in humans. Bone 87:44–56
3. Azariah J, Terranova U (2025) Microgravity and cardiovascular health in astronauts: a narrative review. Health Sci Rep 8:e70316
4. Juhl OJ, Buettmann EG, Friedman MA et al (2021) Update on the effects of microgravity on the musculoskeletal system. NPJ Microgravity 7:28
5. Man J, Graham T, Squires-Donelly G, Laslett AL (2022) The effects of microgravity on bone structure and function. NPJ Microgravity 8:9
6. Lang T, LeBlanc A, Evans H et al (2004) Cortical and trabecular bone mineral loss from the spine and hip in long-duration spaceflight. J Bone Miner Res 19(6):1006–1012
7. Blaber EA, Finkelstein H, Dvorochkin N et al (2015) Microgravity reduces the differentiation and regenerative potential of embryonic stem cells. Stem Cells Dev 24(22):2605–2621
8. Colleran PN, Wilkerson MK, Bloomfield SA et al (2000) Alterations in skeletal perfusion with simulated microgravity: a possible mechanism for bone remodeling. J Appl Physiol 89:1046–1054
9. Baran RA, Wehland M, Schulz H et al (2022) Microgravity-related changes in bone density and treatment options: a systematic review. Int J Mol Sci 23(15):8650
10. Cramer EEA, Ito K, Hofmann S (2021) Ex vivo bone models and their potential in preclinical evaluation. Curr Osteoporos Rep 19:75–87
11. Niculescu AG, Chircov C, Bîrcă AC, Grumezescu AM (2021) Fabrication and applications of microfluidic devices: a review. Int J Mol Sci 22:2011
12. Mansoorifar A, Gordon R, Bergan RC, Bertassoni LE (2021) Bone-on-a-chip: microfluidic technologies and microphysiologic models of bone tissue. Adv Funct Mater 31:2006796
13. Strods A, Narbute K, Movč ana V et al (2024) Development of organ-on-a-chip system with continuous flow in simulated microgravity. Micromachines 15:370
14. Nie HY, Ge J, Liu KG et al (2024) The effects of microgravity on stem cells and the new insights it brings to tissue engineering and regenerative medicine. Life Sci Space Res 41:1–17
15. Paek K, Kim S, Tak S et al (2023) A high-throughput biomimetic bone-on-a-chip platform with artificial intelligence-assisted image analysis for osteoporosis drug testing. Bioeng Transl Med 8:e10313
16. Galván-Chacón VP, Zampouka A, Hesse B et al (2022) Bone-on-a-chip: a microscale 3D biomimetic model to study bone regeneration. Adv Eng Mater 24:2101467
17. Jogdand A, Landolina M, Chen Y (2024) Organs in orbit: how tissue chip technology benefits from microgravity, a perspective. Front Lab Chip Technol 3:1356688
18. Kostina D, Lobov A, Klausen P et al (2022) Isolation of human osteoblast cells capable for mineralization and synthetizing bone-related proteins in vitro from adult bone. Cells 11:3356
19. Dvorakova J, Wiesnerova L, Chocholata P et al (2023) Human cells with osteogenic potential in bone tissue research. Biomed Eng Online 22:33
20. Ceccarelli S, Nodale C, Vescarelli E et al (2018) Neuropilin 1 mediates keratinocyte growth factor signaling in adipose-derived stem cells: potential involvement in adipogenesis. Stem Cells Int 2018:1075156
21. Ceccarelli S, Gerini G, Megiorni F et al (2022) Inhibiting DNA methylation as a strategy to enhance adipose-derived stem cells differentiation: focus on the role of Akt/mTOR and Wnt/β-catenin pathways on adipogenesis. Front Cell Dev Biol 10:926180
22. Gerini G, Traversa A, Cece F et al (2025) Deciphering the transcriptional metabolic

profile of adipose-derived stem cells during osteogenic differentiation and epigenetic drug treatment. Cells 14:135
23. Wagner JM, Reinkemeier F, Wallner C et al (2019) Adipose-derived stromal cells are capable of restoring bone regeneration after post-traumatic osteomyelitis and modulate B-cell response. Stem Cells Transl Med 8:1084–1091
24. Cerneckis J, Cai H, Shi Y (2024) Induced pluripotent stem cells (iPSCs): molecular mechanisms of induction and applications. Signal Transduct Target Ther 9:112
25. Mende W, Götzl R, Kubo Y et al (2021) The role of adipose stem cells in bone regeneration and bone tissue engineering. Cells 10:975
26. Alamán-Díez P, García-Gareta E, Arruebo M, Pérez MÁ (2023) A bone-on-a-chip collagen hydrogel-based model using pre-differentiated adipose-derived stem cells for personalized bone tissue engineering. J Biomed Mater Res A 111:88–105
27. Ceccarelli S, Pontecorvi P, Anastasiadou E et al (2020) Immunomodulatory effect of adipose-derived stem cells: the cutting edge of clinical application. Front Cell Dev Biol 8:236
28. Anastasiadou E, Ceccarelli S, Messina E et al (2021) MiR-200c-3p maintains stemness and proliferative potential in adipose-derived stem cells by counteracting senescence mechanisms. PLoS One 16:e0257070

Part VI

Perspectives

Chapter 15

Integrated Platforms to Further Advance Space Biology Research

Lauren M. Sanders and Sylvain V. Costes

Abstract

Space biology research increasingly requires integrated platforms capable of managing complex datasets. This chapter introduces the NASA Open Science Data Repository (OSDR), designed to unify multiomics, physiological, environmental telemetry, and radiation exposure data from spaceflight and analog experiments. By strictly adhering to FAIR (Findable, Accessible, Interoperable, Reusable) principles, OSDR ensures standardized data curation and facilitates integrated analyses critical for understanding space-induced biological responses. The integration of AI technologies and federated learning frameworks will transform OSDR into a dynamic predictive modeling platform, enhancing discoveries and astronaut health research.

Keywords Space biology, Open science, FAIR, Multiomics, Environmental telemetry, Radiation dosimetry, Federated learning, Artificial intelligence, Digital twin

1 Introduction

The study of biological responses to spaceflight has advanced significantly with the development of integrated data platforms [1]. The complexity of life's response to the space environment requires integration of biological data together with experimental metadata and environmental parameters. Biological datasets span a wide array of data types, requiring domain expertise for analysis. Therefore, integrated data platforms accelerate research through standardizing and harmonizing data and metadata and providing interactive, open-source analysis and modeling tools for users of all levels.

Over the past few decades, space biology data that were originally siloed in domain-specific databases have been integrated into a single research platform: the NASA Open Science Data Repository (osdr.nasa.gov). OSDR serves as a premier FAIR (Findable, Accessible, Interoperable, Reusable) platform [2] for space biology data, distinguishing itself from other databases through rigorous curation,

Mariano Bizzarri (ed.), *Biomedicine and Biotechnology in Space*, Methods in Molecular Biology, vol. 3000,
https://doi.org/10.1007/978-1-0716-5174-2_15,

rich metadata, and standardized data formats. OSDR consolidates multiomics data, phenotypic data including medical imaging and physiological data, and contextual data, such as environmental telemetry and radiation exposure information, enabling researchers to analyze spaceflight effects comprehensively and engage in integrated analysis [3].

As space missions enter a new era engaging commercial entities and civilian astronauts [4], a deeper understanding of the molecular and physiological changes induced by spaceflight is critical. To gain new knowledge, it is essential to gather high-quality datasets that adhere to standardized formats, processing methods, and analytical frameworks. By ensuring data comparability through rigorous standards and metadata completeness, researchers can integrate and analyze information across studies, enabling a clearer understanding of the challenges posed by microgravity, radiation exposure, and environmental stressors, as well as the effectiveness of potential countermeasures. Because of the adherence to principles of FAIR and data reusability since 2017, the OSDR has become the central international hub for integrating biological and environmental datasets, supporting translational research for astronaut health, and beyond [5]. This chapter details the unique contributions of OSDR and its role in advancing space biology research, as well as its integration with other databases and emerging AI-driven technologies.

2 OSDR as a Centralized, Integrated Space Biology Research Platform

In contrast to NASA's repositories only focused on Astronauts such as the NASA Life Sciences Portal and the EXPAND platform maintained by the Translational Research Institute for Space Health (TRISH), OSDR collects the entire space biology portfolio, including space relevant data such as radiation and microgravity analogs on earth, across the full gamut of life from humans, animals, and plants to microbes. In addition to hosting all studies sponsored by NASA Biological and Physical Sciences Division, OSDR also holds data from the Human Research Program focused on radiation analog experiments conducted at the NASA Space Radiation Laboratory (NSRL), located at Brookhaven National Laboratory. The NSRL uses beams of heavy ions extracted from Brookhaven's Booster accelerator to simulate the high-energy cosmic rays found in space [6]. OSDR implements data and metadata curation standardization in alignment with the FAIR principles [2], with structured data processing pipelines facilitating high-quality data ingestion and analysis.

OSDR has played a pivotal role in unifying space biology data from various agencies and organizations worldwide, and integrating these data to enable cross-study comparisons. Notably, the European Space Agency (ESA) has begun federating its omics space biology data into OSDR, further strengthening international collaborations [3]. ESA's integration includes several human analog experiments that provide critical comparative insights for spaceflight studies, such as the Artificial Gravity Bed Rest Study program [7] and the French–Italian Concordia Station in Antarctica [8]. Concordia serves as a valuable spaceflight analog due to its extreme environment, where crew members experience chronic hypoxia, prolonged isolation, and circadian disruption—factors relevant to long-duration missions. These studies further expand OSDR's repository, enabling cross-study comparisons that improve our understanding of spaceflight stressors and potential countermeasures.

Many agencies restrict access to their repositories, requiring extensive approvals that delay research progress. To facilitate data sharing while ensuring security, OSDR has implemented an access model. This model allows principal investigators to keep data private until they are ready for publication, at which point they can generate encrypted URLs for selective sharing with collaborators and reviewers. This mechanism ensures that researchers can confidently share their data without the risk of premature exposure.

Given the constraints of spaceflight experimentation—including high costs, limited crew time, and restricted payload capacity—maximizing the scientific return from available biospecimens is critical. To respond to such a need, OSDR expands biological research through its collaboration with NBISC (NASA Biological Institutional Scientific Collection; visualization.osdr.nasa.gov/nbisc/home) and the GeneLab Sample Processing Lab (SPL). The NBISC biospecimen repository enables investigators to request tissues from space biology experiments through a search tool integrated into the OSDR platform. This functionality allows researchers to validate hypotheses generated from OSDR dataset analyses by conducting new studies using archived animal tissues previously studied in OSDR—highlighting a key feature of an integrated space biology research platform. Through NBISC and SPL, preserved samples from past and current missions, including those from the ISS and legacy Space Shuttle-era experiments, are available for our scientific community [3].

Furthermore, OSDR enhances data accessibility and usability through advanced visualization tools. The Multi-Study Visualization Platform enables simultaneous comparison of combined datasets, allowing researchers to apply filters based on experimental factors, assay technologies, organisms, and tissue types. Advanced normalization schema for differential gene expression developed by the OSDR team [9] is available to further support data exploration, offering interactive visualizations, such as PCA plots, volcano plots, gene set enrichment analysis, and more.

By integrating environmental, radiation, and biological datasets, along with biospecimen repositories, OSDR ensures that data are not only accessible but also comparable through standardized data processing, enabling cross-study insights that would otherwise be impossible. This comprehensive approach accelerates discoveries related to spaceflight stressors, biological adaptation, and potential countermeasures, ultimately advancing human space exploration.

3 Challenges of Data Federation in Space Biology

Federated data systems are increasingly adopted in large-scale biomedical research to enable secure, decentralized data sharing across hospital networks, research institutions, and government agencies, such as the National Cancer Institute (NCI). These systems help address data privacy concerns, regulatory restrictions (e.g., HIPAA, GDPR), and the need for interoperability across diverse datasets. Recent initiatives, such as the NIH's National COVID Cohort Collaborative (N3C) [10], the Cancer Research Data Commons (CRDC) [11], and the Observational Health Data Sciences and Informatics (OHDSI) [12] network, demonstrate how federated architectures facilitate collaborative research while preserving data security and patient confidentiality.

However, space biology research differs in scale and structure. The traditional argument in favor of federation—retaining data at the source for security and scalability—does not apply well to space biology. Given the limited number of spaceflight experiments, federated access can create significant bottlenecks in analysis due to inconsistent data availability, differences in metadata format and different qualities of data curation. By contrast, consolidating data within OSDR ensures that researchers work with harmonized datasets, significantly reducing the challenges of interoperability. Researchers studying rare diseases and cancers use similar approaches, where just like space biology, assembling a database large enough for statistical analysis and modeling is crucial [13].

The primary rationale for currently adopting a single, harmonized platform like OSDR is the complexity of managing access control for sensitive datasets. Although federated databases promote collaboration, they do not eliminate the need for stringent access controls, especially when handling privacy-sensitive data. Managing such permissions across multiple federated sites rapidly complicates the data-sharing landscape. By consolidating all spaceflight relevant data, OSDR provides unified tools for dataset exploration and clear indicators of access restrictions.

For controlled data, such as raw RNAseq from missions like Inspiration4 [14], investigators must submit a Data Access Request (DAR) Form with Institutional Review Board (IRB) documentation and agree to a Data User Agreement. Requests are reviewed by the OSDR Curation Team and a Data Access Committee led by the original data submitter. Once approved, users must follow the agreed terms, provide progress updates, cite the data properly, and delete all files upon project completion. This controlled-access model, adapted from derived from NCBI's database of Genotypes and Phenotypes (dbGaP) framework [15], supports secure sharing of sensitive human data. It promotes timely, inclusive research while protecting participant privacy, helping researchers uncover new insights into the biological effects of spaceflight.

4 Environmental Data and Radiation Exposure

To advance space biology research, it is essential to contextualize biological findings with environmental and physical parameters collected throughout the duration of spaceflight. The Open Science Data Repository (OSDR) addresses this need by integrating longitudinal telemetry and dosimetry data alongside traditional biological datasets, enabling multifactor analyses that reveal how spaceflight conditions influence biological responses.

The Environmental Data Application (EDA; visualization.osdr.nasa.gov/eda) allows researchers to visualize and analyze contextual variables, such as CO_2 levels, humidity, temperature, and radiation exposure, using telemetry from the International Space Station (ISS) and ground analogs. This tool bridges a long-standing gap by making environmental data accessible and interoperable with biological experiment data.

Similarly, the RadLab portal (visualization.osdr.nasa.gov/radlab) consolidates space radiation dosimetry data from ISS sensors and planetary missions, providing a centralized resource to study the biological effects of radiation. By enabling precise correlations between radiation exposure and biological outcomes, RadLab supports critical research on one of the most significant hazards for long-duration missions beyond low Earth orbit.

5 Future Directions: AI and Digital Twins in Space Biology

With the rapid advancements in machine learning and AI technology, our ability to model and predict complex biological effects continues to expand. Machine learning analysis techniques have been used to great effect in recent years for knowledge discovery in space biology and health [16–21]. Further, the growing emphasis on longitudinal data collection, such as in the Space Omics and Medical Atlas (SOMA) and the EXPAND Essential Measures (bcm.edu/academic-centers/space-medicine/expand) enables detailed modeling and simulation of spaceflight effects with AI methods.

The new era in space biology and AI requires traditional database models to evolve. The concept of a static dataset is increasingly being replaced by a dynamic, living version, with platforms, such as Hugging Face and Kaggle providing mechanisms for dataset-model connections, model training hubs, community engagement, and hackathons and challenges in a single space. Classic database queries are giving way to chatbot-driven interactions capable of navigating unstructured data and exploring complex scientific concepts.

It is also now possible to generate realistic synthetic data and simulations, such as digital twins, moving far beyond the original dataset [22, 21]. Knowledge graph technology, coupled with large language models, promises to revolutionize the traditional database search into an end-to-end knowledge discovery platform, facilitating scientific insights through a simple text prompt [23]. This was recently illustrated when our team used this approach on NASA GeneLab transcriptomic data, uncovering human-like physiological responses in space-flown mice. By comparing gene network patterns between space and ground control samples, the analysis highlighted top-ranked nodes, such as space motion sickness, blood vessel regulation, taste receptors, vitamin D metabolism, and the sympathetic nervous system [24]. All these nodes were associated with known human spaceflight effects, demonstrating that even inbred animal models can reveal meaningful insights into human space physiology.

As noted previously, the current Data Access Request (DAR) process on OSDR enables responsible sharing of sensitive astronaut data but would significantly slow scientific analysis in a federated database environment. A promising solution to expand OSDR capabilities through federation with larger databases is federated learning, a privacy-preserving AI approach that trains models without moving raw data. Instead of transferring raw data, only the model's learned parameters (such as weight updates) are shared—making it ideal for sensitive, distributed datasets across agencies and sectors. Our team demonstrated this with FLUID [25], the first federated learning model trained between Earth and the ISS, showing its potential for real-time, privacy-preserving analysis in spaceflight research.

However, it is important to recognize that federated learning only benefits AI-based tools. It does not support classical bioinformatics workflows, which require direct access to raw data for processing and interpretation. This distinction reinforces the need to maintain both access-controlled and AI-enabled pipelines, depending on research needs.

To make federated learning scalable and effective, we must standardize all biomedical spaceflight data—from NASA, ESA, and beyond. AI models rely on structured, harmonized input, and without a shared data format, training across multiple datasets is unfeasible. Standardization is, thus, a foundational step toward AI-readiness.

A critical component of federated learning is the aggregator, which is the central server that collects local model updates from each data node, integrates them (e.g., by averaging), and redistributes the improved global model. The OSDR platform is ideally positioned to serve as this aggregator, enabling secure orchestration of AI model training across a federation of data holders, whether governmental, academic, or commercial. By combining federated learning, standardized data formats, and a centralized aggregator hosted at OSDR, we can move beyond the limitations of current access models and build a scalable, secure, and inclusive AI-driven ecosystem for space health research, which would include many more data resources.

6 Conclusion

OSDR has emerged as the leading platform for integrating and standardizing space biology data, emphasizing consolidation to ensure data quality, accessibility, and analytical consistency. Its security features support controlled access, while tools for environmental telemetry and radiation exposure significantly enhance research capabilities.

Looking ahead, the future of space biology data management is poised to leverage AI-driven digital twins, transitioning OSDR from primarily a data storage solution into a dynamic predictive modeling and knowledge synthesis platform. By fully integrating standardized existing datasets with real-time data streams from ongoing spaceflight missions, OSDR could enable powerful predictive models for identifying health risks and optimizing countermeasures in real-time mission scenarios.

Additionally, extending data federation to broader partner networks using AI technologies like federated learning, OSDR could further accelerate discoveries without compromising data privacy. As human space exploration continues to expand its horizons, OSDR will remain foundational to advancing biological research and ensuring the health and safety of future explorers.

References

1. Costes S, Gentemann C, Platts SH, Carnell LA (2024) Biological horizons: pioneering open science in the cosmos. Nat Commun 15(1):4780
2. Berrios DC, Beheshti A, Costes SV (2018) FAIRness and usability for open-access omics data systems. AMIA Annu Symp Proc 2018:232–241
3. Gebre SG, Scott RT, Saravia-Butler AM, Lopez DK, Sanders LM, Costes S (2024) NASA open science data repository: open science for life in space. Nucleic Acids Res 53(D1):D1697–D1710.
4. Mason CE, Green J, Adamopoulos KI, Afshin EE, Baechle JJ, Basner M, Bailey SM, Bielski L, Borg J, Borg J, et al (2024) A second space age spanning omics, platforms and medicine across orbits. Nature 632(8027):995–1008
5. Overbey EG, Kim J, Tierney BT, Park J, Houerbi N, Lucaci AG, Medina SG, Damle N, Najjar D, Grigorev K et al (2024) The space omics and medical atlas (SOMA) and international astronaut biobank. Nature 632(8027):1145–1154
6. La Tessa C, Sivertz M, Chiang I-H, Lowenstein D, Rusek A (2016) Overview of the NASA space radiation laboratory. Life Sci Space Res 11:18–23
7. Strauch L, von der Wiesche M, Noppe A, Mulder E, Rieger I, Aeschbach D, Elmenhorst E-M (2024) Simulating microgravity with 60 days of 6 degree head-down tilt bed rest compromises sleep. NPJ Microgravity 10(1):109

8. Feuerecker M, Crucian BE, Quintens R, Buchheim J-I, Salam AP, Rybka A, Moreels M, Strewe C, Stowe R, Mehta S et al (2019) Immune sensitization during 1 year in the Antarctic high-altitude Concordia Environment. Allergy 74(1):64–77
9. Sanders LM, Chok H, Samson F, Acuna AU, Polo S-HL, Boyko V, Chen Y-C, Dinh M, Gebre S, Galazka JM et al (2023) Batch effect correction methods for NASA GeneLab transcriptomic datasets. Front Astron Space Sci 10:1200132
10. Haendel MA, Chute CG, Bennett TD, Eichmann DA, Guinney J, Kibbe WA, Payne PRO, Pfaff ER, Robinson PN, Saltz JH et al (2021) The National COVID cohort collaborative (N3C): rationale, design, infrastructure, and deployment. J Am Med Inform Assoc 28(3):427–443
11. Kim E, Davidsen T, Davis-Dusenbery BN, Baumann A, Maggio A, Chen Z, Meerzaman D, Casas-Silva E, Pot D, Pihl T et al (2024) NCI cancer research data commons: lessons learned and future state. Cancer Res 84(9):1404–1409
12. Reinecke I, Zoch M, Reich C, Sedlmayr M, Bathelt F (2021) The usage of OHDSI OMOP–a scoping review. Stud Health Technol Inform 283:95–103
13. Vaske OM, Bjork I, Salama SR, Beale H, Tayi Shah A, Sanders L, Pfeil J, Lam DL, Learned K, Durbin A et al (2019) Comparative tumor RNA sequencing analysis for difficult-to-treat pediatric and young adult patients with cancer. JAMA Netw Open 2(10):e1913968
14. Sanders LM, Grigorev K, Scott RT, Saravia-Butler AM, Polo S-HL, Gilbert R, Overbey EG, Kim J, Mason CE, Costes S (2024) Inspiration4 data access through the NASA Open Science Data Repository. NPJ Microgravity 10(1):56
15. Tryka KA, Hao L, Sturcke A, Jin Y, Wang ZY, Ziyabari L, Lee M, Popova N, Sharopova N, Kimura M et al (2014) NCBI's database of genotypes and phenotypes: dbGaP. Nucleic Acids Res 42(Database issue):D975–9
16. Sanders LM, Scott RT, Yang JH, Qutub AA, Martin HG, Berrios DC, Hastings JJA, Rask J, Mackintosh G, Hoarfrost AL et al (2023) Biological research and self-driving labs in deep space supported by artificial intelligence. Nat Mach Intell 5(3):208-219
17. Scott RT, Sanders LM, Antonsen EL, Hastings JJA, Park S-M, Mackintosh G, Reynolds RJ, Hoarfrost AL, Sawyer A, Greene CS et al (2023) Biomonitoring and precision health in deep space supported by artificial intelligence. Nat Mach Intell 5(3):196-207
18. Li K, Desai R, Scott RT, Steele JR, Machado M, Demharter S, Hoarfrost A, Braun JL, Fajardo VA, Sanders LM et al (2023) Explainable machine learning identifies multi-omics signatures of muscle response to spaceflight in mice. NPJ Microgravity 9(1):90
19. Casaletto JA, Scott RT, Myrick M, Mackintosh G, Chok H, Saravia-Butler A, Hoarfrost A, Galazka JM, Sanders LM, Costes SV (2025) Analyzing the relationship between gene expression and phenotype in space-flown mice using a causal inference machine learning ensemble. Sci Rep 15(1):2363
20. Casaletto JA, Foley P, Fernandez M, Sanders LM, Scott RT, Ranjan S, Jain S, Haynes N, Boerma M, Costes SV et al (2025) Foundational architecture enabling federated learning for training space biomedical machine learning models between the International Space Station and Earth. bioRxiv. 2025.01.14.633017. https://www.biorxiv.org/content/10.1101/2025.01.14.633017v1.abstract. doi:https://doi.org/10.1101/2025.01.14.633017
21. Pantha N, Ramasubramanian M, Gurung I, Maskey M, Sanders LM, Casaletto J, Costes SV (2024) Feature selection in high-dimensional space with applications to gene expression data. In: SoutheastCon 6. IEEE, pp 6–15
22. Guillaudeux M, Rousseau O, Petot J, Bennis Z, Dein C-A, Goronflot T, Vince N, Limou S, Karakachoff M, Wargny M et al (2023) Patient-centric synthetic data generation, no reason to risk re-identification in biomedical data analysis. npj Digit Med 6(1):37
23. Soman K, Rose PW, Morris JH, Akbas RE, Smith B, Peetoom B, Villouta-Reyes C, Cerono G, Shi Y, Rizk-Jackson A et al (2023) Biomedical knowledge graph-optimized prompt generation for large language models. arXiv [cs.CL]. http://arxiv.org/abs/2311.17330
24. Nelson CA, Acuna A, Paul A, Scott RT, Butte A, Cekanaviciute E, Baranzini S, Costes S (2021) Knowledge network embedding of transcriptomic data from spaceflown mice uncovers signs and symptoms associated with terrestrial diseases. Life 11(1):42
25. Casaletto JA, Foley P, Fernandez M, Sanders LM, Scott RT, Ranjan S, Jain S, Haynes N, Boerma M, Costes SV et al (2025) Bridging Earth and space: a flexible and resilient federated learning framework deployed on the International Space Station. Bioinformatics. https://www.biorxiv.org/content/10.1101/2025.01.14.633017v1.full

Chapter 16

Precision Medicine in Astronauts: Principles and Targets in Molecular Assessment

Michael A. Schmidt, Julian C. Schmidt, Caleb M. Schmidt, Jeffrey A. Jones and Christopher E. Mason

Abstract

The field of space medicine has served the population of professional astronauts well over the many decades of human space travel, which is reflected in the considerable success of astronauts executing missions and returning to Earth in good health (rare comorbidities notwithstanding). At the same time, there are considerable gaps in our biomedical knowledge that will only scale as less fit individuals embark upon spaceflight, and we enter more extreme environments such as the Moon and Mars for longer duration missions. One means to partially fill this knowledge gap is to expand the practice of precision medicine, which we define as comprehensive genotyping and molecular phenotyping—coupled with derived countermeasures tailored to individuals and crews—to improve health and mitigate risk of disease. Improving the methods and tools of spaceflight precision medicine (PM) will improve our ability to optimize astronaut health and performance before entering space, while providing additional insight into the needs of specific astronauts while in space and upon return to Earth. A critical feature of PM in human spaceflight rests upon the foundation of complex molecular profiling, as the basis for developing these tailored countermeasures. This chapter provides an introductory overview of one means to use systems thinking to explore principles and molecular targets that are fundamental to PM in space.

Keywords Precision medicine, Personalized medicine, Spaceflight, Astronaut, Genotype, Single nucleotide polymorphism, Pharmacogenomics, Nutrition, Vitamins, Minerals

1 Introduction

Venturing into the spaceflight environment presents myriad unique exposures for humans and other animals. Primary risk exposures include physical forces upon the body (e.g., altered gravitation, loading due to propulsion, fluid shifts, etc.), radiation, isolation and confinement, the possibly hostile closed environment of the spacecraft (e.g., conditions of air, water, and microbial contamination), and distance from earth [1]. These risk factors significantly contribute to overall allostatic load and have a notable influence on astronaut health, safety, and performance. Fortunately, as technology

Mariano Bizzarri (ed.), *Biomedicine and Biotechnology in Space*, Methods in Molecular Biology, vol. 3000,
https://doi.org/10.1007/978-1-0716-5174-2_16,

has advanced, our ability to measure, manage, and counteract spaceflight risk has begun to flourish.

Fundamentally, an astronaut's genotype and molecular phenotype bear heavily upon how each individual will interact with the spaceflight environment. Increasingly, molecular-focused precision medicine has demonstrated its utility in terrestrial medicine. Specifically, performing deep quantitative molecular analysis of an individual's genotype, coupled with assessment of the molecular, physiological, morphological, environmental, and behavioral phenotypes allows for the development of precision countermeasures (though the physiological, morphological, environmental, and behavioral phenotypes will not be discussed herein).

Due to the complexity involved in merely understanding these molecular dynamics, it is useful to outline a systems biology construct that aims to reduce this complexity. This systems approach is based on the principle of inputs, processes, and outputs, whereby the input component is further reduced by characterizing essential, conditionally essential, and nonessential elements. When coupled with a refined approach to pharmacogenomics and targeted genome profiling, including SNPs, indels, CNVs, and SVs (single nucleotide polymorphisms, insertions/deletions, copy-number variants, and structural variants), a better knowledge of the function of molecular network dynamics can be created, leading to a better understanding of biological meaning that informs precision countermeasure development.

2 Precision Medicine: Definitions

For the purpose of the present discussion, precision medicine (PM) is defined as *the science and practice of providing methods of treatment and prevention tailored to an individual's molecular, physiological, morphological, and behavioral characteristics* [2]. Regarding human spaceflight, a related requirement is that the assessment and countermeasures development must be directed to the specific mission architecture and space conditions in which each individual astronaut is operating, including duration, radiation dose, and operational requirements.

Throughout the history of space medicine, clinicians have always given personalized attention to each individual astronaut, including processing medical waivers for certain conditions and preparing personalized medical kits to address specific needs. Precision medicine today takes this to a greater level of granularity, by deepening and expanding the molecular analytical suite, while also carefully incorporating individual genetic variants and polygenic considerations. Another key tenet of PM in human spaceflight incorporates the concepts of *personalization* and *stratification*.

Personalization applies when careful evaluation reveals a pattern of variance that is unique to an individual, whereby countermeasures are then tailored to that individual. Stratification applies to groups of individuals who share certain traits, characteristics, and molecular patterns. An example is a mission crew where three of six astronauts possess the CYP450 2D6 slow metabolizer phenotype for certain drugs. As such, their mission drug list would be shared with respect to medications that are approved, modified, and rejected for this genotype.

Precision medicine on Earth is most commonly focused on molecular analytics to facilitate *diagnosis*, ultimately leading to personalized (precision) intervention and *disease management*. This approach will remain important in professional astronauts and, increasingly, for civilians as civilian space travel becomes more common. Unique to spaceflight, however, is the need for *optimal performance*, which impacts both mission capability and safety. Since performance is not disease associated, there is a specific approach needed beyond diagnosis, which is *pattern analysis*. This consists of characterizing and understanding molecular networks to the greatest extent possible, as a basis to inform the development of individualized (tailored) interventions.

The more nodes within these molecular networks that we can characterize, the better we can describe the functional status of networks. Optimally, this is done with fully quantitative measures, which are used to identify patterns of metabolic deficiency or excess, genetic variability, and describe biological meaning and predicted impact. This allows one, then, to tailor precision countermeasures to the individual astronaut or stratified group within a crew [2]. In general, all assessment of molecular measures should seek to align to the phenotype. In space, we at minimum seek to align the molecular analytics (especially genotype) to the following phenotypes:

- Molecular Phenotype
- Physiological Phenotype
- Behavioral Phenotype
- Morphological Phenotype
- Environmental Phenotype
- Clinical Phenotype
- Performance Phenotype

Of the foundational analytics employed in this systems approach to PM in human spaceflight, the present overview is focused on (1) small molecule essential and conditionally essential inputs, (2) molecular signals that define the process layer (metabolic processes, proteins, and signaling features), (3) pharmacogenomic measures that guide drug prescribing, and (4) selected single nucleotide polymorphisms (SNPs and variants) of potential relevance to human health and performance in space.

Two points warrant consideration as one reviews this summary. First, the scope of this overview is on assessment, rather than intervention (countermeasures). Second, not all genomic elements discussed below may have distinct countermeasures developed as of this writing and even fewer have been directly explored in human spaceflight. However, in order to understand how to develop countermeasures for human spaceflight, it is useful to begin to understand how these SNPs currently utilized on Earth may apply to risk and countermeasure concepts in space.

3 Longitudinal Measures

Essential to PM in spaceflight is to follow the patterns of molecular variability over time, using serial measures over as long a period as possible. Fundamentally, this includes attention to pre-flight, in-flight, and post-flight measures. Determining a baseline as early in the astronaut's career as possible is important. In order to better capture individual variability, multiple baseline levels early in the process of astronaut training are also warranted. While the PM approach should always incorporate research to the extent possible, pre-flight assessment is primarily meant to guide decisions that optimize health and performance of an individual before entering space. Thus, a primary goal should be to identify patterns of variance early in training and follow the astronaut on a temporal basis using the same repeated measures as those collected at the baseline point (though not necessarily limited to those measures).

This analysis should also help determine whether any special attention to specific markers or processes should be given while in space. For example, we have found a high penetrance of the hemochromatosis genotype and phenotype in elite athletes, which has been verified by others [3]. High iron stores in young athletes on Earth appear to benefit performance, though there is a growing cost with age. However, systemic elevation of the oxidative catalyst iron in the radiation of space presents a risk that is not well characterized, though is of theoretical concern. Current indications are that high levels of iron can accumulate reactive oxygen species via ferroptosis. As such, male astronauts in space who possess this genotype may wish to adopt a low iron diet and assess iron status in long-duration missions.

At minimum, baseline levels prior to the start of a mission will begin to help clinicians assess how an individual astronaut responds to stressors over time and as their launch day (LD+0) approaches. To perform optimally in a spaceflight environment, it is important to frequently capture longitudinal measures throughout each phase of the mission, as well as pre/post-flight metrics. During the pre-mission phase, clinicians can begin to identify, characterize, and optimize molecular networks for the individual and the specific demands of the mission. Thus, knowable molecular features of an astronaut's metabolic networks are not left unattended prior to spaceflight. This allows clinicians to develop individual and personalized countermeasures for each astronaut entering space prior to the start of each mission, to optimize pre-flight preparation as well as to enhance in-flight performance.

Due to the inherent difficulty of collecting measures in space (i.e., limited in-flight analytical instrumentation and onboard cold storage space for post-flight sample return), in-flight measures are usually limited. Longitudinal serial measures should continue to be collected to the extent available, however, the primary focus of inflight collection should be on critical measures. Key molecular findings should lead to appropriate countermeasures that optimize astronaut safety, health, performance, and mission success. Looking forward, as analytical techniques and the field of PM in spaceflight advance, additional measures can be added to the suite of molecular analytics in flight.

Upon return to earth, post-mission molecular analytics are central to astronaut health and recovery. The first aim in post-flight analysis should be to repeat the baseline metrics and evaluate which metabolic markers remain perturbed or to reveal new perturbations that have arisen as a result of the stressors of the flight environment, and to also characterize their persistence. Furthermore, it is important to address the symptom presentation and the associated metabolic needs that have resulted from exposure to the space environment. For example, even if a given clinical phenotype is not present, underlying metabolic needs will still require dedicated attention to enhance and optimize recovery once astronauts have returned to the terrestrial environment. Return to Earth exposures and assessment thereof will be critical to astronaut health and longevity, which is especially true following long-duration missions.

The present chapter will introduce a series of molecular markers that provide selected tools to optimize molecular assessment and countermeasures for human health, safety, performance, and survival in space. In this systems-based approach, we begin by addressing the concept of inputs, processes, and outputs.

4 Characterizing Molecular Networks in Precision Medicine

One goal of PM is to use quantitative molecular measures for the purpose of reconstructing molecular pathways and networks relevant to human health and performance. These analyte measures serve as nodes within the network or pathway that help one describe the behavior of that network at a specific point in time. When pathway and network data are aggregated, a greater understanding of the system can be developed, which enhances the ability to assign biological meaning and utility. It is from this understanding of biological meaning that new countermeasure concepts can be developed. While complex, one means to reduce this complexity is to gain an understanding of basic and fundamental processes that drive the system. This begins by looking at *inputs* into the system that shape the system's behavior:

Inputs can be broken down into three categories: (1) *Essential Inputs*, (2) *Conditionally Essential Inputs*, and (3) *Non-essential inputs* (Fig. 1). Essential inputs are substances (molecules, atoms) that cannot be produced by the body and must be acquired through the diet, such as essential amino acids. Each essential input can be involved in hundreds of metabolic processes. The number of processes associated with each essential input can be contextualized by understanding its *participation number* (Table 1), which refers to

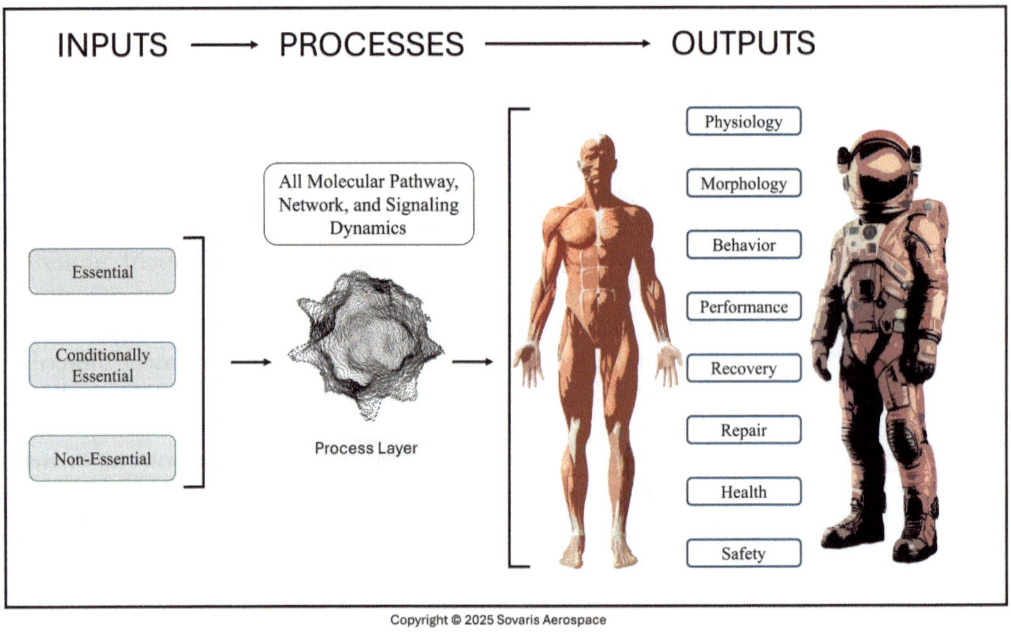

Fig. 1 Foundational Input–Output Framework for Precision Medicine in Space. In this model, inputs (often small molecules and precursors derived from the diet) are categorized as essential, conditionally essential, or nonessential, and flow into the process layer, which encompasses all molecular pathways, networks, and signaling dynamics, with components such as proteins, metabolites, including the gut metagenome and metabolome. The process layer is the metabolic engine that gives rise to a range of outputs (collectively, the phenotype), including physiology, morphology, behavior, performance, recovery, repair, health, and safety in astronauts. Greater detail in identifying and understanding individual components within the input, process, and output layers improves the precision and impact of stratified and personalized medicine strategies Image credit: Sovaris Aerospace

Table 1
Participation number of selected inputs. Reprinted with permission, Sovaris Aerospace

Input	Participation number and associated processes
S-adenosylmethionine	*S*-adenosyl-L-methionine (AdoMet) is the second most commonly used enzyme cofactor after ATP [5–7].
Iron	Roughly 2% of human genes encode an iron protein. It is estimated that 6.5% of all human enzymes are iron-dependent [8].
Magnesium	To date, the enzymatic database lists over 600 enzymes for which Mg^{2+} serves as a cofactor (15%), and another 200 in which Mg^{2+} may act as activator (5%) [9].
Nicotinamide (B3)	The ENZYME database lists over 400 NAD(P)-dependent reactions (10%), giving it the potential to influence almost every area of metabolism [10].
Pyridoxine (B6)	Pyridoxal-5-phosphate (vitamin B6) is utilized by 112 (3%) of the 3870 enzymes cataloged in the ENZYME database [11].
Riboflavin (B2)	A flavin-containing cofactor, FAD or FMN (riboflavin-derived), is utilized by 151 (4%) of the 3870 enzymes cataloged in the ENZYME database [10].
Selenium	Selenium (Se) is critical for the formation of thyroid hormone through its role in deiodinase enzymes. Though the total number of enzymes dependent on Se is low, its reach is substantial, given that thyroid hormone impacts all energy metabolism in cells [12].
Vitamin D 3	Roughly 10% of all genes possess sites for the vitamin D binding protein [13].
Vitamin K	Vitamin K is essential to the blood clotting cascade by serving as a cofactor for the γ-carboxylation of vitamin K-dependent coagulation factors (including factors II, VII, IX, and X). Though the total number of enzymes dependent on vitamin K is low, its influence is vast due to its impact on these clotting dynamics [14].
Zinc	An estimated 3200 genes of the human genome, comprising roughly 15% of the total protein-encoding genes, encode proteins that bind or are predicted to bind zinc ions [15, 16].

the number of enzymatic or other processes in which a given compound is involved.

Conditionally essential inputs are molecular entities that are normally produced in sufficient quantities to meet the body's fundamental needs. However, quantities or synthesis may be insufficient for numerous reasons (e.g., disease, genetics, essential input deficits, cofactor deficits, environmental exposures, or psychosocial stressors) [4]. This can result in inadequate availability for metabolic processes, which can cause a compound the body can make under ordinary circumstances to become essential and require exogenous supplementation.

Non-essential inputs are those that the body does not require but to which the body is intentionally, unwittingly, or accidentally exposed or exposed at levels that lead to metabolic dysregulation (e.g., drugs, xenobiotics, CO_2, radiation, etc.). These exposures can lead to profound, and often complex, molecular alterations that impact metabolic network efficiency [2].

A fundamental approach to the practice of precision medicine in human spaceflight is to first characterize the essential and conditionally essential inputs, so that no astronaut has to enter the space environment with ***knowable, actionable metabolic deficits.*** Further characterization of non-essential inputs solidifies the basic foundation of PM in practice. Understanding specific genetic risk can further inform PM practice in space. This thought process can be exemplified by exploring the convergence of one carbon metabolism, genome stability, and radiation in space.

4.1 Concepts in Molecular Assessment in Space: The Case of One Carbon Networks and Radiation Susceptibility

One can use susceptibility to radiation to illustrate how the convergence of essential inputs, genetics, and radiation dynamics impact individuals and shape interindividual variability. This regards the one-carbon molecular network. In the formation of DNA bases, the conversion of uracil to thymine requires the transfer of a single methyl group, which is provided by N5, N10-methylenetetrahydrofolate (folate). If the methyl pool is sufficiently depleted (e.g., low folate, B12), thymine levels are diminished, and uracil accumulates in the nucleus (Fig. 2). Methyl donor depletion may occur pre-mission if unattended or become depleted over the course of a long duration mission with inadequate diet in space.

As the process of DNA synthesis proceeds, low thymine can lead to misincorporation of uracil (rather than thymine) into DNA, which requires the activity of DNA repair glycosylase enzymes. Sustained insufficient thymine incorporation (expanding uracil pool) may ultimately result in single strand breaks, double strand breaks, and eventual DNA instability. This can be seen as a mutational event (mutational event #1), which occurs in the absence of any exposure to radiation. The DNA repair system will continue its attempts to restore DNA integrity, but if high uracil levels persist, the DNA repair system will be impaired.

Allelic variants in the genes coding for one of several methyl transfer enzymes (MTHFR, MTR, MTRR, SHMT, etc.) can further raise this risk in humans. In particular, the MTHFR 677TT genotype, coupled with folate and/or vitamin B12 deficiency, further impacts DNA stability in adverse ways. This genotype also raises the requirement for riboflavin.

Independently of genotype or nutritional status, space radiation may impact DNA by inducing single or double strand breaks. This represents a separate type of mutational event in space (noted herein as mutational event #2). One who enters the space environment with

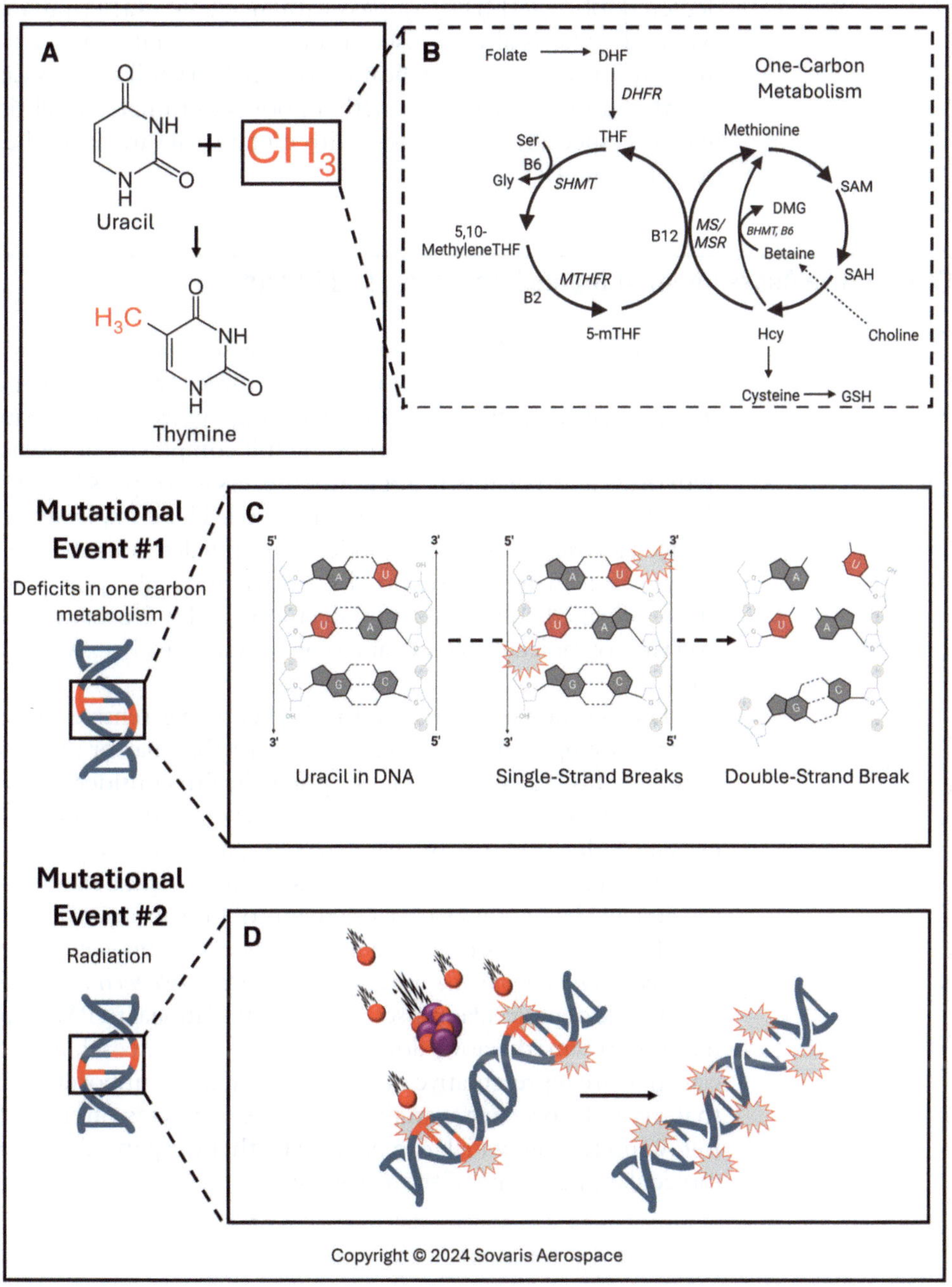

Fig. 2 One carbon metabolism, DNA stability, and DNA repair. The conversion of uracil to thymine requires the transfer of a single methyl group, which is provided by N5, N10-methylenetetrahydrofolate (folate). If the methyl pool is sufficiently depleted, thymine levels are diminished, and uracil accumulates in the nucleus (**A**). This can lead to misincorporation of uracil (rather than thymine) into DNA, resulting in single strand breaks, double strand breaks, and DNA instability. This can be seen as a mutational event (mutational event #1) that does not require radiation (**C**). Genetic mutations in one of several methyl transfer enzymes (MTHFR, MTR, MTRR) can further raise this risk in humans (**B**). Independently in space, radiation may impact DNA by inducing single or double strand breaks (**D**). This represents a separate type of mutational event in space (noted herein

depleted folate and B12, coupled with increased uracil (mutational event #1) who is also exposed to radiation (mutational event #2) may expect to see an amplification of DNA instability. This can be countered by ensuring the methyl pool is optimized at all times, which is especially true in those with mutational variants in the one carbon genes [3].

5 Quantitative Assessment of Inputs, Processes, and Outputs

Beyond the one carbon network, one can easily envision how dysregulation can scale across networks if two, three, or more essential or conditionally essential input deficits are aggregated (a combinatorial event). In practice, assessment of the full complexity of molecular networks in astronauts is not always necessary or possible, due to the substantial number of analytes involved. This complexity can be reduced by using this systems model in making decisions about which analytes and analyte classes provide sufficient details upon which to make informed countermeasure decisions. Table 2 provides a brief overview of molecular classes and representative analytes within each class.

The question often arises as to whether such extensive molecular analysis is simply an exploration for deficits that may or may not be present. This is a fair question for public health considerations that are sensitive to scalability and cost. However, in the context of the extreme vulnerabilities, the harshness of space, and the potential to impact health, safety, performance, and even survival, these measures can actually be seen as essential elements of human space travel. We, therefore, proceed from the premise that *no astronaut should enter the harsh environment of space with knowable, actionable metabolic deficits.*

A table of molecular classes deemed foundational to PM in spaceflight is presented below, along with analytes within each molecular class that are representative of the types of analytes important within that class. In practice, variability in these analytes often leads to follow-up testing, which can provide further deep insights into the molecular phenotype of the astronaut.

Fig. 2 (continued) as mutational event #2). One who enters the space environment with depleted folate, B12, and others and increased uracil (mutational event #1) who is also exposed to radiation (mutational event #2) may expect to see an amplification of DNA instability. This can be countered by ensuring the methyl pool is optimized at all times, which is especially true in those with mutational variants in the one carbon genes. Reprinted with permission, Sovaris Aerospace See also: Building a Space-Faring Civilization. Eds. Schmidt, MA and Legato, MJ. Using Functionally Characterized Networks in the Application of Precision Medicine in Space flight. Schmidt, MA, Schmidt, JC. Elsevier, Academic Press. New York, New York. 2025;223

Table 2

Molecular class assessment for functionally characterized networks. Reprinted with permission, Sovaris Aerospace

Analyte class	Representative analytes
Amino acids	Arginine, histidine, isoleucine, leucine, lysine, methionine, phenylalanine, threonine, tryptophan, valine, (also non-protein amino acids)
Antioxidant status—Nonenzymatic	Ascorbic acid, α-tocopherol, γ-tocopherol, β-carotene, lycopene, selenium, zinc glutathione (GSH), glutathione:glutathione disulfide ratio (GSH:GSSG); uric acid, bilirubin, coenzyme Q10, polyphenols, flavonoids
Antioxidant statusEnzymatic	Superoxide dismutase, catalase, glutathi—one peroxidase, glutathione reductase
Androgens	Testosterone (total), dihydrotestosterone, dehydroepiandrosterone (DHEA), DHEA-S, sex-hormone binding globulin, free androgen index, estrogen
Catecholamines	Epinephrine, norepinephrine, dopamine
Conditionally essential nutrients	GSH, carnitine, CoQ10, taurine, eicosapentaenoic acid, docosahexaenoic acid, etc.
Cortisol	Diurnal salivary cortisol, 24-h urinary cortisol
Cytokines (TH1/TH2)	IL-1β, IL-2, IL-3, IL-4, IL-5, IL-6, IL-10, IL-12, IL-13, IL-17, IFN-γ, TNFα, CXCL-10
Environmental toxicological panel	Phthalates, xylene, toluene, styrene/ethylbenzene, benzene, acrylonitrile, 1-bromopropane, 1,3-butadiene, ethylene oxide, vinyl chloride, organophosphates, nitrogen tetroxide, hydrazine, ammonia
Estrogens	Estradiol (E2), estrone (E1), estriol (E3), estetrol (E4), estrogen receptors, LH, FSH, progesterone, testosterone, sex hormone binding globulin
Fatty acids and phospholipids	Omega-3 Index, AA:EPA ratio, omega-6:omega-3 ratio, omega-3:omega-9 ratio, omega-3:6:9 ratio, α-linolenic acid, eicosapentaenoic acid, docosahexaenoic acid, linoleic acid, γ-linolenic acid, dihomo-γ-linolenic acid, arachidonic acid, docosapentaenoic acid (N6, N3); phospholipids, sphingolipids; acylcarnitines
Fecal metabolome including lipidome, SCFA[a]	Acetate, propionate, butyrate, isobutyrate, isovalerate, lactate, formate, succinate, *p*-cresol, free fatty acids, triglycerides, phospholipids
Fecal metagenome community structure, diversity, keystone species, and functional genes	Community: Firmicutes, Bacteroidetes, Actinobacteria, Proteobacteria; alpha diversity, beta diversity, Shannon diversity, functional diversity
	Keystone spp: *Akkermansia muciniphila, Faecalibacterium prausnitzii, Lactobacillus acidophilus, Bifidobacterium bifidum*
	Functional genes [17]: *p*-cresol genes (e.g., urease, uricase, and tyrosine lyase); butyrate producing genes (e.g., phosphotransbutyrylase and butyrate kinase)

(continued)

Table 2
continued

Analyte class	Representative analytes
Fecal pathogen screen (targeted)	Bacteria: *Campylobacter* spp., *Clostridioides difficile*, *Salmonella* spp., *E. coli*
	Viruses: Norovirus, Rotavirus, etc.
	Parasites: *Giardia lamblia, Cyclospora cayetanensis, cryptosporidium* spp.
Fecal microbial metabolites derived from urine	*p*-Hydroxybenzoate, *p*-hydroxyphenylacetate, 2-hydroxyphenylacetate, 3-indoleacetate, arabinitol, tricarballylate, bile acids (secondary)
Heavy metals	Molybdenum (Mo), cadmium (Cd), arsenic (As), chromium (Cr), mercury (Hg), aluminum (Al), titanium (Ti), lead (Pb)
High resolution lipid panel	Total cholesterol (C), LDL-C, HDL-C, triglycerides, VLDL-C, LDL and HDL particle number and particle size, apolipoprotein A & B, lipoprotein(a), hsCRP
Immunome	Major histocompatibility complex (MHC) class I genes (HLA-A, HLA-B, HLA-C, B2M)
Infectious disease screen (as indicated)	Hep. B, Hep. C, HIV, West Nile Virus, Malaria
Inflammatory panel	hsCRP, rheumatoid Factor (RF), homocysteine, creatine phosphokinase (CPK), lactate dehydrogenase (LDH), fibrinogen, serum amyloid A, IL-1β, IL-2, IL-3, IL-4, IL-5, IL-6, IL-10, IL-12, IL-13, IL-17, IFN-γ, TNFα, CXCL-10, MCP-1, sICAM-1, sCD40L
Oxidative stress panel	Malondialdehyde, F2-isoprostane, 4-hydroxynonenal, 8- hydroxy-2'-deoxyguanosine, protein carbonyls, 3-nitro-tyrosine, glycoxidation adducts, lipoxidation adducts
Neuro panel	BNDF, GFAP, neurofilament light chain, calcium-binding protein S100β, ubiquitin C-terminal hydrolase L1, ENO2 (neuron-specific enolase 2), VEGF, VEGFD
Organic acids	Adipate, suberate, ethylmalonate, methylsuccinate, vanilmandelate, homovanillate, 5-hydroxyindoleacetate, kynurenine, kynurenic acid, quinolinate, pyruvate, lactate, citrate, alpha-ketoglutarate, succinate, fumarate, malate
Pharmacogenomic profiling	CYP450 Genes, UGT
Polygenic risk scores (PRS)	Based on systems or disease risk (e.g., Type II diabetes PRS; breast carcinoma PRS)
Radiation measures	Micronucleated lymphocytes, nuclear buds, nucleoplasmic bridges, chromosomal aberrations, H2AX, total nuclear uracil
Targeted SNP profiling	Selected single nucleotide polymorphisms
Trace elements	Calcium (Ca), magnesium (Mg), selenium (Se), zinc (Zn), manganese (Mn), cobalt (Co), copper (Cu), iron (Fe), sulfur (S), molybdenum (Mo)
Vitamins	A, C, D, E, K, thiamine, riboflavin, niacin, pantothenic acid, pyridoxine, folate, biotin, cobalamin, choline

[a]Representative examples: Note the analytes shown in each representative analyte class are shown only by example, as they represent a small number of analytes from very large analyte classes, which sometimes encompass tens of compounds (e.g., 10–100 or more)

Table 3

Pharmacogenomic panel of selected markers to assess drug-gene dynamics. The following summarizes a comprehensive set of genes that form the foundation of a clinical PGx panel that one would encounter in the commercial laboratory community. Each laboratory will incorporate similar core genes, though labs that specialize in a medical subdiscipline may tailor their panel (e.g., psychiatry, cardiology). The PGx panel can be used to develop a mission drug list that is personalized to each astronaut. From a PGx report, flight surgeons can develop a stratified, or crew-based, drug list, so that the mission formulary can be designed with greater precision. Reprinted with permission, Sovaris Aerospace

Name	Gene	Alleles
Cytochrome P450	CYP1A2	*1C, *1D, *1E, *1F, *1J, *1K, *1L, *1V, *1W
	CYP2B6	*4, *5, *6, *7, *9, *18, *22, *35, *36
	CYP2C Cluster	rs12777823
	CYP2C9	*2, *3, *4, *5, *6, *8, *11
	CYP2C19	*2, *3, *4, *4B, *10, *17
	CYP2D6	*2, *3, *4, *5, *6, *7, *8, *9, *10, *11, *12, *13, *14, *15, *17, *18, *19, *29, *31, *34, *35, *36, *39, *41, *42, *59, *63, *64, *65, *68, *69, *70, *82, *83, *91, *109, *114, *119, copy number variations
	CYP3A4	*1B, *22
	CYP3A5	*3, *6, *7
	CYP4F2	*3
Catechol-*O*-methyltransferase	COMT	Val158Met (rs4680)
Dihydropyrimidine dehydrogenase	DPYD	*2A, c.2846A>T, *13, HapB3, c.557A>G
Dopamine receptor D2	DRD2	-241A>G (rs1799978)
Coagulation factor II (prothrombin)	F2	20210G>A (rs1799963)
Coagulation factor V	F5	Arg534Gln (Leiden, 1691G>A)
Glutamate ionotropic receptor kainate type subunit 4	GRIK4	Intronic T>C variant (rs1954787)
Major histocompatibility complex, class I, A	HLA-A	HLA-A*31:01
Major histocompatibility complex, class I, B	HLA-B	HLA-B*15:02, HLA-B*57:01, HLA-B*58:01
5-Hydroxytryptamine (serotonin) receptor 2A	HTR2A	Intron 2, T>C variant (rs7997012)
5-Hydroxytryptamine (serotonin) receptor 2C	HTR2C	-759C>T (rs3813929)

(continued)

Table 3
continued

Name	Gene	Alleles
Interferon lambda 4 (formerly known as interleukin 28B)	IL28B (IFNL4)	Intronic C>T variant (rs12979860)
Methylenetetrahydrofolate reductase	MTHFR	c.665C>T (rs1801133); c.1298A>C (rs1801131)
Nudix hydrolase 15	NUDT15	Arg139Cys (rs116855232)
Opioid receptor mu 1	OPRM1	Asn40Asp (rs1799971)
Solute carrier family 6 member 4 (serotonin transporter)	SLC6A4	-1936A>G (rs25531); -1917_-1875del43 (rs774676466)
Solute carrier organic anion transporter family member 1B1	SLCO1B1	*5
Thiopurine *S*-methyltransferase	TPMT	*2, *3A, *3B, *3C, *4
UDP glucuronosyltransferase family 1 member A1	UGT1A1	*6, *28 (TA7)
Vitamin K epoxide reductase complex subunit 1	VKORC1	-1639G>A (rs9923231)

6 Molecular Class Assessment

When practicing PM in the spaceflight environment, it is useful to capture a suite of analytes that represents nodes in a series of metabolic networks that can be used to better characterize the astronaut phenotype with greater clarity. While it is beyond the scope of this chapter to outline the full profile of analytes, Table 2 provides a representative list of fully quantitative targeted analytes that is assembled based on molecular classes and can provide clinical decision support. Note that each analyte or analyte class requires consideration of ideal specimen type in order to yield the information most relevant to the clinical or performance question. Those specimens to be considered include but are not limited to saliva, serum, plasma, RBCs, WBCs, urine, 24-h urine, hair, toenails, breath, stool, and others. In many cases, the mission or purpose may determine which tissue, fluid, or compartment is optimal for a given analyte.

For instance, the authors prefer the reliability of red blood cells when assaying for fatty acid status. However, given the proper dietary controls, plasma may be suitable. In general, plasma reflects day-to-day status, while RBC reflects long-term fatty acid status over several months. Hepatic tissue may be optimum for transcriptomics of drug metabolizing enzymes. However, hepatic biopsy is not optimal for space and a less informative tissue such as PBMCs may be used for this purpose. Therefore, CYP450 transcriptome profiling from PBMCs will always have limitations. In general, careful attention should be paid to optimal specimen for a specific research or clinical question.

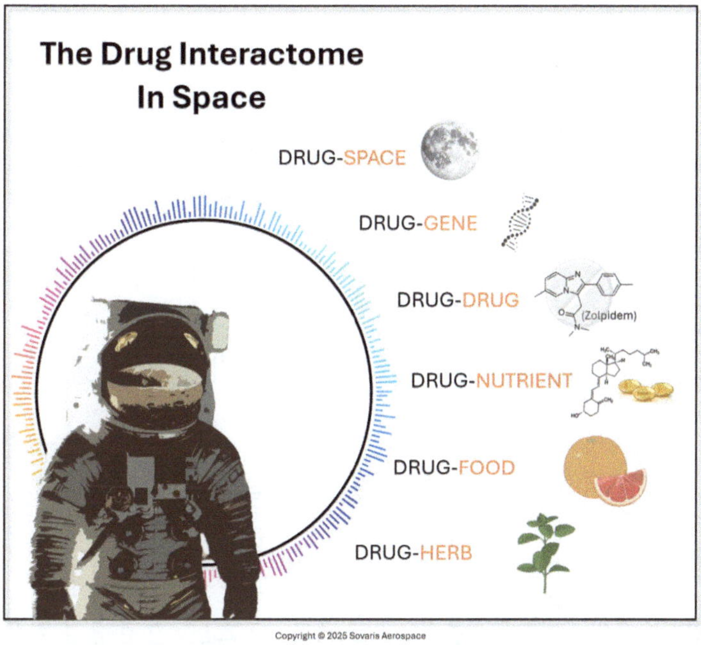

Fig. 3 Drug interactions relevant to human spaceflight. Pharmacogenomics can be seen as one central element in the approach to tailoring medication to individual astronauts in the practice of precision medicine in space. However, pharmacogenomics should be viewed and practiced within the contextual framework of a constrained drug interactome, which consists of drug–gene, drug–drug, drug–nutrient, drug–food, drug–herb, and drug–space interactions. For the purposes of this chapter on molecular analytics, the drug–gene interaction is emphasized Reprinted with permission, Sovaris Aerospace

7 Pharmacogenomic Assessment in Spaceflight

Pharmacogenomics (PGx) is the study of how drugs and genes interact to alter the drug response. It is a branch of pharmacology that uses genomic and genetic information to tailor drug selection, timing, and dosage for individuals. As clinicians develop specific countermeasures for individual astronauts, it is essential to the practice of PM in space to fully understand the detailed interactions of commonly prescribed therapeutics and known drug metabolizing enzymes, based on specific drug-gene dynamics.

Table 3 shows a representative PGx panel of selected genetic variants that affect an individual's drug metabolism. These same measures have been used to assemble a broad set of 608 known drug-gene interactions (to date) that are relevant to clinical practice and have been confirmed by regulatory bodies such as the US FDA [18]. A detailed examination of these genes can provide clinicians and flight surgeons with genotype-predicted interactions for common therapeutics across a range of drugs and indications. This can be used to build a mission drug list for each astronaut and crew.

Table 4
SNP variants affecting drug metabolism of selected medications used in space. The following represents examples of drug–gene dynamics of medications commonly used in human spaceflight, coupled with the clinical impact of each and relevance to the space environment. Reprinted with permission, Sovaris Aerospace

Drug [21]	Gene/enzyme	Key SNP variants	Clinical impact	Spaceflight relevance
Zolpidem (sleep aid)	*CYP3A4*	rs35599367 (*CYP3A422*)	↓ Metabolism → ↑ sedation, residual effects [21]	Sleep disruption is common in space, 44.7% of ISS crew use sleep aids
Promethazine (anti-nausea)	*CYP2D6*	rs3892097 (*CYP2D64*), rs35742686 (*CYP2D65*)	Poor metabolizers (PMs): ↑ sedation, ↓ efficacy [22]	Rescue medication for space motion sickness (SMS)
Ibuprofen (pain/anti-inflammatory)	*CYP2C9*	rs1799853 (*CYP2C92*), rs1057910 (*CYP2C93*)	↓ Metabolism → ↑ GI/ renal toxicity [23]	Used for pain
Metoprolol (heart rate control)	*CYP2D6*	rs3892097 (*CYP2D64*), rs1135840 (*CYP2D63*)	PMs: ↑ bradycardia; ultrarapid metabolizers (UMs): ↓ efficacy [21, 24]	Used for cardiovascular management
Acetaminophen (pain/fever)	*UGT1A6*	rs6759892 (T > G; Ser7Ala) rs2070959 A > G	↓ Glucuronidation → ↑ hepatotoxicity risk [25]	Common analgesic; altered absorption dynamics and bioavailability
Acetaminophen	*CYP2E1*	rs2031920 SNPs in GR-binding regions	Altered CYP2E1 metabolism of acetaminophen may form a toxic metabolite, N-acetyl-pbenzoquinone imine (NAPQI), which can lead to liver toxicity	May render drugs such as acetaminophen more toxic in space
Oxycodone (pain)	*CYP2D6*	rs3892097 (*CYP2D64*), rs1135840 (*CYP2D63*)	PMs: ↓ analgesia; UMs: ↑ opioid toxicity [27, 28]	Available, but less used in space due to side effects. May be given for severe pain (e.g., obstructing urinary calculus)

While the present discussion is focused on direct molecular analytics (e.g., SNP profiling and drug–gene interactions), pharmaceutical intervention in space should also examine a *constrained* portion of the drug-interactome, which does not require further molecular analytics. The term *interactome* is used to describe the whole set of molecular interactions taking place in a cell, such as methylation and chromatin state of a PGx gene. We use the term *constrained* interactome to delineate a specific set of molecular influences that are usually concomitant with drug administration in space. Included are drug–gene interactions, drug–drug interactions, drug–nutrient interactions, drug-food interactions, drug–herb interactions, and drug–space interactions. For our purposes, the term 'drug-space' interaction is used to note that numerous genes governing drug metabolism in space may be altered in their expression or behavior. While the drug does not interact directly with space, it is the dynamic of a drug interacting with space-altered genes that is referenced by this term. The constrained drug interactome in space is represented in Fig. 3, which is discussed in more detail in [2] and its direct relation to space in [19].

Beyond the foundational PGx panel, Table 4 introduces selected medications commonly available to astronauts aboard the ISS and examines key genes responsible for metabolizing these medications, key genetic variants (i.e., SNPs) that have a known effect on drug-specific metabolism, the clinical manifestations of the selected SNP variants (e.g., poor metabolizers vs ultra-rapid metabolizers), and their relevance to spaceflight. An expanded overview of pharmacogenomics in space is presented by Schmidt et al. [20].

8 Selected Single Nucleotide Polymorphism Assessment for Spaceflight Genomics

When astronauts embark upon short-, medium-, and long-duration spaceflight, complex intersecting dynamics exist between the genome, epigenome, and space, including structural changes in telomere length [29, 30]. Moving beyond PGx, one of the next key areas of genome assessment is centered on single nucleotide polymorphisms that impact nutrition and metabolism, the musculoskeletal system, cardiovascular system, nervous system, digestive system, and others. Deriving meaning from these measures requires a brief examination of some of the features of SNP profiling, its utility, its limitations, and basic concepts about SNP interaction with the space environment.

9 Targeted vs. Untargeted SNP Profiling

In the application of SNP profiling for human spaceflight, there are three common approaches that are employed. These are whole-genome sequencing, whole-exome sequencing, and targeted SNP profiling. Whole *genome* sequencing (WGS) covers the entirety of an astronaut's genome, including non-protein encoding regions. This data-heavy approach is more useful in research applications than in clinical decision support. The approach is called *untargeted* because the individual genes (analytes) are not preselected (though the degree of coverage is typically known).

Whole *exome* sequencing (WES) covers exons, which are regions of the genome transcribed into RNA (the primary focus being on the protein-coding regions). The exome is only about 2% of the total DNA, but it is the region that accounts for some 80% of disease-associated mutations [31, 32]. This approach is also called *untargeted* because the genes (analytes) are not preselected (other than the fact that the analysis is limited to the exome). Similar to WGS, WES is more commonly applied in research, though the clinical value is increased in areas such as spaceflight, because there are few astronauts on missions and such data are useful to better understand the human spaceflight response. In addition, specific genetic features (e.g., SNPs) can be extracted from the untargeted exome sequence data and used for clinical decision support, since genes of interest can be reliably quantified from the exome analysis. Both WGS and WES are considered important disciplines in spaceflight genomics. When coupled with multiomic assessment, such analysis can be highly informative [33–35].

Targeted SNP profiling refers to an analysis that is focused on gene variants that are preselected because of their known function, known utility, likelihood of actionability, and where sufficient levels of evidence may exist to warrant their use in a given setting. Ideally, the genes of choice are useful in understanding risk or directing one to an intervention (or countermeasure strategy). Presently, targeted SNP profiling is the more common application for precision medicine in human spaceflight, because it limits the number of genes selected and those selected genes that are more likely to have known clinical implications or clear paths of action.

When considering targeted SNP profiling for space, it is important to first consider that the potential predictive value of a given SNP should always be examined in relation to how it may or may impact the phenotype. It should have some predictive value or be useful in developing a strategy that favorably impacts health, safety, performance, or survival. Second, a given SNP should, where indicated, be viewed in the context of the actual protein, metabolites, or molecular network associated with a given SNP, or the phase of one variant with another in the same genome. In this context, measurement of the SNP or SNPs is most informative when they are used to characterize molecular network behavior that can be used to guide an action.

For example, to properly characterize the one carbon molecular network using SNPs, one should understand:

- SNPs: MTHFR, MTR, MTRR, SHMT, CBS, PEMT, etc.
- Nutrient Inputs: Vitamin B2 (riboflavin), B6 (pyridoxine), B12, choline, trimethylglycine, and folate
- Analytes: Homocysteine, cystathionine, cysteine, taurine, and glutathione

In practice, we always strive to incorporate untargeted genomics and targeted molecular analytics in our spaceflight applications. This provides the ability to study the human response across all missions (e.g., using WGS and WES), while using targeted approaches to develop personalized countermeasures in our precision medicine applications in space. This coupling of targeted and untargeted methods are valuable in advancing both research and clinical decision support. It is also useful when the desire is to compare the human response across different mission designs.

For the purposes of this overview, the focus will be on targeted SNP profiling. In this regard, it should be noted that there are varying levels of evidence supporting the SNP summaries that follow. However, the authors see value in outlining a range of SNPs that are useful on Earth today, along with those that may have utility in the future. Finally, the present overview is focused on assessment, rather than treatment. The reader is invited to explore the individual SNPs outlined herein and independently explore the levels of evidence, the association with clinical phenotypes, and the actionable elements associated with a given SNP. We also direct readers to our previous work in Rutter et al. for a discussion of protective and risk alleles in human spaceflight [36].

10 Contextualizing the Use of SNP Profiling in Space PM

The use of SNP profiling in PM in human spaceflight should be contextualized in relation to both its utility and its limitations. In one consideration, selected gene variants may have *directly actionable steps, based on metabolic network dynamics*. For example, homozygous fatty acid desaturase 1 (FADS1) adversely impacts the desaturation of polyunsaturated fatty acids to their more highly unsaturated metabolites [e.g., conversion of alpha-linolenic acid to eicosapentaenoic acid (EPA) and docosahexaenoic acid (DHA) by sequential addition of a double bond]. Variants in the FADS1 [37] gene (particularly rs174537) and the attendant change in the FADS1 enzyme have been consistently associated with lower serum/plasma EPA and DHA. For those with this genotype, this knowledge argues for a greater need for preformed EPA/DHA via the diet or supplementation as a spaceflight countermeasure.

SNP profiling may also reveal *risk phenotypes* relevant to the *clinical response to space*. For instance, those with the APOE4/4 genotype who undergo chemotherapy have been shown to be more susceptible to cognitive decline (e.g., visual memory, working memory, processing speed, etc.) than those not possessing the e4 allele [38–40]. Those with the APOE4/4 genotype have also been shown to have greater decline in learning and memory, executive function, and the Clinical Trial Battery Composite score after treatment with whole brain radiation therapy compared with APOE e4 noncarriers [41–43]. This increased susceptibility to cognitive decline when exposed to an exogenous insult raises questions as to whether space radiation exposure in APOE4 carriers confers a greater risk of cognitive decline in space (colloquially referred to as *space stupids*) compared to noncarriers. It also raises the issue of whether APOE genotyping should be used for astronaut screening and selection, or be used to develop countermeasures in those who choose to proceed into space with this trait. It further raises the question, "might APOE2/2 one day be considered a protective genotype in space?"

Adding further to the complexity and to the potential value of genotyping, one can consider the *converging variables* of the APOE genotype and fatty acid metabolism, just noted above. For example, Tomaszewski et al. studied APOE genotype status in adults ingesting 2 g DHA daily for 18 months. Non-APOE4 carriers experienced a greater increase in plasma and cerebrospinal fluid DHA/AA. *APOE* $\varepsilon 2/\varepsilon 3$ carriers ingesting DHA also exhibited a 10% less decline in left hippocampal volume than *APOE* $\varepsilon 4/\varepsilon 4$ carriers. Meanwhile, *APOE* $\varepsilon 2/\varepsilon 3$ carriers and *APOE* $\varepsilon 2/\varepsilon 4$ carriers experienced significantly less decline in the right hippocampal volume than *APOE* $\varepsilon 4/\varepsilon 4$ carriers [44].

In contemplating the use of molecular profiling methods, it merits consideration as to how single variants may impact the phenotype, but also how two, three, several, or many converging variants begin to scale across molecular networks. When these scaling dynamics occur in the context of the environmental extremes of space, they may grow further in their collective impact.

11 The Nascent Field of Transcriptomic Profiling in Space PM

While SNP profiling often yields useful information that provides insights for use on Earth, it does not account for the gene expression changes that occur over time in the space environment. In recent years, the application of transcriptomic profiling (RNA sequencing) has yielded greater insight into gene expression changes in space. However, the limited number of subjects, limited number of tissue types sequenced, and the variable mission durations yield a complex picture that has not, yet, led to clear path in how the transcriptome should be converted into clinical action. The field of epigenetics in space is subject to similar limitations in its clinical translation. Given that both transcriptomic and epigenetic changes modify the impact of a given SNP in space, the utility of SNP profiling for application of PM on Earth should be viewed with attention to the degree of unknowns introduced once the astronaut is in space.

One can illustrate this dynamic by examining transcriptome changes in recent long-duration human space missions. Moreno-Villanueva et al. found that, in comparison to pre-flight, a total of 2030 genes were differentially expressed in PBMCs collected between 135 and 210 days in orbit. Not unexpectedly, the degree of RNA changes was highly variable between crew members, though the degree to which individual SNPs contributed to this variability is not known (since, SNP profiling was not reported) [45]. In the NASA Twins Study of One Year in Space (340 days), there were 8564 differentially expressed genes observed (in PBMCs) in the 6–12-month inflight period compared with the Earth control (though the number varied by cell type) [29]. This suggests that longer space missions may result in even larger changes in gene expression patterns.

Using the FADS1 example, as above, we explored the changes in FADS1 gene expression in PBMCs, during the 1 year NASA Twins Study [46]. While SNP data on the individual astronauts were not released due to privacy concerns, there were only slight increases in FADS1 RNA (log2 change of 0.5) in astronaut PBMCs, and no significant differences in CD4, CD8, or CD19 cells. It warrants noting that this study also bears a limitation, in that PBMCs reflect only one tissue. While far more difficult to obtain, a hepatic transcriptome may have been more representative of the true changes that impact fatty acid metabolism in space. Future studies will be needed with greater numbers of subjects where SNP profiling is analyzed in conjunction with transcriptomics in the most relevant tissue to better understand the extent to which a given SNP is diminished or enhanced in the space environment.

Such findings merely illustrate the point that SNP profiling for clinical decision support in spaceflight PM has utility, but we are not yet sufficiently advanced in the field to the point where we can assert with confidence how a given genotype might behave while in space. However, we can assert that there is great utility in using SNP profiling on Earth in an effort to optimize astronauts in their preparation for entering space and in addressing unique needs following return to Earth.

12 A Sample of SNPs Relevant to Precision Medicine on Earth: Considerations for Space PM

The following tables represent SNPs that have been employed in terrestrial precision medicine that are associated with health and performance phenotypes that we assert may have relevance to human

Table 5
SNPs associated with nutrient status and metabolism. The following demonstrates selected genes that impact vitamin and trace element metabolism with relevance to human health and performance in space. Reprinted with permission, Sovaris Aerospace

Nutrient	Gene	SNP	Variant impact on metabolism	Key findings
Vitamin D	DHCR7/NADSYN1	rs12785878	Lower 25(OH)D levels, ↑ risk of insufficiency	GG genotype linked to ↓ vitamin D [47]
	CYP2R1	rs10741657 (Caucasian and Asian), rs12794714 (African Americans)	Impaired hydroxylation of vitamin D	GG genotype (GA in African Americans) associated with ↓ 25(OH)D [47]
	GC	rs2282679	↓ Vitamin D binding protein (DBP)	Risk allele linked to ↓ 25(OH)D and ↑ deficiency risk [47]
Vitamin E	TTPA	rs6994076 (980A/T)	↓ α-tocopherol levels	TT genotype associated with ↓ enzyme activity and ↓ response to supplementation [47]
	CYP4F2	rs2108622	Impaired vitamin E catabolism	TT genotype linked to ↑ α-tocopherol retention [47]
Vitamin C	SLC23A1	rs33972313	↓ Vitamin C absorption	Minor allele linked to ↓ plasma vitamin C [47]
	SLC23A2	rs1279683 (A>G)	↓ Plasma vitamin C levels	GG genotype associated with ↓ vitamin C [47]
	Hp (Haptoglobin)	Hp2-2 phenotype	↓ vitamin C due to instability	Hp2-2 individuals have ↓ serum vitamin C [47]
Folate	MTHFR	rs1801133 (C677T)	↓ Folate metabolism, ↑ homocysteine	TT genotype linked to reduced enzyme activity and ↓ folate absorption [47]
Vitamin B12	TCN2	rs1801198	↓ Vitamin B12 transport	C allele (CC genotype) linked to ↓ B12 levels [47]
Calcium	VDR	rs4516035	↓ Calcium absorption	Associated with vitamin D receptor dysfunction [48]
	COL1A1	rs1800012	↓ Bone mineral density	Linked to calcium deficiency and osteoporosis risk [48]

Iron	TFR2	rs7385804	↓ Serum iron	C/C genotype associated with ↓ iron [48]
	TF	rs3811647	Altered iron transport	A/A genotype linked to ↑ ferritin [48]
	HFE	rs1800562	Disrupts sulfide bond formation and intracellular transport of HFE, and decreases hepcidin expression	Systemic iron overload. May be advantageous in terrestrial performance but infer greater biomedical risk in space [3].
		rs1799945	Altered cell surface signaling and cellular uptake	Moderate reduction in signaling and increased iron accumulation (exacerbated by cofactors). May be advantageous in terrestrial performance but infer greater biomedical risk in space [3].
		rs1800730	Less pronounced functional impacts compared to other HFE mutations.	Mild iron accumulation. Considered nonclinically significant in terrestrial medicine. May infer greater biomedical risk in space [49].
Fatty acid desaturation	FADS1	rs174537	↓ Endogenous synthesis of EPA and DHA	Minor alleles seem to have lower EPA and DHA levels [37]
Magnesium	MUC1	rs4072037	Impaired magnesium reabsorption	Associated with ↓ magnesium levels [50]
Selenium	SEPP1	rs3877899	↓ Selenoprotein synthesis	Linked to ↓ selenium status [48]
Zinc	SLC39A8	rs13107325	↓ Zinc transport	Associated with zinc deficiency and immune dysfunction [48]
	SLC30A8	rs13266634	Impaired insulin secretion, ↑ diabetes risk	Linked to ↓ zinc levels [48]

Table 6

Key SNP variants relevant in performance environments. The following demonstrates selected genes that impact human performance in space. Reprinted with permission, Sovaris Aerospace

Gene	SNP	Performance trait	Biological mechanism	Key findings
ACTN3	rs1815739 (R577X)	Power/strength	α-Actinin-3 expression in fast-twitch muscle fibers	RR genotype associated with sprint/power athletes. XX genotype linked to endurance athletes [51, 52]
ACE		Endurance	Regulates angiotensin-converting enzyme (ACE)	II genotype associated with endurance; DD genotype linked to power [52]
PPARGC1A (PGC-1α)	rs8192678	Aerobic capacity	Mitochondrial biogenesis, energy metabolism	CC genotype linked to higher VO_2max, endurance performance, more slow-twitch muscle fibers, and increased mitochondria activity [53]
AQP1	rs1049305	Endurance running	Aquaporin-1 water channel in erythrocytes	C allele carriers show faster running times [54, 55]
IL6	rs1474347	VO_2max gains	Interleukin-6 signaling during inflammation	G allele associated with increased VO_2 max [56]
COL5A1	rs12722	Injury risk	Collagen V synthesis in tendons/ligaments	Genotypes are associated with varying levels of injury risk in differing populations. [57, 58]
ADRB3	rs4994 (Trp64Arg)	Energy metabolism	β3-adrenergic receptor in fat oxidation	Arg64 allele associated with endurance performance [59]
HFE	rs1799945 (C282Y) rs1800562 (H63D)	Iron metabolism	Regulates iron absorption/storage	High risk allele linked to improved endurance and increased VO_{2peak} via iron homeostasis [60]
GABPB1	rs12594956	Mitochondrial function	Regulates mitochondrial gene expression	A allele associated with endurance performance [61]
MMP3	rs591058	Injury susceptibility	Matrix metalloproteinase-3 in tendon remodeling	Associated with higher risk of tendon injuries [62]

FTO	rs9939609 (T>A)	Body composition	Fat mass and obesity-associated gene	AA genotype associated with lower BMI and lean muscle mass [62]
BDKRB2	rs5810761	Improved oxygen regulation	Vasodilation/vasoconstriction, increases glucose uptake in skeletal muscle and blood flow in muscles during exercise	+9/+9 genotype is associated with endurance performance (Ipekoglu, 2024), −9/−9 genotype is associated with greater response in exercise induced cytokines (e.g., IL-10), BDNF, and muscle recovery [63]
EPOR	W439X	High oxygen transport	Increased red blood cell production, hemoglobin and hematocrit, increased mitochondrial biogenesis, and increased nitric oxide production.	W439X causes beneficial symptoms of erythrocytosis [36]
PDE10A	rs3008052	Improved oxygen regulation	Increased erythropoiesis through thyroid hormone (T4) regulation and enhanced oxygen storage	rs3008052 is associated with increased spleen size and erythrocytic cell volume [64]

Table 7

Clinically relevant SNP variants across physiological systems. The following contains representative genes that impact various systems that may be impacted in the space environment. Reprinted with permission, Sovaris Aerospace

System	Gene	SNP	Associated trait/disorder	Biological mechanism	Source
Cognition/behavior	COMT	rs4680 (Val158Met)	Executive function, working memory, schizophrenia, ADHD	Alters dopamine degradation in prefrontal cortex, Met allele reduces enzyme activity	[65, 66]
	BDNF	rs6265 (Val66Met)	Memory, depression, neural plasticity	Reduces BDNF secretion; impacts hippocampal function and synaptic plasticity.	[65, 67]
	MTHFR	rs1801133 (C677T)	Cognitive impairment, increased homocysteine levels	Reduces folate metabolism, increasing homocysteine (neurotoxic).	[68]
	IL-8	rs4073	Suicidal ideation, inflammation	Alters IL-8 expression; linked to neuroinflammatory pathways.	[69]
Cardiac/circulation	PHACTR1	rs12526453	High LDL, coronary artery disease risk, blood pressure, Increased survival following myocardial infarction (Szpakowicz, 2015)	Modulates endothelial function vascular health	[70]
	VEGFA	rs2010963	Angiogenesis, CVD risk pathologies	Alters VEGF-A levels; impacts vascular repair and lipid metabolism.	[71]
	Factor V Leiden	rs6025	Clotting risk (venous)	Risk to deep vein thrombosis (DVT) and pulmonary embolism (PE); Heterozygotes (Het): 3–8 increased risk; Homozygotes (Hm): 50–-100× increased risk (Note: women on oral contraceptives: Het = 38× and Hm = 100× increased risk)	[72, 73]
	ANGPTL3	rs1748195	Less atherosclerosis	LOF SNP leads to reduced ANGPTL3, and thus, increased LPL and EL activity → protective for atherosclerosis and coronary artery disease and MI	[74]

	APOA1	rs670	Less cardiovascular disease (CVD)	Leads to increased ApoA1, especially in HDL particles, which facilitates reverse cholesterol transport	[75]
	PCSK9	rs11591147	Less CVD, known harmful effect	Lowers LDLR degradation enhances LDL clearance	[76]
	CCR5		Less heart disease and infections, known harmful effect	Immune modulation, reduced inflammation, and higher HDL	[77]
Connective tissue	SUMF1	rs3864051	COPD, lung ECM remodeling	Disrupts sulfation balance in extracellular matrix proteins	[78]
	COL5A1	rs12722	Tendon/ligament integrity	Affects collagen V synthesis; CC genotype lowers injury risk.	[79, 80]
Respiratory/lung	DNAH11	rs75679995	Lung function decline	Impacts ciliary function in airway epithelia; linked to COPD progression.	[81]
	SUMF1	rs793391	COPD, airway obstruction	Reduces sulfatase activity, impairing ECM stability in lung tissue.	[78]
Inflammation	TNFα	rs1800629 (G-308A)	Inflammatory diseases, neurological deficits	Increases TNFα production; linked to chronic inflammation and neuropsychiatric disorders.	[82]
	IFNG-AS1	rs7134599	IBD, autoimmune disorders	Regulates IFNG expression; enhances proinflammatory cytokine production.	[83, 84]
	IL-6	rs1474347	Chronic inflammation, muscle recovery	Modulates IL-6 signaling; G allele enhances post-exercise VO2max gains.	[56]
Ocular	SHMT1	rs1979277	Ocular edema in space	SHMT1 1420 CC associated with all but one case of disc edema after flight. Spaceflight Associated Neuro-ocular Syndrome	[85]
	MTRR	rs1801394	Choroidal folds, cotton wool spots in space	All of the astronauts with the homozygous MTRR 66 GG genotype exhibited choroidal folds and cotton wool spots. Spaceflight Associated Neuro-ocular Syndrome	[85]

Table 8

SNP variants relevant to muscle and bone. The following contains representative genes that impact muscle and bone, which may be relevant in the space environment. Reprinted with permission, Sovaris Aerospace

Gene	SNP	Performance trait	Biological echanism	References
ACTN3	rs1815739 (R577X)	Power/Strength	α-Actinin-3 expression in fast-twitch muscle fibers. RR genotype associated with sprint/power athletes. XX genotype linked to endurance athletes	[51, 52]
SPTBN1	rs13024191	Bone mineral density (BMD), muscle health	Regulates TGFβ/Smad3 signaling; critical for osteoblast function.	[86]
TRIM63	rs3008232	Muscle atrophy (sarcopenia)	Encodes MuRF1, a key E3 ubiquitin ligase in protein degradation.	[87]
LRP5	rs4988321	Bone retention; rheumatoid arthritis susceptibility	Alters Wnt/β-catenin activity, impacting osteoblast proliferation and bone formation.	[88, 89]
SOST	rs2741856	Bone retention, fracture risk	Impacts Wnt signaling, leading to altered osteoblast differentiation. Loss of Wnt inhibition leads to increased bone formation	[90, 91]
MAFBX	n/a	Muscle retention, muscle atrophy	Muscle-specific E3 ubiquitin ligase, enhances proteolysis and suppresses protein synthesis.	[92]
MURF1	rs2275950	Muscle retention, muscle protein degradation	Muscle specific E3 ubiquitin ligase involved in protein homeostasis. AA homozygotes are associated with increased muscle strength and reduced muscle soreness.	[93, 94]
MSTN	rs1805086	Muscle growth and cardiac output	MSTN modulates proliferation and differentiation of myoblasts. Increased frequency of R allele in strength-oriented athletes. A55T SNP (rs180565) with the AT genotype is associated with increased hypertrophy and faster recovery.	[95]

Table 9

SNP variants linked to performance and recovery. The following contains representative genes that may influence injury and recovery in the space environment. Reprinted with permission, Sovaris Aerospace

Gene	SNP	Recovery trait	Biological mechanism	Key findings
GABPB1	rs7181866	Mitochondrial adaptation	Regulates mitochondrial biogenesis and function	G allele associated with combat-sport athletic status, especially in world-class athletes [96]
	rs8031031	Mitochondrial adaptation	Regulates mitochondrial biogenesis and function	T allele possibly linked to enhanced intermittent exercise capacity [96]
COL1A1	rs1800012	Muscle strength recovery	Impacts type I collagen synthesis in tendons/muscles	T allele carriers showed slower recovery of muscle strength TT genotype linked to substantially lower risk of musculoskeletal soft tissue injuries [97]
	rs2249492	Muscle strength recovery	Impacts type I collagen synthesis in tendons/muscles	TT genotype linked to weaker muscle strength and muscle soreness, indicative of postexercise-induced damage [93]
COL2A1	rs2070739	Muscle soreness	Affects type II collagen in cartilage/joints	A allele carriers reported greater postexercise soreness [93]
COL5A1	rs12722	Tendon/ligament recovery	Modulates type V collagen in connective tissues	C allele decreases injury risk [80]

(continued)

Table 9
continued

Gene	SNP	Recovery trait	Biological mechanism	Key findings
DSG1	rs10502567	Injury resilience	Desmoglein-1 in epithelial/cell adhesion	Linked to elite performance, may enhance tissue integrity [98]
PAPPA2	rs11580456	Significantly associated with injury	PAPPA2 in ECM remodeling; regulates bioavailability of IGF	Involved in diseases and development; increased Injury risk [98]
MAS1	rs220735	Muscle/joint injury susceptibility	Angiotensin receptor signaling; involved in smooth muscle relaxation, cardioprotection, and regulation of blood pressure	Associated with injury risk [98]
COL22A1	rs3924862	Muscle Injury Risk	Responsible for producing collagen, stabilizing muscle-tendon junctions, and skeletal muscle attachments during contractions	Increased injury risk [98]

spaceflight. These should be viewed as practical examples, as there are likely to be varying opinions between clinicians and researchers on the ideal SNPs most appropriate for application in space PM. Of further note, clinical laboratories often present "levels of evidence" analysis for each of the SNPs in a patient report. These should be taken into consideration when making clinical decisions about countermeasures derived from a given SNP status. As noted previously, metabolite and sometimes protein levels add further clarity to SNP results.

Note that the series of tables that follow (Tables 5–9) are generally not preceded by introductory narrative remarks. As such, this section is intended as a set of reference tables for SNPs to be considered for the specific systems or indications as noted in each respective table heading.

13 Discussion

The present discussion represents a brief overview of a highly complex subject, which is to consider selected principles and molecular targets important in the practice of precision medicine in human spaceflight. While there are several approaches one may take to PM in space, the approach described, herein, is focused on a systems biology construct based upon inputs, processes, and outputs. The inputs are further refined by considering essential, conditionally essential, and nonessential inputs.

Precision (and personalized) medicine in space balances the dynamic between diagnosis and pattern analysis. While the focus on pattern analysis does not in any way preclude diagnosis, the analysis of patterns of deficiency, excess, and dysregulation in molecular networks provides a clinician with a fundamental tool for "gap analysis" that naturally lends itself to a more precise description of biological meaning, which leads to knowledge generation suitable for informed countermeasure development. In practice, this approach typically takes the form of addressing substrates, precursors, cofactors, and intermediates, which are often in the form of micronutrients and macronutrients.

This "entry point" into the PM process adds further refinement to the use of drugs in space. It is not uncommon to observe that, when molecular network balance is restored, the need for drugs is reduced or eliminated. When drugs remain indicated, their administration can be done with greater precision by applying the *constrained drug interactome* approach of precision medicine described herein. At its core, this guides a formal approach to drug prescribing, based on attention to drug–gene, drug–drug, drug–nutrient, drug–food, drug–herb, and drug-space interactions. At the foundation of this approach is pharmacogenomics, which provides for the tailoring of drugs to crews, missions, and individuals. Beyond PGx, selected (targeted) SNP profiling can be used to further understand risk and guide countermeasure development.

It is the opinion of the authors that PM assessment should be at the foundation of astronaut training and preparation for all space agencies and spaceflight organizations. Longitudinal measures provide a means of following the trajectory of astronaut progress during training and preparation for missions, while affording the ability to make adjustments based on laboratory analytics over time. This allows for the refinement of countermeasure packages at selected intervals. A central objective is to optimize the astronaut mission package, so that their unique needs can be addressed while in space. When astronauts return to Earth, comprehensive assessment and longitudinal postmission measures can be conducted, so that incremental recommendations aimed at optimized recovery can be carefully directed and tailored.

In practice and in principle, we proceed from the axiom that *no astronaut should have to enter the space operating environment with knowable, actionable metabolic deficits* that can adversely impact performance, recovery, health, safety, and survival.

References

1. Afshinnekoo E, Scott RT, MacKay MJ et al (2020) Fundamental biological features of spaceflight: advancing the field to enable deep-space exploration. Cell 183:1162–1184. https://doi.org/10.1016/j.cell.2020.10.050
2. Schmidt MA, Jones JA, Mason CE (2023) Optimizing human performance in extreme environments through precision medicine: from spaceflight to high-performance operations on Earth. Camb Prism Precis Med 1:e27. https://doi.org/10.1017/pcm.2023.16
3. Schmidt MA, Schmidt JC (2025) Using functionally characterized networks in the application of precision medicine in spaceflight. In: Schmidt MA, Legato MJ (eds) Building a space-faring civilization. Elsevier, New York, NY, pp 217–243
4. Kendler BS (2006) Supplemental conditionally essential nutrients in cardiovascular disease therapy. J Cardiovasc Nurs 21:9–16. https://doi.org/10.1097/00005082-200601000-00004
5. Cheng X, Blumenthal RM (1999) S-adenosylmethionine-dependent methyltransferases. World Scientific, Hackensack, NJ
6. Zhang C, Sultan SA, Rehka T, Chen X (2021) Biotechnological applications of S-adenosyl-methionine-dependent methyltransferases for natural products biosynthesis and diversification. Bioresour Bioprocess 8:72. https://doi.org/10.1186/s40643-021-00425-y
7. Luo H, Hansen ASL, Yang L et al (2019) Coupling S-adenosylmethionine–dependent methylation to growth: design and uses. PLoS Biol 17(3):e2007050. https://doi.org/10.1371/journal.pbio.2007050
8. Andreini C, Putignano V, Rosato A, Banci L (2018) The human iron-proteome. Metallomics 10:1223–1231. https://doi.org/10.1039/c8mt00146d
9. de Baaij JHF, Hoenderop JGJ, Bindels RJM (2015) Magnesium in man: implications for health and disease. Physiol Rev 95:1–46. https://doi.org/10.1152/physrev.00012.2014
10. Ames BN, Elson-Schwab I, Silver EA (2002) High-dose vitamin therapy stimulates variant enzymes with decreased coenzyme binding affinity (increased Km): relevance to genetic disease and polymorphisms. Am J Clin Nutr 75:616–658. https://doi.org/10.1093/ajcn/75.4.616
11. Kirkland JB (2012) Niacin requirements for genomic stability. Mutat Res 733:14–20. https://doi.org/10.1016/j.mrfmmm.2011.11.008
12. Schomburg L (2012) Selenium, selenoproteins and the thyroid gland: interactions in health and disease. Nat Rev Endocrinol 8:160–171. https://doi.org/10.1038/nrendo.2011.174
13. Ramagopalan SV, Heger A, Berlanga AJ et al (2010) A ChIP-seq defined genome-wide map of vitamin D receptor binding: associations with disease and evolution. Genome Res 20:1352–1360. https://doi.org/10.1101/gr.107920.110
14. Mishima E, Wahida A, Seibt T, Conrad M (2023) Diverse biological functions of vitamin K: from coagulation to ferroptosis. Nat Metab 5:924–932. https://doi.org/10.1038/s42255-023-00821-y
15. Andreini C, Banci L, Bertini I, Rosato A (2006) Counting the zinc-proteins encoded in the human genome. J Proteome Res 5:196–201. https://doi.org/10.1021/pr050361j
16. Maret W (2013) Zinc biochemistry: from a single zinc enzyme to a key element of life. Adv Nutr 4:82–91
17. Saito Y, Sato T, Nomoto K, Tsuji H (2018) Identification of phenol- and p-cresol-producing intestinal bacteria by using media supplemented with tyrosine and its metabolites. FEMS Microbiol Ecol 94:1–11. https://doi.org/10.1093/femsec/fiy125
18. U.S. Food & Drug Administration (2024) Table of pharmacogenomic biomarkers in drug labeling. https://www.fda.gov/drugs/science-and-research-drugs/table-pharmacogenomic-biomarkers-drug-labeling
19. Schmidt MA, Schmidt JC, Schmidt CM (2026) Precision medicine in human spaceflight: applying systems thinking to prinicples, concepts, and methods assessments. In: Waisberg E (ed) Fundamentals of space medicine and clinical technology. Elsevier, London,UK
20. Schmidt MA, Schmidt CM, Goodwin TJ (2019) Pharmacogenomics in spaceflight. In: Pathak Y, Araújo dos Santos M, Zea L (eds) Handbook of space pharmaceuticals. Springer International Publishing, Cham, pp 1–39
21. (2023) NASA-STD-3001 Technical Brief OCHMO-TB-006: Pharmaceutical Care. Washington, D.C

22. Foster A, Mobley E, Wang Z (2007) Complicated pain management in a CYP450 2D6 poor metabolizer. Pain Pract 7:352–356. https://doi.org/10.1111/j.1533-2500.2007.00153.x
23. Krasniqi V, Dimovski A, Domjanović IK et al (2016) How polymorphisms of the cytochrome P450 genes affect ibuprofen and diclofenac metabolism and toxicity/Kako polimorfizmi gena citokroma P450 utječu na metabolizam i toksičnost ibuprofena i diklofenaka. Arh Hig Rada Toksikol 67:1–8. https://doi.org/10.1515/aiht-2016-67-2754
24. Collett S, Massmann A, Petry NJ et al (2023) Metoprolol and CYP2D6: a retrospective cohort study evaluating genotype-based outcomes. J Pers Med 13:416. https://doi.org/10.3390/jpm13030416
25. Cerezo-Arias M, de las O, Gómez-Tabales J, Martí M et al (2022) Common UGT1A6 variant alleles determine acetaminophen pharmacokinetics in man. J Pers Med 12:720. https://doi.org/10.3390/jpm12050720
26. Hu W, Jiang C, Kim M et al (2021) Individual-specific functional epigenomics reveals genetic determinants of adverse metabolic effects of glucocorticoids. Cell Metab 33:1592–1609.e7. https://doi.org/10.1016/j.cmet.2021.06.004
27. Huddart R, Clarke M, Altman RB, Klein TE (2018) PharmGKB summary: oxycodone pathway, pharmacokinetics. Pharmacogenet Genomics 28:230–237. https://doi.org/10.1097/FPC.0000000000000351
28. Stamer UM, Zhang L, Book M et al (2013) CYP2D6 genotype dependent oxycodone metabolism in postoperative patients. PLoS One 8:e60239. https://doi.org/10.1371/journal.pone.0060239
29. Garrett-Bakelman FE, Darshi M, Green SJ et al (2019) The NASA Twins study: a multidimensional analysis of a year-long human spaceflight. Science (1979) 364(6436):eaau8650. https://doi.org/10.1126/science.aau8650
30. Garcia-Medina JS, Sienkiewicz K, Narayanan SA et al (2024) Genome and clonal hematopoiesis stability contrasts with immune, cfDNA, mitochondrial, and telomere length changes during short duration spaceflight. Precis Clin Med 7(1):pbae007. https://doi.org/10.1093/pcmedi/pbae007
31. Aspden JL, Wallace EWJ, Whiffin N (2023) Not all exons are protein coding: addressing a common misconception. Cell Genomics 3:100296. https://doi.org/10.1016/j.xgen.2023.100296
32. Choi M, Scholl UI, Ji W et al (2009) Genetic diagnosis by whole exome capture and massively parallel DNA sequencing. Proc Natl Acad Sci U S A 106:19096–19101. https://doi.org/10.1073/pnas.0910672106
33. Kim J, Tierney BT, Overbey EG et al (2024) Single-cell multi-ome and immune profiles of the Inspiration4 crew reveal conserved, cell-type, and sex-specific responses to spaceflight. Nat Commun 15:4954. https://doi.org/10.1038/s41467-024-49211-2
34. Tierney BT, Kim J, Overbey EG et al (2024) Longitudinal multi-omics analysis of host microbiome architecture and immune responses during short-term spaceflight. Nat Microbiol 9:1661–1675. https://doi.org/10.1038/s41564-024-01635-8
35. Park J, Overbey EG, Narayanan SA et al (2024) Spatial multi-omics of human skin reveals KRAS and inflammatory responses to spaceflight. Nat Commun 15:4773. https://doi.org/10.1038/s41467-024-48625-2
36. Rutter LA, MacKay MJ, Cope H et al (2024) Protective alleles and precision healthcare in crewed spaceflight. Nat Commun 15:6158. https://doi.org/10.1038/s41467-024-49423-6
37. Loukil I, Mutch DM, Plourde M (2024) Genetic association between FADS and ELOVL polymorphisms and the circulating levels of EPA/DHA in humans: a scoping review. Genes Nutr 19:11. https://doi.org/10.1186/s12263-024-00747-4
38. Fernandez HR, Varma A, Flowers SA, Rebeck GW (2020) Cancer chemotherapy related cognitive impairment and the impact of the Alzheimer's disease risk factor APOE. Cancers (Basel) 12:3842. https://doi.org/10.3390/cancers12123842
39. Ahles TA, Li Y, McDonald BC et al (2014) Longitudinal assessment of cognitive changes associated with adjuvant treatment for breast cancer: the impact of *APOE* and smoking. Psychooncology 23:1382–1390. https://doi.org/10.1002/pon.3545
40. Mandelblatt JS, Small BJ, Luta G et al (2018) Cancer-related cognitive outcomes among older breast cancer survivors in the thinking and living with cancer study. J Clin Oncol 36:3211–3222. https://doi.org/10.1200/JCO.18.00140
41. Cramer CK, Cummings TL, Andrews RN et al (2019) Treatment of radiation-induced cognitive decline in adult brain tumor patients. Curr Treat Options Oncol 20:42. https://doi.org/10.1007/s11864-019-0641-6
42. Makale MT, McDonald CR, Hattangadi-Gluth JA, Kesari S (2017) Mechanisms of radiotherapy-associated cognitive disability in patients with brain tumours. Nat Rev Neurol 13:52–64. https://doi.org/10.1038/nrneurol.2016.185
43. Wefel JS, Deshmukh S, Brown PD et al (2024) Impact of apolipoprotein E genotype on neurocognitive function in patients with brain metastases: an analysis of NRG oncology's RTOG 0614. Int J Radiat Oncol Biol Phys

119:846–857. https://doi.org/10.1016/j.ijrobp.2023.12.004
44. Tomaszewski N, He X, Solomon V et al (2020) Effect of APOE genotype on plasma docosahexaenoic acid (DHA), eicosapentaenoic acid, arachidonic acid, and hippocampal volume in the Alzheimer's disease cooperative study-sponsored DHA clinical trial. J Alzheimer's Dis 74:975–990. https://doi.org/10.3233/JAD-191017
45. Moreno-Villanueva M, Jimenez-Chavez LE, Krieger S et al (2025) Transcriptomics analysis reveals potential mechanisms underlying mitochondrial dysfunction and T cell exhaustion in astronauts' blood cells in space. Front Immunol 15:1512578. https://doi.org/10.3389/fimmu.2024.1512578
46. Schmidt MA, Meydan C, Schmidt CM et al (2020) The NASA Twins Study: the effect of one year in space on long-chain fatty acid desaturases and elongases. Lifestyle Genom 13:107–121. https://doi.org/10.1159/000506769
47. Niforou A, Konstantinidou V, Naska A (2020) Genetic variants shaping inter-individual differences in response to dietary intakes—a narrative review of the case of vitamins. Front Nutr 7:558598. https://doi.org/10.3389/fnut.2020.558598
48. Dhuli K, Medori MC, Bonetti G et al (2023) Nutrigenomics: SNPs correlated to minerals' deficiencies. Clin Ter 174:193–199. https://doi.org/10.7417/CT.2023.2487
49. Holmstrom P, Marmur J, Eggersten G et al (2002) Mild iron overload in patients carrying the HFE S65C gene mutation: a retrospective study in patients with suspected iron overload and healthy controls. Gut 51:723–730. https://doi.org/10.1136/gut.51.5.723
50. Meyer TE, Verwoert GC, Hwang S-J et al (2010) Genome-wide association studies of serum magnesium, potassium, and sodium concentrations identify six loci influencing serum magnesium levels. PLoS Genet 6:e1001045. https://doi.org/10.1371/journal.pgen.1001045
51. El Ouali EM, Barthelemy B, Del Coso J et al (2024) A systematic review and meta-analysis of the association between ACTN3 R577X genotypes and performance in endurance versus power athletes and non-athletes. Sports Med Open 10:37. https://doi.org/10.1186/s40798-024-00711-x
52. Guth LM, Roth SM (2013) Genetic influence on athletic performance. Curr Opin Pediatr 25:653–658. https://doi.org/10.1097/MOP.0b013e3283659087
53. Petr M, Maciejewska-Skrendo A, Zajac A et al (2019) Association of elite sports status with gene variants of peroxisome proliferator activated receptors and their transcriptional coactivator. Int J Mol Sci 21:162. https://doi.org/10.3390/ijms21010162
54. de la Iglesia R, Espinosa-Salinas I, Lopez-Silvarrey FJ et al (2020) A potential endurance algorithm prediction in the field of sports performance. Front Genet 11:711. https://doi.org/10.3389/fgene.2020.00711
55. Saunders CJ, Posthumus M, O'Connell K et al (2015) A variant within the *AQP1* 3′-untranslated region is associated with running performance, but not weight changes, during an Ironman Triathlon. J Sports Sci 33:1342–1348. https://doi.org/10.1080/02640414.2014.989535
56. Harvey NR, Voisin S, Dunn PJ et al (2020) Genetic variants associated with exercise performance in both moderately trained and highly trained individuals. Mol Genet Genomics 295:515–523. https://doi.org/10.1007/s00438-019-01639-8
57. Alvarez-Romero J, Laguette MN, Seale K et al (2023) Genetic variants within the *COL5A1* gene are associated with ligament injuries in physically active populations from Australia, South Africa, and Japan. Eur J Sport Sci 23:284–293. https://doi.org/10.1080/17461391.2021.2011426
58. Heffernan SM, Kilduff LP, Erskine RM et al (2017) COL5A1 gene variants previously associated with reduced soft tissue injury risk are associated with elite athlete status in rugby. BMC Genomics 18, 820. https://doi.org/10.1186/s12864-017-4187-3
59. Santiago C, Ruiz JR, Buxens A et al (2011) Trp64Arg polymorphism in *ADRB3* gene is associated with elite endurance performance: table 1. Br J Sports Med 45:147–149. https://doi.org/10.1136/bjsm.2009.061366
60. Thakkar D, Sicova M, Guest NS et al (2021) HFE genotype and endurance performance in competitive male athletes. Med Sci Sports Exerc 53:1385–1390. https://doi.org/10.1249/MSS.0000000000002595
61. Eynon N, Ruiz JR, Bishop DJ et al (2013) The rs12594956 polymorphism in the NRF-2 gene is associated with top-level Spanish athlete's performance status. J Sci Med Sport 16:135–139. https://doi.org/10.1016/j.jsams.2012.05.004
62. Appel M, Zentgraf K, Krüger K, Alack K (2021) Effects of genetic variation on endurance performance, muscle strength, and injury susceptibility in sports: a systematic review. Front Physiol 12:694411. https://doi.org/10.3389/fphys.2021.694411
63. Sierra APR, Martínez Galán BS, de Sousa CAZ et al (2022) Exercise induced-cytokines response in marathon runners: role of ACE I/D and BDKRB2 +9/-9 polymorphisms.

Front Physiol 13:919544. https://doi.org/10.3389/fphys.2022.919544
64. Ilardo MA, Moltke I, Korneliussen TS et al (2018) Physiological and genetic adaptations to diving in sea nomads. Cell 173:569–580.e15. https://doi.org/10.1016/j.cell.2018.03.054
65. Spoto G, Di Rosa G, Nicotera AG (2024) The impact of genetics on cognition: insights into cognitive disorders and single nucleotide polymorphisms. J Pers Med 14:156. https://doi.org/10.3390/jpm14020156
66. Chen J, Lipska BK, Halim N et al (2004) Functional analysis of genetic variation in catechol-O-methyltransferase (COMT): effects on mRNA, protein, and enzyme activity in postmortem human brain. Am J Hum Genet 75:807–821. https://doi.org/10.1086/425589
67. Baj G, Carlino D, Gardossi L, Tongiorgi E (2013) Toward a unified biological hypothesis for the BDNF Val66Met-associated memory deficits in humans: a model of impaired dendritic mRNA trafficking. Front Neurosci 7:188. https://doi.org/10.3389/fnins.2013.00188
68. Colson NJ, Naug HL, Nikbakht E et al (2017) The impact of MTHFR 677 C/T genotypes on folate status markers: a meta-analysis of folic acid intervention studies. Eur J Nutr 56:247–260. https://doi.org/10.1007/s00394-015-1076-x
69. Janelidze S, Suchankova P, Ekman A et al (2015) Low IL-8 is associated with anxiety in suicidal patients: genetic variation and decreased protein levels. Acta Psychiatr Scand 131:269–278. https://doi.org/10.1111/acps.12339
70. Krishnamurthy HK, Balaguru UM, Pereira M et al (2023) Influence of genetic polymorphisms on serum biomarkers of cardiac health. Medicine 102:e33953. https://doi.org/10.1097/MD.0000000000033953
71. Meza-Alvarado JC, Page RA, Mallard B et al (2023) VEGF-A related SNPs: a cardiovascular context. Front Cardiovasc Med 10:1190513. https://doi.org/10.3389/fcvm.2023.1190513
72. Dave R, Rosero LRD, Pamugo C et al (2024) The role of factor V leiden deficiency in venous thromboembolism: a comprehensive review. Ann Rev Res 12(1):555829. https://doi.org/10.19080/ARR.2024.12.555829
73. Rayamajhi S, Sayonara Lopez Capa G, Carolina Flores Reyes L et al (2024) Massive pulmonary embolism associated with factor V leiden mutation in a young female on oral contraceptive pills: a case report. Cureus 16(6):e62451. https://doi.org/10.7759/cureus.62451
74. Stitziel NO, Khera AV, Wang X et al (2017) ANGPTL3 deficiency and protection against coronary artery disease. J Am Coll Cardiol 69:2054–2063. https://doi.org/10.1016/j.jacc.2017.02.030
75. Xu L-B, Zhou Y-F, Yao J-L et al (2017) Apolipoprotein A1 polymorphisms and risk of coronary artery disease: a meta-analysis. Arch Med Sci 4:813–819. https://doi.org/10.5114/aoms.2017.65233
76. Benn M, Tybjærg-Hansen A, Nordestgaard BG (2019) Low LDL cholesterol by PCSK9 variation reduces cardiovascular mortality. J Am Coll Cardiol 73:3102–3114. https://doi.org/10.1016/j.jacc.2019.03.517
77. Afzal AR, Kiechl S, Daryani YP et al (2008) Common CCR5-del32 frameshift mutation associated with serum levels of inflammatory markers and cardiovascular disease risk in the Bruneck population. Stroke 39:1972–1978. https://doi.org/10.1161/STROKEAHA.107.504381
78. Jarenbäck L, Frantz S, Weidner J et al (2022) Single-nucleotide polymorphisms in the sulfatase-modifying factor 1 gene are associated with lung function and COPD. ERJ Open Res 8(2):00668. https://doi.org/10.1183/23120541.00668-2021
79. Lv Z-T, Gao S-T, Cheng P et al (2018) Association between polymorphism rs12722 in COL5A1 and musculoskeletal soft tissue injuries: a systematic review and meta-analysis. Oncotarget 9:15365–15374. https://doi.org/10.18632/oncotarget.23805
80. Gkritzali E, Akam E, Mastana S (2024) Role of COL5A1 gene polymorphism (rs12722) in lower limb musculoskeletal injuries among adults—a systematic review and meta-analysis. J Res Pract Musculoskelet Syst 08:60–70. https://doi.org/10.22540/JRPMS-08-060
81. Kim CY, Park B, Jung JY et al (2025) Genome wide interaction study of genetic variants associated with lung function decline. Sci Rep 15:9824. https://doi.org/10.1038/s41598-025-93147-6
82. Duan R, Wang N, Shang Y et al (2022) TNF-α (G-308A) polymorphism, circulating levels of TNF-α and IGF-1: risk factors for ischemic stroke—an updated meta-analysis. Front Aging Neurosci 14:831910. https://doi.org/10.3389/fnagi.2022.831910
83. Hu Y, Lu Y, Fang Y et al (2024) Role of long non-coding RNA in inflammatory bowel disease. Front Immunol 15:1406538. https://doi.org/10.3389/fimmu.2024.1406538
84. Rankin CR, Shao L, Elliott J et al (2020) The IBD-associated long noncoding RNA *IFNG-AS1* regulates the balance between inflammatory and anti-inflammatory cytokine production after T-cell stimulation. Am J Physiol Gastrointest Liver Physiol 318:G34–G40. https://doi.org/10.1152/ajpgi.00232.2019
85. Zwart SR, Gregory JF, Zeisel SH et al (2016) Genotype, B-vitamin status, and androgens affect spaceflight-induced ophthalmic

changes. FASEB J 30:141–148. https://doi.org/10.1096/fj.15-278457
86. Jung J, Wu Q (2024) Identification of bone mineral density associated genes with shared genetic architectures across multiple tissues: functional insights for EPDR1, PKDCC, and SPTBN1. PLoS One 19:e0300535. https://doi.org/10.1371/journal.pone.0300535
87. Peris-Moreno D, Taillandier D, Polge C (2020) MuRF1/TRIM63, master regulator of muscle mass. Int J Mol Sci 21:6663. https://doi.org/10.3390/ijms21186663
88. Zehra S, Ghouri M, Jafari H et al (2023) Single nucleotide polymorphisms (rs3736228 and rs4988321) in low-density lipoprotein receptor-related protein-5 gene with predisposition to rheumatoid arthritis. Gene 851:147025. https://doi.org/10.1016/j.gene.2022.147025
89. Ferrari SL, Deutsch S, Choudhury U et al (2004) Polymorphisms in the low-density lipoprotein receptor–related protein 5 (LRP5) gene are associated with variation in vertebral bone mass, vertebral bone size, and stature in whites. Am J Hum Genet 74:866–875. https://doi.org/10.1086/420771
90. Ortiz M, Jauset-Rubio M, Trummer O et al (2023) Generic platform for the multiplexed targeted electrochemical detection of osteoporosis-associated single nucleotide polymorphisms using recombinase polymerase solid-phase primer elongation and ferrocene-modified nucleoside triphosphates. ACS Cent Sci 9:1591–1602. https://doi.org/10.1021/acscentsci.3c00243
91. Estrada K, Styrkarsdottir U, Evangelou E et al (2012) Genome-wide meta-analysis identifies 56 bone mineral density loci and reveals 14 loci associated with risk of fracture. Nat Genet 44:491–501. https://doi.org/10.1038/ng.2249
92. Ruparelia AA, Montandon M, Merriner J et al (2024) Atrogin-1 promotes muscle homeostasis by regulating levels of endoplasmic reticulum chaperone BiP. JCI Insight 9(8):e167578. https://doi.org/10.1172/jci.insight.167578
93. Baumert P, Stewart CE, Lake MJ et al (2018) Variations of collagen-encoding genes are associated with exercise-induced muscle damage. Physiol Genomics 50:691–693. https://doi.org/10.1152/physiolgenomics.00145.2017
94. Bodine SC, Baehr LM (2014) Skeletal muscle atrophy and the E3 ubiquitin ligases MuRF1 and MAFbx/atrogin-1. Am J Physiol Endocrinol Metab 307:E469–E484. https://doi.org/10.1152/ajpendo.00204.2014
95. Kruszewski M, Aksenov MO (2022) Association of myostatin gene polymorphisms with strength and muscle mass in athletes: a systematic review and meta-analysis of the MSTN rs1805086 mutation. Genes (Basel) 13:2055. https://doi.org/10.3390/genes13112055
96. Guilherme JPLF, Souza-Junior TP, Lancha Junior AH (2021) Association study of performance-related polymorphisms in Brazilian combat-sport athletes highlights variants in the *GABPB1* gene. Physiol Genomics 53:47–50. https://doi.org/10.1152/physiolgenomics.00118.2020
97. Guo R, Gao S, Shaxika N et al (2024) Associations of collagen type 1 α1 gene polymorphisms and musculoskeletal soft tissue injuries: a meta-analysis with trial sequential analysis. Aging 16:8866–8879. https://doi.org/10.18632/aging.205846
98. Ebert JR, Magi A, Unt E et al (2023) Genome-wide association study identifying variants related to performance and injury in high-performance athletes. Exp Biol Med 248:1799–1805. https://doi.org/10.1177/15353702231198068

Index

Mariano Bizzarri (ed.), *Biomedicine and Biotechnology in Space*, Methods in Molecular Biology, vol. 3000, https://doi.org/10.1007/978-1-0716-5174-2,

MIX
Papier aus verantwortungsvollen Quellen
Paper from responsible sources
FSC® C105338

If you have any concerns about our products,
you can contact us on
ProductSafety@springernature.com

In case Publisher is established outside the EU,
the EU authorized representative is:
Springer Nature Customer Service Center GmbH
Europaplatz 3, 69115 Heidelberg, Germany

Printed by Libri Plureos GmbH
in Hamburg, Germany